finite element methods

PURE AND APPLIED MATHEMATICS

A Program of Monographs, Textbooks, and Lecture Notes

LECTURE NOTES IN PURE AND APPLIED MATHEMATICS

1. *N. Jacobson,* Exceptional Lie Algebras
2. *L.-Å. Lindahl and F. Poulsen,* Thin Sets in Harmonic Analysis
3. *I. Satake,* Classification Theory of Semi-Simple Algebraic Groups
4. *F. Hirzebruch et al., Differentiable Manifolds and Quadratic Forms*
5. *I. Chavel,* Riemannian Symmetric Spaces of Rank One
6. *R. B. Burckel,* Characterization of C(X) Among Its Subalgebras
7. *B. R. McDonald et al.,* Ring Theory
8. *Y.-T. Siu,* Techniques of Extension on Analytic Objects
9. *S. R. Caradus et al.,* Calkin Algebras and Algebras of Operators on Banach Spaces
10. *E. O. Roxin et al.,* Differential Games and Control Theory
11. *M. Orzech and C. Small,* The Brauer Group of Commutative Rings
12. *S. Thomier,* Topology and Its Applications
13. *J. M. Lopez and K. A. Ross,* Sidon Sets
14. *W. W. Comfort and S. Negrepontis,* Continuous Pseudometrics
15. *K. McKennon and J. M. Robertson,* Locally Convex Spaces
16. *M. Carmeli and S. Malin,* Representations of the Rotation and Lorentz Groups
17. *G. B. Seligman,* Rational Methods in Lie Algebras
18. *D. G. de Figueiredo,* Functional Analysis
19. *L. Cesari et al.,* Nonlinear Functional Analysis and Differential Equations
20. *J. J. Schäffer,* Geometry of Spheres in Normed Spaces
21. *K. Yano and M. Kon,* Anti-Invariant Submanifolds
22. *W. V. Vasconcelos,* The Rings of Dimension Two
23. *R. E. Chandler,* Hausdorff Compactifications
24. *S. P. Franklin and B. V. S. Thomas,* Topology
25. *S. K. Jain,* Ring Theory
26. *B. R. McDonald and R. A. Morris,* Ring Theory II
27. *R. B. Mura and A. Rhemtulla,* Orderable Groups
28. *J. R. Graef,* Stability of Dynamical Systems
29. *H.-C. Wang,* Homogeneous Branch Algebras
30. *E. O. Roxin et al.,* Differential Games and Control Theory II
31. *R. D. Porter,* Introduction to Fibre Bundles
32. *M. Altman,* Contractors and Contractor Directions Theory and Applications
33. *J. S. Golan,* Decomposition and Dimension in Module Categories
34. *G. Fairweather,* Finite Element Galerkin Methods for Differential Equations
35. *J. D. Sally,* Numbers of Generators of Ideals in Local Rings
36. *S. S. Miller,* Complex Analysis
37. *R. Gordon,* Representation Theory of Algebras
38. *M. Goto and F. D. Grosshans,* Semisimple Lie Algebras
39. *A. I. Arruda et al.,* Mathematical Logic
40. *F. Van Oystaeyen,* Ring Theory
41. *F. Van Oystaeyen and A. Verschoren,* Reflectors and Localization
42. *M. Satyanarayana,* Positively Ordered Semigroups
43. *D. L Russell,* Mathematics of Finite-Dimensional Control Systems
44. *P.-T. Liu and E. Roxin,* Differential Games and Control Theory III
45. *A. Geramita and J. Seberry,* Orthogonal Designs
46. *J. Cigler, V. Losert, and P. Michor,* Banach Modules and Functors on Categories of Banach Spaces
47. *P.-T. Liu and J. G. Sutinen,* Control Theory in Mathematical Economics
48. *C. Byrnes,* Partial Differential Equations and Geometry
49. *G. Klambauer,* Problems and Propositions in Analysis
50. *J. Knopfmacher,* Analytic Arithmetic of Algebraic Function Fields
51. *F. Van Oystaeyen,* Ring Theory
52. *B. Kadem,* Binary Time Series
53. *J. Barros-Neto and R. A. Artino,* Hypoelliptic Boundary-Value Problems
54. *R. L. Sternberg et al.,* Nonlinear Partial Differential Equations in Engineering and Applied Science
55. *B. R. McDonald,* Ring Theory and Algebra III
56. *J. S. Golan,* Structure Sheaves Over a Noncommutative Ring
57. *T. V. Narayana et al.,* Combinatorics, Representation Theory and Statistical Methods in Groups

Additional Volumes in Preparation

finite element methods
superconvergence, post-processing, and a posteriori estimates

edited by

M. Křížek
Mathematical Institute of the Czech Academy of Sciences
Prague, Czech Republic

P. Neittaanmäki
University of Jyväskylä
Jyväskylä, Finland

R. Stenberg
Helsinki University of Technology
Espoo, Finland

MARCEL DEKKER, INC.　　NEW YORK · BASEL · HONG KONG

ISBN: 0-8247-0148-8

The publisher offers discounts on this book when ordered in bulk quantities. For more information, write to Special Sales/Professional Marketing at the address below.

This book is printed on acid-free paper.

MARCEL DEKKER, INC.
270 Madison Avenue, New York, New York 10016
http://www.dekker.com

Current printing (last digit):
10 9 8 7 6 5 4 3 2 1

PRINTED IN THE UNITED STATES OF AMERICA

"When I am thinking of A, I am saying B, I am writing C, and it is, in fact, D."

VLADIMÍR KOŘÍNEK
(1899–1981)

Preface

The conference Finite Element Methods: Superconvergence, Post-processing, and A Posteriori Error Estimates was held at the University of Jyväskylä, Finland. It was, in fact, the first conference devoted to superconvergence phenomena in the finite element method. Its aim was to bring together numerical analysts, specialists in computational mathematics, and software developers who work in superconvergence, post-processing and a posteriori error estimates for finite element solutions of differential equations and other problems of mathematical physics. These proceedings contain selected reviewed papers presented at the conference.

During the development of the finite element method it was found that the rate of convergence of FE-approximations exceeds the optimal global rate at some exceptional points. This phenomenon has come to be known as "superconvergence." During the last 25 years hundreds of papers and several books on superconvergence phenomena have been published. The notion of superconvergence was first analyzed by J. Douglas and T. Dupont in *Some superconvergence results for Galerkin methods for the approximate solution of two-point boundary value problems*, Topics in Numerical Analysis II (Proc. Royal Irish Acad. Conf., Dublin, 1972), Academic Press, London, 1973, 89-93. However, Oganesjan and Ruchovec had already in 1969 proved that the linear interpolation on uniform triangular meshes is superclose to the finite element solution of second order elliptic problems. Later it was found that this property plays a key role in proving many superconvergence phenomena. In the 1970s great strides were made in understanding superconvergence at nodal points and also at Gauss–Legendre, Jacobi, and Lobatto points. Later various post-processing techniques were developed that increase the accuracy of finite element solutions and get higher-order a priori error estimates. Nowadays superconvergence of the finite element method is a rapidly developing field of research. The phenomenon of superconvergence has been observed on meshes that are: (1) locally symmetric with respect to one point (Schatz, Sloan, Wahlbin), e.g., uniform or piecewise uniform, (2) quasiuniform (Chen, Levine, Lin, Zhu), (3) locally periodic (Babuška, Strouboulis), (4) self-similar (Schatz). In the conference special attention was also paid to a posteriori error estimates, which yield reliable bounds for the error in the computer solutions. Some of these estimates are based on various superconvergence techniques.

The scientific committe of the conference consisted of I. Babuška (US), C.M. Chen (China), I. Hlaváček (Czech Republic), R. D. Lazarov (US), Q. Lin (China), T. Regińska (Poland), T. Strouboulis (US), M. Stynes (Ireland), A. Zhou (China) and M. Zlámal (Czech Republic). The scientific program was comprised of invited lectures and contributions. The invited lectures were delivered by G. F. Carey (US), C. M. Chen (China), J. Chleboun (Czech Republic), Y. Q. Huang (China), Q. Lin (China), P. Monk (US), A. Schatz (US), E. Süli (UK), and L. Wahlbin (US).

During the conference a social program was organized for the participants, including a Finnish sauna, a get-together party, a conference dinner, and a boat cruise along the second biggest lake in Finland.

We plan to organize the conference Finite Element Methods every three years. The first conference, Finite Element Methods: Fifty Years of the Courant Element was also held at the University of Jyväskylä. Its proceedings were published as Lecture Notes in Pure and Applied Mathematics, Vol. 164, Marcel Dekker, New York, 1994.

We would like to thank Ms. Marja-Leena Rantalainen for her invaluable help in the preparation of this book. Our great thanks go also to the secretary of the Conference Office, Mrs. Heidi Laaksonen. The financial support of the Academy of Finland and the grant of the Academy of Sciences of the Czech Republic no. A1019601 are gratefully acknowledged.

Michal Křížek
Pekka Neittaanmäki
Rolf Stenberg

Contents

Contents

Contributors

J. Aalto   Department of Civil Engineering, University of Oulu, Oulu, Finland

Jan H. Brandts   Laboratory of Scientific Computing, University of Jyväskylä, Jyväskylä, Finland

P. Burda   Department of Mathematics, Faculty of Mechanical Engineering, Czech Technical University, Prague, Czech Republic

Graham F. Carey   ASE/EM Dept., The University of Texas at Austin, Austin, Texas

Chuanmiao Chen   Institute of Computer Science, Hunan Normal University, Changsha, Hunan, People's Republic of China

Jan Chleboun   Mathematical Institute, Academy of Sciences of the Czech Republic, Prague, Czech Republic

Josef Dalík   Department of Mathematics, Technical University of Brno, Brno, Czech Republic

P. Destuynder   CNAM/IAT, Saint-Cyr-l'École, France

Pedro Díez   Departamento de Matemática Aplicada III, E.T.S. de Ingenieros de Caminos, Universitat Politècnica de Catalunya, Barcelona, Spain

Juan José Egozcue   Departamento de Matemática Aplicada III, E.T.S. de Ingenieros de Caminos, Universitat Politècnica de Catalunya, Barcelona, Spain

István Faragó   Department of Applied Analysis, Eötvös Loránd University, Budapest, Hungary

Ivan Hlaváček   Mathematical Institute, Academy of Sciences of the Czech Republic, Prague, Czech Republic

R. H. W. Hoppe   Mathematical Institute, University of Augsburg, Augsburg, Germany

Yun-Qing Huang   Department of Mathematics, Xiangtan University, Hunan, People's Republic of China

Antonio Huerta   Departamento de Matemática Aplicada III, E.T.S. de Ingenieros de Caminos, Universitat Politècnica de Catalunya, Barcelona, Spain

Michal Křížek   Mathematical Institute, Academy of Sciences of the Czech Republic, Prague, Czech Republic

A. M. Lakhany   BICOM, Institute of Computational Mathematics, Brunel University, Uxbridge, Middlesex, United Kingdom

Qun Lin   Institute of Systems Science, Academia Sinica, Beijing, People's Republic of China

B. Métivet   EDF/DER, Clamart, France

Pekka Neittaanmäki   Department of Mathematics, University of Jyväskylä, Jyväskylä, Finland

M. Perälä   Department of Civil Engineering, University of Oulu, Oulu, Finland

Robert Sandboge   Department of Mathematics, Chalmers University of Technology, Göteborg, Sweden

Alfred H. Schatz   Department of Mathematics, Cornell University, Ithaca, New York

Anton A. Smolianski   Laboratory of Scientific Computing, University of Jyväskylä, Jyväskylä, Finland

T. Tiihonen   Department of Mathematics, University of Jyväskylä, Jyväskylä, Finland

Lars B. Wahlbin   Department of Mathematics, Cornell University, Ithaca, New York

J. R. Whiteman   BICOM, Institute of Computational Mathematics, Brunel University, Uxbridge, Middlesex, United Kingdom

N.-E. Wiberg   Department of Structural Mechanics, Chalmers University of Technology, Göteborg, Sweden

B. Wohlmuth   Mathematical Institute, University of Augsburg, Augsburg, Germany

Qiding Zhu   Department of Mathematics and Mechanics, Southeast University, Nanjing, People's Republic of China

S. Ziukas   Department of Structural Mechanics, Chalmers University of Technology, Göteborg, Sweden

Two robust patch recovery methods with built-in field equations and boundary conditions

J. AALTO and M. PERÄLÄ   University of Oulu, Department of Civil Engineering, P.O. Box 191, FIN–90101 Oulu, Finland

Abstract   Two systematic procedures of obtaining a local polynomial representation for the basic unknown function of a boundary value problem and the corresponding derivative quantities of interest, which contains "built-in" information from the field equations, are presented. The first procedure is based on power series method and the second one is based on weighted residual method. Based on these local representations two simple and robust patch recovery methods, in which the unknown parameters are determined by simultaneous discrete least squares fitting of both the basic unknown and the corresponding derivatives, are proposed. A computationally simple way of improving the results of these patch recovery methods by adding information from the boundary conditions is further proposed. Two numerical examples dealing with potential flow and plane elasticity are finally given.

1   INTRODUCTION

An efficient way of getting a posteriori error estimates for a finite element analysis has been proposed by Zienkiewicz and Zhu [1]. In this procedure original, consistent derivative quantities are first improved using a suitable smoothing procedure. The difference between the smoothed and the original derivatives is then regarded to represent the error.

The usefulness and reliability of such error estimation is highly dependent on the efficiency and accuracy of the smoothing technique. The rate of convergence of the smoothed derivatives should be higher than that of the consistent derivatives. So called patch recovery procedures recently proposed by Zienkiewicz and Zhu [2], seem to be most efficient and accurate derivative smoothing procedures.

Wiberg and Abdulwahab [3], and independently Blacker and Belytchko [4], were the first who used information from the field equations (equilibrium) to improve the accuracy of a patch recovery procedure. This improvement, however, demands more local equations, which must be solved within each patch, and thus reduces the efficiency of the technique. Aalto [5], [6] has recently proposed, a patch recovery procedure, which uses local polynomials containing "built-in" information from the field equations. In this technique, the number of local equations is smaller and in a sense ideal.

This paper is a continuation of the work of references [5], [6], which considered in detail only field equations with constant coefficients. Here two approaches [7], [8] of constructing local polynomial representations containing the information from the field equations, which are more generally applicable, are presented. Based on these local representations two simple and robust patch recovery methods, in which the unknown parameters are determined by simultaneous discrete least squares fitting both at sampling points of the basic unknown and sampling points of the derivatives. Further an efficient and computationally simple way of improving the results of both patch recovery methods by using information from the boundary conditions is presented.

2   LOCAL POLYNOMIAL REPRESENTATION WITH "BUILT-IN" FIELD EQUATIONS

2.1   Field equations

Consider a boundary value problem in two dimensions governed by n second order linear partial differential equations

$$\mathcal{R}(\mathbf{u}) \equiv \mathbf{A}_{xx}\frac{\partial^2 \mathbf{u}}{\partial x^2} + 2\mathbf{A}_{xy}\frac{\partial^2 \mathbf{u}}{\partial x \partial y} + \mathbf{A}_{yy}\frac{\partial^2 \mathbf{u}}{\partial y^2} \tag{1}$$
$$+ \mathbf{A}_x\frac{\partial \mathbf{u}}{\partial x} + \mathbf{A}_y\frac{\partial \mathbf{u}}{\partial y} + \mathbf{A}\mathbf{u} + \mathbf{f} = \mathbf{0} \qquad \text{in } \Omega,$$

where $\mathbf{u}(x,y)$ is $n \times 1$ vector of unknown functions, $\mathbf{A}_{xx}(x,y), ..., \mathbf{A}(x,y)$ are $n \times n$ matrices and $\mathbf{f}(x,y)$ is $n \times 1$ vector of known functions.

2.2   On representing the unknown function

Let us first represent the unknown $\mathbf{u}(x,y)$ locally within the recovery patch using a complete polynomial of degree p as

$$\tilde{\mathbf{u}} = \sum_{i=0}^{p}\sum_{j=0}^{i} \lambda^{i-j}\mu^j \mathbf{u}_{ij}, \tag{2}$$

where $\mathbf{u}_{ij}$ are $n \times 1$ vectors of unknown parameters,

$$\lambda = \frac{x - x_0}{h}, \quad \mu = \frac{y - y_0}{h} \tag{3}$$

are dimensionless coordinates, x_0 and y_0 are coordinates of the patch assembly point and h is a characteristic length of the patch. The dimensionless coordinates λ and

μ have been adopted here in order to shorten the notation of the paper and to keep the parameters $\mathbf{u}_{ij}$ dimensionally homogeneous. A suitable value for the length h is

$$h = \frac{1}{2}\left[(x_{\max} - x_{\min})(y_{\max} - y_{\min})\right]^{\frac{1}{2}},\tag{4}$$

where $x_{\max}$, $x_{\min}$, $y_{\max}$ and $y_{\min}$ are maximum and minimum values of the coordinates of the nodes of the patch considered. Equation (2) can also be expressed in matrix form as

$$\tilde{\mathbf{u}} = \mathbf{P}^p\mathbf{U}^p,\tag{5}$$

where

$$\mathbf{P}^p = [\mathbf{I}, \lambda\mathbf{I}, \mu\mathbf{I}, \lambda^2\mathbf{I}, \lambda\mu\mathbf{I}, \mu^2\mathbf{I}, ..., \lambda^p\mathbf{I}, \lambda^{p-1}\mu\mathbf{I}, ..., \lambda\mu^{p-1}\mathbf{I}, \mu^p\mathbf{I}]\tag{6}$$

and

$$\mathbf{U}^p = [\mathbf{u}_{00}^T, \mathbf{u}_{10}^T, \mathbf{u}_{11}^T, \mathbf{u}_{20}^T, \mathbf{u}_{21}^T, \mathbf{u}_{22}^T, ..., \mathbf{u}_{p0}^T, \mathbf{u}_{p1}^T, ..., \mathbf{u}_{p(p-1)}^T, \mathbf{u}_{pp}^T]^T.\tag{7}$$

2.3 Constraint equations between the parameters $\mathbf{U}^p$ based on field equations

To form constraint equations between the parameters $\mathbf{U}^p$, we demand that the right-hand-side (residual) $\mathcal{R}(\mathbf{u})$ of the field equation (1), with $\mathbf{u} = \tilde{\mathbf{u}}$, should approximately vanish. This can be done using either (i) a power series method or (ii) a weighted residual method. These methods are briefly described in the following.

(i) Power series method

We construct a complete polynomial representation

$$\widetilde{\mathcal{R}} = \sum_{i=0}^{p-2}\sum_{j=0}^{i}\lambda^{i-j}\mu^j\mathcal{R}_{ij}\tag{8}$$

of degree $p-2$ of the residual $\mathcal{R}(x, y)$. Based on two dimensional Taylor series of the function $\mathcal{R}(x, y)$, the corresponding coefficients have the expression

$$\mathcal{R}_{ij} = \frac{h^i}{(i-j)!j!}\frac{\partial^i\mathcal{R}}{\partial x^{i-j}\partial y^j}(x_0, y_0).\tag{9}$$

In order to develop the coefficients $\mathcal{R}_{ij}$ further, in addition to the unknown function $\mathbf{u}(x, y)$, we have to express the known functions $\mathbf{A}_{xx}(x, y), ..., \mathbf{f}(x, y)$ of the field equation (which are not constants) using similar polynomials of degree p. For example, for $\mathbf{A}_{xx}(x, y)$ we get

$$\widetilde{\mathbf{A}}_{xx} = \sum_{i=0}^{p}\sum_{j=0}^{i}\lambda^{i-j}\mu^j\mathbf{A}_{xxij},\tag{10}$$

where $\mathbf{A}_{xxij}$ are known coefficients. Based on two dimensional Taylor series of function $\mathbf{A}_{xx}(x, y)$, they can be evaluated from

$$\mathbf{A}_{xxij} = \frac{h^i}{(i-j)!j!}\frac{\partial^i\mathbf{A}_{xx}}{\partial x^{i-j}\partial y^j}(x_0, y_0).\tag{11}$$

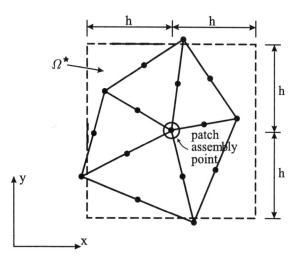

Figure 1   Domain Ω^* of integration.

Further we need similar equation

$$\mathbf{u}_{ij} = \frac{h^i}{(i-j)!j!} \frac{\partial^i \mathbf{u}}{\partial x^{i-j} \partial y^j}(x_0, y_0) \tag{12}$$

for the unknown parameters $\mathbf{u}_{ij}$.

With the help of equations (9) and (1), using rules of differentiation, using further equation (12), (11) and similar equations for the coefficients of the other known functions, the coefficients $\mathcal{R}_{ij}$ of the residual can be expressed as (linear) functions $\mathcal{R}_{ij}(\mathbf{U}^p)$ of the unknown parameters $\mathbf{U}^p$ (see Appendix A). Demanding that the polynomial representation (8) should vanish with all values of x and y (or λ and μ) results to linear equations $\mathcal{R}_{ij}(\mathbf{U}^p) = \mathbf{0}$ ($j = 0, ..., i$, $i = 0, ..., p-2$), which can be expressed in matrix form as

$$\mathbf{C}^P \mathbf{U}^p + \mathbf{d}^P = \mathbf{0}, \tag{13}$$

where the superscript P refers to power series method.

(ii) Weighted residual method

We write weighted residual equations of form

$$\int_{\Omega^*} (\mathbf{P}^{p-2})^T \mathcal{R}(\tilde{\mathbf{u}}) \, d\Omega = \mathbf{0}, \tag{14}$$

where the integration domain Ω^* is simply $x_0 - h \le x \le x_0 + h$ and $y_0 - h \le y \le y_0 + h$ and it is demonstrated in Figure 1. The number of weighted residual equations (14) has been chosen to be $n(p-1)p/2$, which is exactly equal to the number of similar equations in the power series method. It should be noticed here, that, in general, $\widetilde{\mathcal{R}}$ of equation (8) and $\mathcal{R}(\tilde{\mathbf{u}})$ of equations (14) are not equal. The latter is obtained by simply substituting $\tilde{\mathbf{u}}(x, y)$ of equation (2) into the residual $\mathcal{R}(u)$ of equation (1).

With the help of equations (1) and (5) equations (14) can be reduced to

$$\mathbf{C}^W \mathbf{U}^p + \mathbf{d}^W = \mathbf{0}, \tag{15}$$

where

$$\mathbf{C}^W = \int_{\Omega^*} (\mathbf{P}^{p-2})^T \left(\mathbf{A}_{xx} \frac{\partial^2 \mathbf{P}^p}{\partial x^2} + 2\mathbf{A}_{xy} \frac{\partial^2 \mathbf{P}^p}{\partial x \partial y} + \mathbf{A}_{yy} \frac{\partial^2 \mathbf{P}^p}{\partial y^2} \right. \tag{16a}$$

$$\left. + \mathbf{A}_x \frac{\partial \mathbf{P}^p}{\partial x} + \mathbf{A}_y \frac{\partial \mathbf{P}^p}{\partial y} + \mathbf{A}\mathbf{P}^p \right) d\Omega,$$

$$\mathbf{d}^W = \int_{\Omega^*} (\mathbf{P}^{p-2})^T \mathbf{f} \, d\Omega \tag{16b}$$

and the superscript W refers to weighted residual method.

2.4   Original parameters $\mathbf{U}^p$ in terms of the new independent parameters $\mathbf{a}^p$

Equations (13) or (15) are $n(p-1)p/2$ linear constraint equations between the $n(p+1)(p+2)/2$ unknown parameters $\mathbf{U}^p$. They demand that the representation (2) or (5) satisfies the field equation (1) in an average sense. It is thus possible to choose $n(p+1)(p+2)/2 - n(p-1)p/2 = n(2p+1)$ of these parameters as independent parameters

$$\mathbf{a}^p = [\mathbf{a}_1^T, ..., \mathbf{a}_{2p+1}^T]^T. \tag{17}$$

It is reasonable to do the choice of these independent ($n \times 1$ vectors of) parameters $\mathbf{a}_i$ so that first $\mathbf{a}_1 = \mathbf{u}_{00}$ and then corresponding to each degree i (≥ 1) independent parameters $\mathbf{a}_{2i}$ and $\mathbf{a}_{2i+1}$ are equal to two of the original parameters $\mathbf{u}_{i0}, ..., \mathbf{u}_{ii}$.

By substituting the independent parameters $\mathbf{a}^p$ into the constraint equations (13) or (15) we get

$$\mathbf{C}_I \mathcal{U}^p + \mathcal{C}_{II} \mathbf{a}^p + \mathbf{d} = \mathbf{0}, \tag{18}$$

where $\mathcal{U}^p$ is vector of the remaining dependent parameters $\mathbf{u}_{ij}$, $\mathbf{C}_I$ and $\mathbf{C}_{II}$ are matrices containing those columns of matrix $\mathbf{C}^P$ or $\mathbf{C}^W$, which correspond to $\mathcal{U}^p$ and $\mathbf{a}^p$, respectively. Solving equation (18) for $\mathcal{U}^p$ results in

$$\mathcal{U}^p = \mathcal{S}\mathbf{a}^p + \mathcal{T}, \tag{19}$$

where

$$\mathcal{S} = -\mathbf{C}_I^{-1}\mathbf{C}_{II}, \quad \mathcal{T} = -\mathbf{C}_I^{-1}\mathbf{d}. \tag{20}$$

The chosen relations between the dependent and independent parameters together with equations (19) can be written as

$$\mathbf{U}^p = \mathbf{S}\mathbf{a}^p + \mathbf{T}. \tag{21}$$

These equations express the original unknown parameters $\mathbf{U}^p$ in terms of the new independent parameters $\mathbf{a}^p$.

2.5 Polynomial representation with "built-in" field equations

Substituting the result (21) into the original polynomial representation (5) finally gives

$$\tilde{\mathbf{u}} = \mathbf{N}^p \mathbf{a}^p + \mathbf{u}_0^p, \tag{22}$$

where

$$\mathbf{N}^p = \mathbf{P}^p \mathbf{S}, \quad \mathbf{u}_0^p = \mathbf{P}^p \mathbf{T}. \tag{23}$$

Equation (22) is local polynomial representation of the unknown function $\mathbf{u}(x, y)$, which contains "built-in" approximate solution of the field equations (1). The degree of this representation is p and the number of unknown parameters $\mathbf{a}^p$ is $n(2p + 1)$.

Corresponding polynomial representation for the derivative quantities $\gamma \equiv \nabla \mathbf{u}$ is obtained straightforwardly and is

$$\tilde{\gamma} = \mathbf{B}^p \mathbf{a}^p + \gamma_0^p, \tag{24}$$

where

$$\mathbf{B}^p = \nabla \mathbf{N}^p, \quad \gamma_0^p = \nabla \mathbf{u}_0^p. \tag{25}$$

The degree of this representation is $p - 1$.

3 PATCH RECOVERY PROCEDURE

A simple patch recovery procedure, which uses polynomial representations (22) and (24) can now developed. Each internal corner node of the finite element grid is a patch assembly point in this procedure.

Within each patch the unknown parameters $\mathbf{a}^p$ are obtained by least squares fitting of the representations $\tilde{\mathbf{u}}(x, y)$ and $\tilde{\gamma}(x, y)$ to the corresponding consistent (original finite element) values $\hat{\mathbf{u}}(x, y)$ and $\hat{\gamma}(x, y)$ at certain (if possible superconvergent) sampling points of the basic unknown and the derivatives, respectively. The corresponding least squares function gets the form

$$s = \sum_i w_1 [\tilde{\mathbf{u}}(x_i, y_i) - \hat{\mathbf{u}}(x_i, y_i)]^T [\tilde{\mathbf{u}}(x_i, y_i) - \hat{\mathbf{u}}(x_i, y_i)] \tag{26}$$

$$+ \sum_j w_2 h^2 [\tilde{\gamma}(x_j, y_j) - \hat{\gamma}(x_j, y_j)]^T [\tilde{\gamma}(x_j, y_j) - \hat{\gamma}(x_j, y_j)],$$

where w_1 and w_2 are dimensionless weights (the multiplier h^2 has been added in order to keep the function dimensionally homogeneous). The first summation is performed over the sampling points (i) of the basic unknown and the second summation is performed over the sampling points (j) of the derivatives.

The derivative quantities γ are then extrapolated to the appropriate system nodes of the patch using the representation (24). Final nodal values of the derivative quantities γ are obtained by averaging the obtained values at the system nodes. More details of this procedure are given in reference [2].

4 ADDING INFORMATION FROM THE BOUNDARY CONDITIONS TO THE PATCH RECOVERY PROCEDURE

4.1 Patch assembly points on the boundary

In the proposed patch recovery procedure only internal corner nodes of the finite element grid are used as patch assembly points. It is, however, possible to improve the procedure so that also those corner nodes, which are located on the boundary of the domain, are taken as patch assembly points. In order to do this successfully information from the boundary conditions of the problem must be used.

4.2 Boundary conditions

The n Neumann and n Dirichlet boundary conditions of our second order boundary value problem are expressed as

$$\mathcal{R}_N(\mathbf{u}) \equiv \mathbf{B}_x \frac{\partial \mathbf{u}}{\partial x} + \mathbf{B}_y \frac{\partial \mathbf{u}}{\partial y} + \mathbf{g} = 0 \quad \text{on } \Gamma_N,$$

$$\mathcal{R}_D(\mathbf{u}) \equiv \mathbf{u} = \bar{\mathbf{u}} \qquad\qquad \text{on } \Gamma_D. \tag{27}$$

In equations (27) $\mathbf{B}_x(s)$ and $\mathbf{B}_y(s)$ are $n \times n$ matrices of known coefficients and $\mathbf{g}(s)$ and $\bar{\mathbf{u}}(s)$ are $n \times 1$ vectors of known functions.

4.3 Constraint equations between the parameters $\mathbf{U}^p$ based on boundary conditions

We want to form constraint equations between the parameters $\mathbf{U}^p$ by demanding that the right-hand-sides (residuals) $\mathcal{R}_N(\mathbf{u})$ and $\mathcal{R}_D(\mathbf{u})$ of the boundary conditions (27), with $\mathbf{u} = \tilde{\mathbf{u}}$, should approximately vanish. This is done by using either (i) power series method or (ii) weighted residual method.

(i) Power series method

We construct polynomial representations

$$\widetilde{\mathcal{R}}_N = \sum_{i=0}^{p-1} \sigma^i \mathcal{R}_{Ni}, \quad \widetilde{\mathcal{R}}_D = \sum_{i=0}^{p} \sigma^i \mathcal{R}_{Di}, \tag{28}$$

of degrees $p-1$ and p in the boundary coordinate s for the residuals of the boundary conditions. In equations (28)

$$\sigma = \frac{s - s_0}{h} \tag{29}$$

is corresponding dimensionless coordinate and s_0 is the coordinate s of the patch assembly point. Based on one dimensional Taylor series of functions $\mathcal{R}_N(s)$ and $\mathcal{R}_D(s)$, the corresponding coefficients have expression

$$\mathcal{R}_{Ni} = \frac{h^i}{i!} \frac{d^i \mathcal{R}_N}{ds^i}(s_0), \quad \mathcal{R}_{Di} = \frac{h^i}{i!} \frac{d^i \mathcal{R}_D}{ds^i}(s_0). \tag{30}$$

In order to develop the coefficients $\mathcal{R}_{Ni}$ and $\mathcal{R}_{Di}$ further, we have to express the known functions $\mathbf{B}_x(s), ..., \overline{\mathbf{u}}(s)$ using similar polynomials in the boundary coordinate s. For example, for $\mathbf{B}_x(s)$ we get

$$\tilde{\mathbf{B}}_x = \sum_{i=0}^{p} \sigma^i \mathbf{B}_{xi}, \tag{31}$$

where the parameters $\mathbf{B}_{xi}$ are known and can be evaluated from

$$\mathbf{B}_{xi} = \frac{h^i}{i!} \frac{d^i \mathbf{B}_x}{ds^i}(s_0). \tag{32}$$

We also have to express the coordinates $x = x(s)$ and $y = y(s)$ of the boundary curve using polynomials

$$\tilde{x} = \sum_{i=0}^{p} \sigma^i x_i, \quad \tilde{y} = \sum_{i=0}^{p} \sigma^i y_i, \tag{33}$$

where

$$x_i = \frac{h^i}{i!} \frac{d^i x}{ds^i}(s_0), \quad y_i = \frac{h^i}{i!} \frac{d^i y}{ds^i}(s_0). \tag{34}$$

With the help of equations (30) and (27), using the chain rule and other rules of differentiation, using further equations (12), (34), (32) and similar equations for the coefficients of the other known functions, the coefficients of the residuals can be expressed as (linear) functions $\mathcal{R}_{Ni}(\mathbf{U}^p)$ and $\mathcal{R}_{Di}(\mathbf{U}^p)$ of the unknown parameters $\mathbf{U}^p$ (see Appendix B). Demanding that $\tilde{\mathcal{R}}_N$ and $\tilde{\mathcal{R}}_D$ should vanish with all values of s (or σ) results in Neumann case to np equations $\mathcal{R}_{Ni}(\mathbf{U}^p) = \mathbf{0}$, $i = 0, ..., p-1$, and in Dirichlet case to $n(p+1)$ equations $\mathcal{R}_{Di}(\mathbf{U}^p) = \mathbf{0}$, $i = 0, ..., p$. These equations can be written in matrix form as

$$\mathbf{C}_N^P \mathbf{U}^p + \mathbf{d}_N^P = \mathbf{0}, \quad \mathbf{C}_D^P \mathbf{U}^p + \mathbf{d}_D^P = \mathbf{0}. \tag{35}$$

(ii) Weighted residual method

We write weighted residual equations of form

$$\int_{\Gamma^*} (\mathbf{Q}^{p-1})^T \mathcal{R}_N(\tilde{\mathbf{u}}) \, d\Gamma = \mathbf{0}, \quad \int_{\Gamma^*} (\mathbf{Q}^p)^T \mathcal{R}_D(\tilde{\mathbf{u}}) \, d\Gamma = \mathbf{0}, \tag{36}$$

where

$$\mathbf{Q}^p = [\mathbf{I}, \sigma \mathbf{I}, \sigma^2 \mathbf{I}, ..., \sigma^p \mathbf{I}]. \tag{37}$$

The integration domain Γ^* on the boundary curve is simply $s_0 - h \le s \le s_0 + h$ and it is demonstrated in Figure 2. The number of weighted residual equations (36) has been chosen to be np and $n(p+1)$ corresponding to the Neumann and Dirichlet

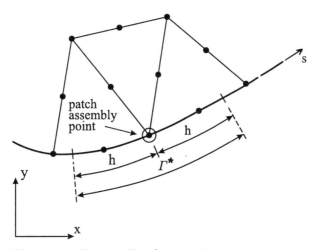

Figure 2 Domain Γ^* of integration.

case, respectively. This is exactly equal to the number of similar equations in the power series method.

With the help of equations (27) and (5) equations (36) reduce to

$$\mathbf{C}_N^W \mathbf{U}^p + \mathbf{d}_N^W = \mathbf{0}, \quad \mathbf{C}_D^W \mathbf{U}^p + \mathbf{d}_D^W = \mathbf{0}, \tag{38}$$

where

$$\mathbf{C}_N^W = \int_{\Gamma^*} (\mathbf{Q}^{p-1})^T \left(\mathbf{B}_x \frac{\partial \mathbf{P}^p}{\partial x} + \mathbf{B}_y \frac{\partial \mathbf{P}^p}{\partial y} \right) d\Gamma, \quad \mathbf{d}_N^W = \int_{\Gamma^*} (\mathbf{Q}^{p-1})^T \mathbf{g} \, d\Gamma \tag{39a}$$

and

$$\mathbf{C}_D^W = \int_{\Gamma^*} (\mathbf{Q}^p)^T \mathbf{P}^p \, d\Gamma, \quad \mathbf{d}_D^W = -\int_{\Gamma^*} (\mathbf{Q}^p)^T \overline{\mathbf{u}} \, d\Gamma. \tag{39b}$$

Equations (35) or (38) are linear constraint equations between the $n(p+1)(p+2)/2$ unknown parameters $\mathbf{U}^p$, which demand that the representation (2) or (5) should satisfy the boundary conditions (27) in an average sense.

4.4 Constraint equations between the new parameters $\mathbf{a}^p$ based on boundary conditions

Equations (35) or (38) are np or $n(p+1)$ additional constraint equations (in addition to equations (13) or (15)) between the original parameters $\mathbf{U}^p$, which should hold, if the patch assembly point is on the boundary of the domain. The corresponding equations between the new parameters $\mathbf{a}^p$ are obtained with the help of equation (21) and they can be written as

$$\mathbf{G}\mathbf{a}^p = \mathbf{H}, \tag{40}$$

where

$$G = \begin{cases} \mathbf{C}_N \mathbf{S} & \text{(Neumann boundary)} \\ \mathbf{C}_D \mathbf{S} & \text{(Dirichlet boundary),} \end{cases}$$

$$H = \begin{cases} -\mathbf{d}_N - \mathbf{C}_N \mathbf{T} & \text{(Neumann boundary)} \\ -\mathbf{d}_D - \mathbf{C}_D \mathbf{T} & \text{(Dirichlet boundary)} \end{cases} \tag{41}$$

and the superscripts P or W have been dropped.

4.5 Lagrange multiplier method for constraint equations based on boundary conditions

One computationally simple way of taking into account the information from the boundary conditions in the patch recovery procedure is to use Lagrange multiplier method in connection with least squares function (26). The corresponding modified least squares function is now

$$s' = \sum_i w_1 [\tilde{\mathbf{u}}(x_i, y_i) - \hat{\mathbf{u}}(x_i, y_i)]^T [\tilde{\mathbf{u}}(x_i, y_i) - \hat{\mathbf{u}}(x_i, y_i)]$$

$$+ \sum_j w_2 h^2 [\tilde{\gamma}(x_j, y_j) - \hat{\gamma}(x_j, y_j)]^T [\tilde{\gamma}(x_j, y_j) - \hat{\gamma}(x_j, y_j)] \tag{42}$$

$$+ (\boldsymbol{\lambda})^T (\mathbf{G} \mathbf{a}^p - \mathbf{H}),$$

where $\boldsymbol{\lambda}$ is vector of corresponding Lagrange multipliers.

5   NUMERICAL EXAMPLES

5.1   Example problems and typical grids

Two example problems with known, nonpolynomial, analytical solutions were used to demonstrate the presented methods:

Problem A: Potential flow around a circular cylinder   Figure 3 shows the domain and demonstrates the problem. It is governed by equation (1) with $n = 1$, $\mathbf{u} = \phi$, $\mathbf{A}_{xx} = \mathbf{A}_{yy} = 1$, $\mathbf{A}_{xy} = \mathbf{A}_x = \mathbf{A}_y = \mathbf{A} = 0$, and $\mathbf{f} = 0$, where ϕ is the unknown potential.

Problem B: Stretching of an infinite plate with a circular hole   Figure 4 shows the domain and demonstrates the problem. It is governed by equation (1) with $n = 2$,

$$\mathbf{u} = \begin{Bmatrix} u \\ v \end{Bmatrix}, \quad \mathbf{A}_{xx} = \frac{G}{\eta - 1} \begin{bmatrix} \eta + 1 & 0 \\ 0 & \eta - 1 \end{bmatrix}, \quad \mathbf{A}_{xy} = \frac{G}{\eta - 1} \begin{bmatrix} 0 & 1 \\ 1 & 0 \end{bmatrix},$$

$$\mathbf{A}_{yy} = \frac{G}{\eta - 1} \begin{bmatrix} \eta - 1 & 0 \\ 0 & \eta + 1 \end{bmatrix}, \quad \mathbf{f} = \begin{Bmatrix} f_x \\ f_y \end{Bmatrix} \tag{43}$$

and $\mathbf{A}_x = \mathbf{A}_y = \mathbf{A} = 0$, where u and v are the unknown displacements, G is shear modulus, $\eta = (3 - \nu)/(1 + \nu)$, ν is Poisson's ratio and f_x and f_y are volume forces. Here $f_x = f_y = 0$ and thus $\mathbf{f} = 0$.

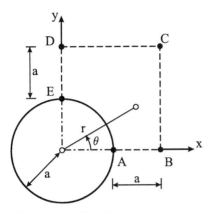

Figure 3   Problem A.

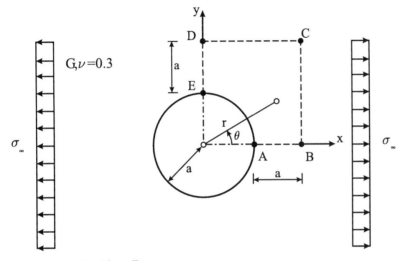

Figure 4   Problem B.

In both examples the sampling points of the basic unknown were simply chosen to be the nodes of the recovery patch and no sampling points of the derivatives were used. Correspondingly the values $w_1 = 1$ and $w_2 = 0$ for the weights were used.

Figure 5 shows typical finite element grids ($h/a = 0.5$) used in experimental convergence study of the two example problems.

5.2   Numerical results

Figures 6 and 7 present experimental convergence study of the relative error in energy (η_E) of the two example problems. Each of these figures show a comparison of the smoothed solution obtained with the present technique based on power series method (BUI_P), the present technique based on weighted residual method (BUI_W), technique (WA) of reference [3], technique (SPR) of reference [2] and the error of the consistent (original finite element) solution. Information from the

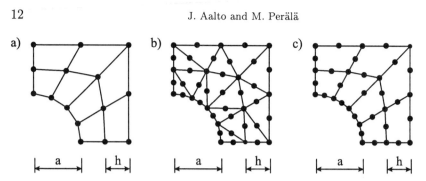

Figure 5 Typical grids of (a) bilinear quadrilaterals, (b) quadratic triangles and (c) quadratic Serendip quadrilaterals.

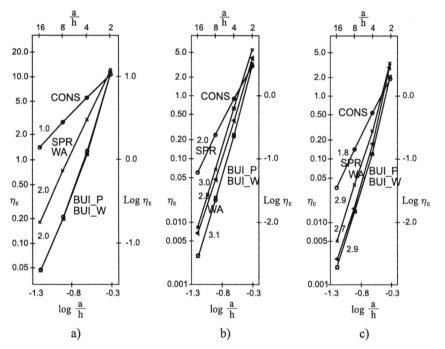

Figure 6 Problem A: (a) $p = 3$; bilinear quadrilaterals (b) $p = 4$; quadratic triangles (c) $p = 5$; quadratic quadrilaterals.

boundary conditions has not been used in these results. Figure 8 present study of problem A. They show a comparison of the smoothed solutions of the present methods obtained without (BUI_P and BUI_W) and with (BUI_P+BC and BUI_W+BC) boundary conditions and the consistent solution. The presented two patch recovery methods seem to work well and give results, which are comparable and even better than those of the two other well known methods (SPR and WA). Adding information from the boundary conditions to the recovery procedure seems to improve the results, especially in connection with coarse grids.

Tables 1 and 2 show a comparison of the effectivity indices

$$\theta = \eta_E^{\mathrm{esti}}/\eta_E, \tag{44}$$

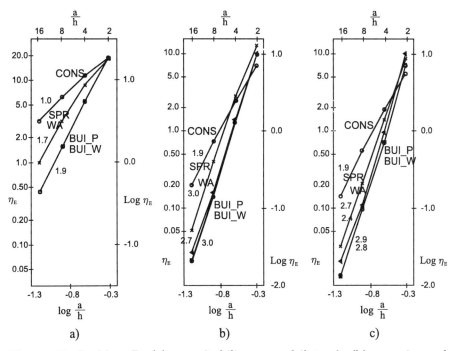

Figure 7 Problem B: (a) $p = 3$; bilinear quadrilaterals (b) $p = 4$; quadratic triangles (c) $p = 5$; quadratic quadrilaterals.

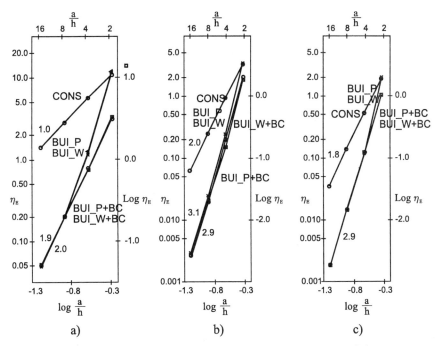

Figure 8 Problem A: (a) $p = 3$; bilinear quadrilaterals (b) $p = 4$; quadratic triangles (c) $p = 5$; quadratic quadrilaterals.

Table 1 Effectivity indices θ with coarse grid ($h/a = 0.5$); Problem A.

Element type	SPR	WA	BUI_P	BUI_W	BUI_P+BC	BUI_W+BC
Bilinear quadrilaterals	1.26	1.24	1.23	1.23	1.06	1.06
Quadratic triangles	1.95	1.55	1.40	1.40	1.22	1.26
Quadratic Serendip quadrilaterals	1.86	1.45	1.45	1.44	1.10	1.11

Table 2 Effectivity indices θ with dense grid ($h/a = 0.0625$); Problem A.

Element type	SPR	WA	BUI_P	BUI_W	BUI_P+BC	BUI_W+BC
Bilinear quadrilaterals	0.991	0.991	0.998	0.998	1.002	1.002
Quadratic triangles	0.989	0.988	0.999	0.998	1.001	1.001
Quadratic Serendip quadrilaterals	1.015	1.003	1.003	1.005	1.004	1.003

where η_E is relative error in energy and η_E^{esti} is the corresponding Zienkiewicz–Zhu [1] estimate, obtained for problem A using different patch recovery methods. The present (BUI_P and BUI_W) methods give better error estimates than the other two methods (SPR and WA) both with coarse and dense grid. If information from the boundary conditions (BUI_P+BC and BUI_W+BC) is taken into account, the effectivity of error estimation of the present methods is further improved.

5.3 A note on equivalence of the two patch recovery methods

It can be shown [8], that in a special case of the field equations (1), the present two patch recovery methods (with boundary conditions not included) are equivalent. This is when the coefficient matrices $\mathbf{A}_{xx}$, $\mathbf{A}_{xy}$ and $\mathbf{A}_{yy}$ are constants, the coefficient matrices $\mathbf{A}_x$, $\mathbf{A}_y$, $\mathbf{A}$ vanish and the source term vector $\mathbf{f}(x,y)$ is a polynomial (of degree less than $p-2$) of x and y. The field equations of the two example problems A and B of this paper belong to this special case. This explains why the corresponding numerical results are so close to each other.

6 CONCLUSIONS

The presented two patch recovery methods have properties, which are common to both of them: They have similar strategy of "building-in" the solution of the field equations, similar local least squares fitting technique and similar way of taking into account the boundary conditions. They are even equivalent in certain circumstances.

The two methods, however, also have essential differences: Different way of forming constrain equations based on both field equations and boundary conditions. Power series method (BUI_P) is fast, especially in problems with constant coefficients. Weighted residual method (BUI_W) is slower, but its efficiency does not depend on the type of the coefficients. Power series method needs Taylor type coefficients of the known functions in the domain and on the boundaries. Thus the implementation is more problem dependent. Weighted residual method is less problem dependent and easier to implement.

Based on the theoretical basis and numerical results obtained this far, the authors claim that the presented two patch recovery procedures can be classified as robust ones.

REFERENCES

1. O. C. Zienkiewicz and J. Z. Zhu, *A simple error estimator and adaptive procedure for practical engineering analysis*, Int. J. Numer. Meth. Engng. **24** (1987), 337–357.
2. O. C. Zienkiewicz and J. Z. Zhu, *The superconvergent patch recovery and a posteriori error estimates. Part 1: The recovery technique'*, Int. J. Numer. Meth. Engng. **33** (1992), 1331–1364.
3. N.-E. Wiberg and F. Abdulwahab, *An efficient postprocessing technique for stress problems based on superconvergent derivatives and equilibrium*, in "Numerical Methods in Engineering '92," (Edited by Hirch et. al.), Elsevier Science, Amsterdam, 1992, pp. 25–32.
4. T. Blacker and T. Belytchko, *Superconvergent patch recovery with equilibrium and conjoint interpolant enhancements*, Int. J. numer. Meth. Engng. **37** (1994), 1517–1536.
5. J. Aalto, *Built-in field equations for recovery procedures*, in "Proc. Second International Conference on Computational Structures Technology: Advances in post and preprocessing for finite element technology," (Edited by M. Papadrakakis and R. H. V. Topping), Civil-Comp Press, 1994, pp. 125–135.
6. J. Aalto, *Built-in field equations for recovery procedures*, Computers and Structures, (Accepted 16 November 1995).
7. J. Aalto and M. Åman, *Polynomial representations with built-in field equations for patch recovery procedures*, in "Proc. Third International Conference on Computational Structures Technology: Advances in Finite Element Technology," (Edited by R. H. V. Topping), Civil-Comp Press, 1996, pp. 109–125.
8. J. Aalto and M. Perälä, *Built-in weighted residual field equations for patch recovery procedures*, in "Proc. Third International Conference on Computational Structures Technology: Advances in Finite Element Technology," (Edited by R. H. V. Topping), Civil-Comp Press, 1996, pp. 135–150.

APPENDIX A

In connection with the power series method the paper explains only very briefly, how the coefficients $\mathcal{R}_{ij}$ of the polynomial representation (8) of the residual of the field equations can be expressed as (linear) function of the unknown parameters $\mathbf{U}^p$. This appendix demonstrates in detail how this is done in connection with first three coefficients $\mathcal{R}_{00}$, $\mathcal{R}_{10}$ and $\mathcal{R}_{11}$. The same idea can be continued for subsequent coefficients. (A general formula for $\mathcal{R}_{ij}$ is given in equation (19) of reference [7].)

Equations (9) and (1) give for the first two coefficients

$$
\begin{aligned}
\mathcal{R}_{00} &= \mathcal{R}(x_0, y_0) \\
&= \mathbf{A}(x_0, y_0)\mathbf{u}(x_0, y_0) + \mathbf{A}_x(x_0, y_0)\frac{\partial \mathbf{u}}{\partial x}(x_0, y_0) + \mathbf{A}_y(x_0, y_0)\frac{\partial \mathbf{u}}{\partial y}(x_0, y_0) \\
&\quad + \mathbf{A}_{xx}(x_0, y_0)\frac{\partial^2 \mathbf{u}}{\partial x^2}(x_0, y_0) + 2\mathbf{A}_{xy}(x_0, y_0)\frac{\partial^2 \mathbf{u}}{\partial x \partial y}(x_0, y_0) \\
&\quad + \mathbf{A}_{yy}(x_0, y_0)\frac{\partial^2 \mathbf{u}}{\partial y^2}(x_0, y_0) + \mathbf{f}(x_0, y_0)
\end{aligned}
\tag{A.1}
$$

and

$$
\begin{aligned}
\mathcal{R}_{10} &= h\frac{\partial \mathcal{R}}{\partial x}(x_0, y_0) \\
&= h\Bigg[\frac{\partial \mathbf{A}}{\partial x}(x_0, y_0)\mathbf{u}(x_0, y_0) \\
&\quad + \left(\mathbf{A}(x_0, y_0) + \frac{\partial \mathbf{A}_x}{\partial x}(x_0, y_0)\right)\frac{\partial \mathbf{u}}{\partial x}(x_0, y_0) + \frac{\partial \mathbf{A}_y}{\partial x}(x_0, y_0)\frac{\partial \mathbf{u}}{\partial y}(x_0, y_0) \\
&\quad + \left(\mathbf{A}_x(x_0, y_0) + \frac{\partial \mathbf{A}_{xx}}{\partial x}(x_0, y_0)\right)\frac{\partial^2 \mathbf{u}}{\partial x^2}(x_0, y_0) \\
&\quad + \left(\mathbf{A}_y(x_0, y_0) + 2\frac{\partial \mathbf{A}_{xy}}{\partial x}(x_0, y_0)\right)\frac{\partial^2 \mathbf{u}}{\partial x \partial y}(x_0, y_0) + \frac{\partial \mathbf{A}_{yy}}{\partial x}(x_0, y_0)\frac{\partial^2 \mathbf{u}}{\partial y^2}(x_0, y_0) \\
&\quad + \mathbf{A}_{xx}(x_0, y_0)\frac{\partial^3 \mathbf{u}}{\partial x^3}(x_0, y_0) + 2\mathbf{A}_{xy}(x_0, y_0)\frac{\partial^3 \mathbf{u}}{\partial x^2 \partial y}(x_0, y_0) \\
&\quad + \mathbf{A}_{yy}(x_0, y_0)\frac{\partial^3 \mathbf{u}}{\partial x \partial y^2}(x_0, y_0) + \frac{\partial \mathbf{f}}{\partial x}(x_0, y_0)\Bigg].
\end{aligned}
\tag{A.2}
$$

Using now equation (11) and similar equations for the known coefficients $\mathbf{A}_{xxij}, ..., \mathbf{f}_{ij}$ and equation (12) for the unknown coefficients $\mathbf{u}_{ij}$ we get

$$
\begin{aligned}
\mathcal{R}_{00} &= \mathbf{A}_{00}\mathbf{u}_{00} + \frac{1}{h}\mathbf{A}_{x00}\mathbf{u}_{10} + \frac{1}{h}\mathbf{A}_{y00}\mathbf{u}_{11} \\
&\quad + \frac{2}{h^2}\mathbf{A}_{xx00}\mathbf{u}_{20} + \frac{2}{h^2}\mathbf{A}_{xy00}\mathbf{u}_{21} + \frac{2}{h^2}\mathbf{A}_{yy00}\mathbf{u}_{22} + \mathbf{f}_{00}
\end{aligned}
\tag{A.3}
$$

and

$$
\begin{aligned}
\mathcal{R}_{10} &= \mathbf{A}_{10}\mathbf{u}_{00} + \left(\mathbf{A}_{00} + \frac{1}{h}\mathbf{A}_{x10}\right)\mathbf{u}_{10} + \frac{1}{h}\mathbf{A}_{y10}\mathbf{u}_{11} \\
&\quad + \frac{2}{h}\left(\mathbf{A}_{x00} + \frac{1}{h}\mathbf{A}_{xx10}\right)\mathbf{u}_{20} + \frac{1}{h}\left(\mathbf{A}_{y00} + \frac{2}{h}\mathbf{A}_{xy10}\right)\mathbf{u}_{21} + \frac{2}{h^2}\mathbf{A}_{yy10}\mathbf{u}_{22} \\
&\quad + \frac{6}{h^2}\mathbf{A}_{xx00}\mathbf{u}_{30} + \frac{4}{h^2}\mathbf{A}_{xy00}\mathbf{u}_{31} + \frac{2}{h^2}\mathbf{A}_{yy00}\mathbf{u}_{32} + \mathbf{f}_{10}.
\end{aligned}
\tag{A.4}
$$

Similarly we get for the third coefficient

$$\mathcal{R}_{11} = \mathbf{A}_{11}\mathbf{u}_{00} + \frac{1}{h}\mathbf{A}_{x11}\mathbf{u}_{10} + \left(\mathbf{A}_{00} + \frac{1}{h}\mathbf{A}_{y11}\right)\mathbf{u}_{11}$$

$$+ \frac{2}{h^2}\mathbf{A}_{xx11}\mathbf{u}_{20} + \frac{1}{h}\left(\mathbf{A}_{x00} + \frac{2}{h}\mathbf{A}_{xy11}\right)\mathbf{u}_{21} + \frac{2}{h}\left(\mathbf{A}_{y00} + \frac{1}{h}\mathbf{A}_{yy11}\right)\mathbf{u}_{22}$$

$$+ \frac{2}{h^2}\mathbf{A}_{xx00}\mathbf{u}_{31} + \frac{4}{h^2}\mathbf{A}_{xy00}\mathbf{u}_{32} + \frac{6}{h^2}\mathbf{A}_{yy00}\mathbf{u}_{33} + \mathbf{f}_{11}.$$

$$(A.5)$$

APPENDIX B

In connection with the power series method the paper explains only very briefly, how the coefficients $\mathcal{R}_{Ni}$ and $\mathcal{R}_{Di}$ of the polynomial representations (28) of the residuals of the boundary conditions can be expressed as (linear) functions of the unknown parameters $\mathbf{U}^p$. This appendix demonstrates in detail how this is done in connection Neumann boundary condition (N) and first two coefficients $\mathcal{R}_{N0}$ and $\mathcal{R}_{N1}$. The same idea can be used in connection with Dirichlet boundary condition (D) and for subsequent coefficients. (Reference [7] presents more details of the subject.)

Equations (28a) and (27a) give for the first two coefficients

$$\mathcal{R}_{N0} = \mathcal{R}_N(s_0) = \mathbf{B}_x(s_0)\frac{\partial \mathbf{u}}{\partial x}(x_0, y_0) + \mathbf{B}_y(s_0)\frac{\partial \mathbf{u}}{\partial y}(x_0, y_0) + \mathbf{g}(s_0) \qquad (B.1)$$

and

$$\mathcal{R}_{N1} = h\frac{d\mathcal{R}_N}{ds}(s_0)$$

$$= h\left\{ \frac{d\mathbf{B}_x}{ds}(s_0)\frac{\partial \mathbf{u}}{\partial x}(x_0, y_0) + \frac{d\mathbf{B}_y}{ds}(s_0)\frac{\partial \mathbf{u}}{\partial y}(x_0, y_0) + \mathbf{B}_x(s_0)\frac{dx}{ds}(s_0)\frac{\partial^2 \mathbf{u}}{\partial x^2}(x_0, y_0) \right.$$

$$+ \left[\mathbf{B}_x(s_0)\frac{dy}{ds}(s_0) + \mathbf{B}_y(s_0)\frac{dx}{ds}(s_0)\right]\frac{\partial^2 \mathbf{u}}{\partial x\partial y}(x_0, y_0) + \mathbf{B}_y(s_0)\frac{dy}{ds}(s_0)\frac{\partial^2 \mathbf{u}}{\partial y^2}(x_0, y_0)$$

$$\left. + \frac{d\mathbf{g}}{ds}(s_0) \right\}.$$

$$(B.2)$$

Using now equation (32) and similar equations for the known coefficients $\mathbf{B}_{xi}, ..., \mathbf{g}_i$ and equation (12) for the unknown coefficients $\mathbf{u}_{ij}$ we get

$$\mathcal{R}_{N0} = \mathcal{R}_N(s_0) = \frac{1}{h}\mathbf{B}_{x0}\mathbf{u}_{10} + \frac{1}{h}\mathbf{B}_{y0}\mathbf{u}_{11} + \mathbf{g}_0 \qquad (B.3)$$

and

$$\mathcal{R}_{N1} = h\frac{d\mathcal{R}_N}{ds}(s_0)$$

$$= \frac{1}{h}\mathbf{B}_{x1}\mathbf{u}_{10} + \frac{1}{h}\mathbf{B}_{y1}\mathbf{u}_{11} + \frac{2}{h^2}\mathbf{B}_{x0}x_1\mathbf{u}_{20} \qquad (B.4)$$

$$+ \frac{1}{h^2}\left[\mathbf{B}_{x0}y_1 + \mathbf{B}_{y0}x_1\right]\mathbf{u}_{21} + \frac{2}{h^2}\mathbf{B}_{y0}y_1\mathbf{u}_{22} + \mathbf{g}_1.$$

Superconvergence similarities in standard and mixed finite element methods

JAN H. BRANDTS   University of Jyväskylä, Laboratory of Scientific Computing, P.O. Box 35, FIN–40351 Jyväskylä, Finland, e-mail: brandts@math.jyu.fi

Abstract   In this presentation, we will show that it is possible to tackle some superconvergence questions in finite elements for the standard and mixed setting simultaneously. The similarities can already seen to be present in the simple one-dimensional case. Essentially, they become clear if in the standard setting, the finite element solution is not (as is traditionally done) compared to Lagrangian interpolants of the exact solution, but to the local interpolation operator of which the curl is equal to the Fortin interpolation of the curl. In two space dimensions, this approach results in relatively simple proofs for superconvergence results for the mixed and standard elements in one go, for example, the linear and quadratic triangular elements and all order rectangular elements.

1   INTRODUCTION

Traditionally, superconvergence for the standard finite element method has been tackled as follows. First, one defines a simple, local interpolation scheme for the exact solution into the finite element space. Then, one proves that the difference between the finite element solution and the interpolant of the exact solution is of higher order than both are to the exact solution (in some norm). This is often the most difficult part, and it can be expressed as:

$$\|\mathrm{FE}(u) - \mathrm{I}(u)\| = \sigma\left(\|u - \mathrm{I}(u)\|\right) \quad (h \to 0). \tag{1}$$

Then one studies the possibility to post-process this local interpolant, which is the more easy task because of this local character:

$$\|u - \mathrm{PP} \circ \mathrm{I}(u)\| = \sigma\left(\|u - \mathrm{I}(u)\|\right) \quad (h \to 0). \tag{2}$$

Finally, one proves that this post-processing scheme is bounded in the norm used above. This is sufficient to transfer the post-processability of the local interpolation scheme to the finite element method, resulting in higher order accuracy and superconvergence-based a posteriori error estimation:

$$\|u-\mathrm{PP}\circ\mathrm{FE}(u)\| \leq \|u-\mathrm{PP}\circ\mathrm{I}(u)\|+\|\mathrm{PP}\|\|\mathrm{FE}(u)-\mathrm{I}(u)\| = \sigma\left(\|u-\mathrm{I}(u)\|\right) \quad (h \to 0).$$

Apart from this view on superconvergence, there is also the "pointwise" view on superconvergence. Research in this direction aims to identify points in the domain in which the rate of convergence is exceptional compared to most of the other points. We refer to [10], [13], [24], [23]. In spite of the fact that much progress has been made in this direction (cf. [23]), we will, in this paper, concentrate on the global-norm superconvergence.

For simple elliptic problems and on rectangular or uniform triangular grids, the global-norm approach has been successful in several cases. We refer to [20], [11], [15], [16], [14], [12], [2], [18]. Although at first sight it might seem that the techniques used are different for the standard finite element and the mixed finite element settings, there are close relations between the two, and indeed, for some cases, the results follow simultaneously or are equivalent.

First, we will give an introduction in one dimension in Sect. 2, which will already show some of the connections, then we move on to the two-dimensional case. After some preliminaries in Sect. 3 we will give an overview of the results and possibilities of the combined approach in Sect. 4, supported by some numerical experiments.

2 ONE-DIMENSIONAL CONSIDERATIONS

Before we turn to the more interesting two-dimensional case, we will briefly study standard and mixed methods in one dimension. We recall the well-known fact that for linear elements, these methods reduce to simple Lagrange interpolation. It is also well-known that for the same problem in *two* space dimensions, the continuous piecewise linear finite element approximation is, in general, *not* equal to the continuous piecewise linear interpolant. We will – informally – explain why.

2.1 Continuous piecewise harmonic approximation

Consider the two point boundary value problem $-u'' = f$ with homogeneous Dirichlet boundary conditions on the unit interval I. Its (equivalent) weak formulation is to find a function $u \in H_0^1(I)$ such that:

$$\forall v \in H_0^1(I) : \quad \left(\frac{\partial}{\partial x}u, \frac{\partial}{\partial x}v\right)_I = (f, v)_I, \qquad (3)$$

and its continuous piecewise linear finite element discretization to find a function $u_h \in V_{0h}^1$ (the continuous piecewise linears that are zero on the boundary) such that:

$$\forall v_h \in V_{0h}^1 : \quad \left(\frac{\partial}{\partial x}u_h, \frac{\partial}{\partial x}v_h\right)_I = (f, v_h)_I. \qquad (4)$$

Now, denote by $\pi_h^1(u) \in V_{0h}^1$ the continuous piecewise linear interpolant of u interpolating at the grid points. Then, using integration by parts, we have that on each subinterval $K =]p, q[$ and for each $v_h \in V_{0h}^1$:

$$\left(\frac{\partial}{\partial x} \pi_h^1(u), \frac{\partial}{\partial x} v_h \right)_K = \pi_h^1(u) \frac{\partial}{\partial x} v_h \bigg|_p^q = u \frac{\partial}{\partial x} v_h \bigg|_p^q = \left(\frac{\partial}{\partial x} u, \frac{\partial}{\partial x} v_h \right)_K. \quad (5)$$

Summing over all subintervals K and combination of eqs. (3) and (4) yields that $\pi_h^1(u)$ satisfies the finite element equation. Uniqueness of the finite element solution results in *perfect* superconvergence: $\pi_h^1(u) = u_h$.

OBSERVATION 2.1 The property essential for perfect superconvergence is not the piecewise linearity, but the fact that the approximating functions are locally harmonic, or, they locally satisfy the homogeneous differential equation. Of course, for our simple model problem these three properties coincide. $\quad\square$

Consider for example the following generalization of our model problem for higher space dimensions, this is, the Poisson problem

$$-\Delta u = f \quad \text{in } \Omega, \qquad u = 0 \quad \text{on } \partial\Omega. \quad (6)$$

Introduce the space $\mathcal{H}_h(\mathcal{T}_h)$ of continuous, piecewise harmonic functions relative to the partitioning $\mathcal{T}_h$ of the domain. Given some function u smooth enough, define its continuous piecewise harmonic interpolant as the function $H_h(u) \in \mathcal{H}_h$ satisfying the following uniquely solvable local Laplace equations:

$$\forall K \in \mathcal{T}_h : \quad -\Delta H_h(u) = 0 \quad \text{on } K, \quad H_h(u) = u \quad \text{on } \partial K. \quad (7)$$

Entirely similar to the one-dimensional case we obtain, by the Green identity, that for arbitrary $K \in \mathcal{T}_h$ and $v_{\mathcal{H}} \in \mathcal{H}_h$ that:

$$\begin{aligned} (\mathbf{grad}\, H_h(u), \mathbf{grad}\, v_{\mathcal{H}})_K &= \langle H_h(u), (\mathbf{grad}\, v_{\mathcal{H}})^{\mathrm{T}} \nu \rangle_{\partial K} \\ &= \langle u, (\mathbf{grad}\, v_{\mathcal{H}})^{\mathrm{T}} \nu \rangle_{\partial K} = (\mathbf{grad}\, u, \mathbf{grad}\, v_{\mathcal{H}})_K = (f, v_{\mathcal{H}})_\Omega. \end{aligned} \quad (8)$$

So, this continuous piecewise harmonic interpolant satisfies the finite element formulation with approximating space $\mathcal{H}_h$) of the Poisson problem (6).

Of course, our misfortune is that the space of continuous piecewise harmonics is not *finite dimensional* for the dimension of the domain greater than one. Since the space of continuous piecewise linears is only a (finite dimensional) subspace of $\mathcal{H}_h$, we do not expect this space to have all the favorable properties of its one-dimensional counterpart.

2.2 Relation with Green's function

Instead of solving a large linear system to obtain the values of u in, say, N points of the domain, we prefer of course to solve a one-by-one system N times and take therefore as approximating space the space generated by the one base function $v_y (y \in]0, 1[)$ defined by

$$v_y(x) = \begin{cases} y^{-1} x & (0 \le x \le y), \\ (y-1)^{-1} x + (1-y)^{-1} & (y \le x \le 0). \end{cases} \quad (9)$$

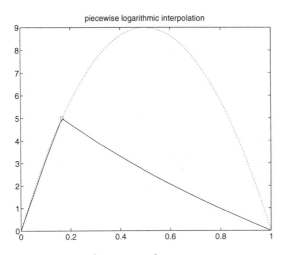

Figure 1 $-\frac{\partial}{\partial x}((1+x)\frac{\partial}{\partial x}u) = 36(1+4x)$. Approximating the solution of this equation using continuous functions that are piecewise of the form $c_1 \log(1+x) + c_2$ yields equality in the grid points. In this picture we only have one interior grid point.

Then the finite element system reduces to the calculation of the one coefficient that equals the value of u_h at $x = y$, since $v_y(y) = 1$: set $u_h = u_h(y)v_y(x)$ and solve:

$$\left(u_h(y)\frac{\partial}{\partial x}v_y, \frac{\partial}{\partial x}v_y\right)_I = (f, v_y)_I. \tag{10}$$

By equality of u and u_h in the grid point $x = y$ we find a familiar formula:

$$u(y) = u_h(y) = \left(f, v_y\left\|\frac{\partial}{\partial x}v_y\right\|_{0,I}^{-2}\right)_I = \int_I f(x)K(x,y)dy,$$

with

$$K(x,y) = \left\|\frac{\partial}{\partial x}v_y\right\|_{0,I}^{-2} \cdot v_y(x). \tag{11}$$

Therefore, in dimension one we can often force equality in the nodes by adapting the finite element space, since Green's function is readily available. An example is shown in Figure 1.

For the more complicated two-dimensional problems, in particular those defined on geometrically complicated domains, we will not have Green's function at hand, and it is not realistic to expect equality in nodes or on edges.

2.3   Mixed elements in one dimension

Obviously, we can rewrite the boundary value problem $-u'' = f$ into a system of two first order equations by introducing a second variable $\mathbf{p}$:

$$-\frac{\partial}{\partial x}u = \mathbf{p}, \qquad \frac{\partial}{\partial x}\mathbf{p} = f \quad \text{in } I, \qquad u = 0 \quad \text{on } \partial I. \tag{12}$$

Testing the first equation in $H^1(I)$ and the second in $L^2(I)$ results in a weak formulation suitable for discretization by mixed finite elements, which aims to find $(u, \mathbf{p}) \in L^2(I) \times H^1(I)$ such that:

$$\forall \mathbf{q} \in H^1(I): \quad (\mathbf{p}, \mathbf{q}) - \left(u, \frac{\partial}{\partial x}\mathbf{q}\right) = 0, \tag{13}$$

$$\forall w \in L^2(I): \quad \left(\frac{\partial}{\partial x}\mathbf{p}, w\right) = (f, w). \tag{14}$$

Note that the first equation (13) imposes the homogeneous Dirichlet boundary conditions on u. Now, take a subspace $V_{0h} \subset H_0^1(I)$ and set $V_h = \mathcal{H}(I) + V_{0h}$ and $W_h = \frac{\partial}{\partial x}V_h$, where $\mathcal{H}(I) = \mathcal{P}^1(I)$ is the space of linear functions on I, for the search for approximations $u_h \in W_h$ and $\mathbf{p}_h \in V_h$ of u and $\mathbf{p}$. Then one can verify the unique solvability of the following discrete system, which aims to find $(u_h, \mathbf{p}_h) \in W_h \times V_h$ such that:

$$\forall \mathbf{q}_h \in V_h: \quad (\mathbf{p}_h, \mathbf{q}_h) - \left(u_h, \frac{\partial}{\partial x}\mathbf{q}_h\right) = 0, \tag{15}$$

$$\forall w \in W_h: \quad \left(\frac{\partial}{\partial x}\mathbf{p}_h, w_h\right) = (f, w_h). \tag{16}$$

Now, for the analysis to follow we will define an operator π_h that represents the standard finite element method for $-u'' = f$ of which we saw that it yields equality on the edges of the elements, which turns it into a *local* operator.

DEFINITION 2.2   Denote by π_h the operator $H^1(I) \to V_h : v \mapsto \pi_h v$ satisfying:

$$\frac{\partial}{\partial x}(\pi_h v) = P_h\left(\frac{\partial}{\partial x}v\right), \quad \pi_h v = v \quad \text{on } \partial I. \tag{17}$$

where P_h is L^2-orthogonal projection on W_h.   □

By equation (16) above and the continuity of $\mathbf{p}_h$ and $\pi_h\mathbf{p}$ it follows that there exists a constant function $C(\mathbf{p}, h)$ such that:

$$\frac{\partial}{\partial x}\mathbf{p}_h = P_h\left(\frac{\partial}{\partial x}\mathbf{p}\right) \quad \text{and} \quad \mathbf{p}_h - \pi_h\mathbf{p} = C(\mathbf{p}, h). \tag{18}$$

REMARK 2.3   Note that if we would have imposed homogeneous Neumann boundary conditions, i.e., $\mathbf{p} = 0$ on ∂I, we would have searched and tested in $H_0^1(I)$ and V_{0h} for $\mathbf{p}$ and $\mathbf{p}_h$, resulting in $C(\mathbf{p}, h) = 0$, immediately giving perfect superconvergence for $\mathbf{p}_h$ with respect to $\pi_h\mathbf{p}$.   □

Subtraction of equation (15) from (13) for a test function $\mathbf{q}_h \in V_h$ gives the error equation:

$$\forall \mathbf{q}_h \in V_h: \quad (\mathbf{p} - \mathbf{p}_h, \mathbf{q}_h) - \left(u - u_h, \frac{\partial}{\partial x}\mathbf{q}_h\right) = 0. \tag{19}$$

Substituting $\mathbf{q}_h = \mathbf{p}_h - \pi_h \mathbf{p} \in V_h$ gives $(\mathbf{p} - \mathbf{p}_h, \mathbf{p}_h - \pi_h \mathbf{p}) = 0$ and therefore:

$$\|\mathbf{p}_h - \pi_h \mathbf{p}\|_{0,I}^2 = (\mathbf{p}_h - \pi_h \mathbf{p}, \mathbf{p} - \pi_h \mathbf{p}). \tag{20}$$

Since $\mathbf{p} - \pi_h \mathbf{p} \in H_0^1(I)$, integration by parts and the "commuting diagram property" of π_h and P_h (cf. Def. 2.2) result in:

$$\|\mathbf{p}_h - \pi_h \mathbf{p}\|_{0,I}^2 = -\left(\eta, \frac{\partial}{\partial x}\mathbf{p} - \mathrm{P}_h\left(\frac{\partial}{\partial x}\mathbf{p}\right)\right), \tag{21}$$

which is obviously zero whenever $\mathcal{H}(I) \subset W_h$ resulting in perfect superconvergence. If $\mathcal{H}(I) \not\subset W_h$, application of the Schwarz inequality after approximating η by its projection on W_h results in:

$$
\begin{aligned}
&\|\mathbf{p}_h - \pi_h \mathbf{p}\|_{0,I}^2 \\
&= \left|\left(\eta - \mathrm{P}_h\eta, \frac{\partial}{\partial x}\mathbf{p} - \mathrm{P}_h\left(\frac{\partial}{\partial x}\mathbf{p}\right)\right)\right| \\
&\leq \underline{\mathrm{gap}}(\mathcal{H}(I), W_h, \|\cdot\|_{0,I}, |\cdot|_{1,I})\|\mathbf{p}_h - \pi_h \mathbf{p}\|_{0,I}\left\|\frac{\partial}{\partial x}\mathbf{p} - \mathrm{P}_h\left(\frac{\partial}{\partial x}\mathbf{p}\right)\right\|_{0,I}.
\end{aligned}
\tag{22}
$$

THEOREM 2.4   Suppose we approximate the solution of $-u'' = f$ in one space dimension using mixed finite elements with approximating spaces W_h and V_{0h} (corresponding to homogeneous Neumann boundary conditions). Then we have that

$$\mathbf{p}_h = \pi_h \mathbf{p}. \tag{23}$$

If homogeneous Dirichlet boundary conditions are imposed, i.e., we use the space V_h instead of V_{0h}, we only have that:

$$\|\mathbf{p}_h - \pi_h \mathbf{p}\|_{0,I} \leq \underline{\mathrm{gap}}(\mathcal{H}(I), W_h, \|\cdot\|_{0,I}, |\cdot|_{1,I})\left\|\frac{\partial}{\partial x}\mathbf{p} - \mathrm{P}_h^k\left(\frac{\partial}{\partial x}\mathbf{p}\right)\right\|_{0,I} \tag{24}$$

which results in (23) if $\mathcal{H}(I) \subset W_h$.   □

For the usual choice of finite element spaces, the continuous piecewise polynomials V_h^{k+1} of degree $k + 1$ for V_h and all the piecewise degree k polynomials W_h^k for W_h, this results in the remarkable fact that only for the lowest order method, this approach does *not* yield superconvergence.

COROLLARY 2.5   Taking $W_h = W_h^k$, $V_h = V_h^{k+1}$, and defining $\pi_h = \pi_h^{k+1}$ and $\mathrm{P}_h = \mathrm{P}_h^k$ according to Definition 2.2 for some $k \geq 1$, is sufficient to obtain $\mathcal{H}(I) \subset W_h$, resulting in (23). For $k = 0$ however, it only gives that:

$$\|\mathbf{p}_h - \pi_h^{k+1}\mathbf{p}\|_{0,I} \leq Ch\left\|\frac{\partial}{\partial x}\mathbf{p} - \mathrm{P}_h^0\left(\frac{\partial}{\partial x}\mathbf{p}\right)\right\|_{0,I} \leq Ch^2|\mathbf{p}|_{2,I}, \tag{25}$$

which is of the same order of h as the a priori estimates for $\mathbf{p}_h$ and $\pi_h^{k+1}\mathbf{p}$ with respect to $\mathbf{p}$, and thus not superconvergent.   □

In fact, according to numerical experiments, the bound (25) can not be improved in general. We refer to [6] for those experiments and for more experimental results on superconvergence and a posteriori error estimation in this situation, also for u.

2.4 Conclusions and remarks for dimension one

Below we will list some of the characteristic features of our one-dimensional analysis that will also return in the two-dimensional analysis.

- In MFE, the homogeneous Neumann boundary conditions were he easiest to analyze, as opposed to the SFE, where the homogeneous Dirichlet boundary conditions are the easiest to analyze.

- For the usual piecewise polynomials, the estimate (22) leads to superconvergence for the mixed Dirichlet problem except for the lowest order method (in this case for $k = 0$).

- Local averaging/projection techniques can be applied to post-process the superconvergent approximations in order to get asymptotically exact a posteriori error estimators. The techniques do not depend on the origin of the approximation, i.e., from SFE or MFE.

- Semi-discrete parabolic and hyperbolic systems can be proved to inherit (most of) the superconvergence properties from the elliptic operator in space by means of (mixed-) elliptic projection and the Gronwall inequality.

- The operator π_h can be considered as an elliptic projection, but then as a special one in the sense that it has become *local* in character because of the equality in the grid points.

- Similar two point boundary value problems, also with non-constant coefficients, will not always give equality in the gridpoints, but will stay superconvergent. Using Green's function, the equality can probably be forced back in most cases.

3 TWO-DIMENSIONAL PRELIMINARIES

3.1 Sobolev spaces

Suppose Ω is a bounded open convex polygonal domain in $\mathbb{R}^2$. Let $\mathbf{grad} = (\frac{\partial}{\partial x}, \frac{\partial}{\partial y})^{\mathrm{T}}$, $\mathbf{curl} = (-\frac{\partial}{\partial y}, \frac{\partial}{\partial x})^{\mathrm{T}}$, $\mathrm{div}(q_1, q_2)^{\mathrm{T}} = \frac{\partial}{\partial x} q_1 + \frac{\partial}{\partial y} q_2$ and $\mathrm{rot}(q_1, q_2)^{\mathrm{T}} = \frac{\partial}{\partial y} q_1 - \frac{\partial}{\partial x} q_2$. Denote by $H^k(\Omega)$ the usual Sobolev spaces of order k, normed and seminormed by $\| \cdot \|_{k,\Omega}$ and $| \cdot |_{k,\Omega}$. Write $L^2(\Omega) = H^0(\Omega)$. Denote by $\mathbf{H}(\mathrm{div}; \Omega)$ the space of $L^2(\Omega)$-vector fields of which the divergence exists in the weak sense and is also in $L^2(\Omega)$. Let $\mathbf{H}_0(\mathrm{div}; \Omega)$ be the subspace of $\mathbf{H}(\mathrm{div}; \Omega)$ of fields having – essentially – zero normal component on $\partial\Omega$.

3.2 Discrete spaces

We will use the Raviart–Thomas spaces $\mathbf{\Gamma}_h^k \subset \mathbf{H}(\mathrm{div}; \Omega)$ (for details, cf. [22]) relative to a family $(\mathcal{T}_h)_h$ of triangulations of Ω, the parameter h indicating the grid size. We remind the reader that the Raviart–Thomas spaces are piecewise polynomial vector fields with continuous components normal to the edges of the triangles across those edges, which ensures they belong to $\mathbf{H}(\mathrm{div};\Omega)$. The space $W_h^k = \mathrm{div}(\boldsymbol{\Gamma}_h^k)$ is the space of all piecewise degree k polynomials relative to the triangulation (without any continuity constraint across the boundaries of the elements). Define the commonly used standard finite element spaces $V_h^k = W_h^k \cap H^1(\Omega)$ and $V_{0h}^k = W_h^k \cap H_0^1(\Omega)$. Finally, set $\boldsymbol{\Gamma}_{0h}^k = \boldsymbol{\Gamma}_h^k \cap \mathbf{H}_0(\mathrm{div};\Omega)$.

THEOREM 3.1 ([21], Discrete Helmholtz decomposition.) The space $\boldsymbol{\Gamma}_{0h}^k$ satisfies the following complete and orthogonal decomposition:

$$\boldsymbol{\Gamma}_{0h}^k = \mathbf{grad}_{0h} W_h^k \oplus \mathbf{curl}\, V_{0h}^{k+1}. \tag{26}$$

where $\mathbf{grad}_{0h} : W_h^k \to \boldsymbol{\Gamma}_{0h}^k : w_h \mapsto \mathbf{grad}_{0h}\, w_h$ is defined by

$$\forall \mathbf{q}_h \in \boldsymbol{\Gamma}_{0h}^k : \quad (\mathbf{grad}_{0h}\, w_h, \mathbf{q}_h)_\Omega = -(w_h, \mathrm{div}\, \mathbf{q}_h)_\Omega. \tag{27}$$

For the proof and details on the operator $\mathbf{grad}_{0h}$ we refer to [4], [21].   □

REMARK 3.2 A useful ingredient in the proof of Th. 3.1 (and elsewhere) is the following property of Raviart–Thomas space:

$$\forall \mathbf{q}_h \in \boldsymbol{\Gamma}_{0h}^k : \mathrm{div}\, \mathbf{q}_h = 0 \Rightarrow \mathbf{q}_h \in \mathbf{curl}\, V_{0h}^{k+1} \tag{28}$$

3.3 Interpolation and projection operators

We will now give the definition of Fortin's interpolation $\boldsymbol{\Pi}_h^k : [H^1(\Omega)]^2 \to \boldsymbol{\Gamma}_h^k$, which always turns out to be a useful tool in the analysis of the mixed finite element method (cf. [9], [12], [22]) because it satisfies the so called *commuting diagram property*, i.e., it is valid that for all vector fields $\mathbf{q} \in [H^1(\Omega)]^2$:

$$\mathrm{div}\, \boldsymbol{\Pi}_h^k \mathbf{q} = \mathrm{P}_h^k\, \mathrm{div}\, \mathbf{q}, \tag{29}$$

where by P_h^k we mean the $L^2(\Omega)$-orthogonal projection on the space W_h^k.

DEFINITION 3.3 (Fortin interpolation.) For $\mathbf{q} \in [H^1(\Omega)]^2$ we define $\boldsymbol{\Pi}_h^k \mathbf{q}$ to be the unique vector field in $\boldsymbol{\Gamma}_h^k$ for which on each triangle K with edges $\partial K_1, \partial K_2, \partial K_3$ (with the convention $\mathcal{P}^{-1}(K) = \emptyset$):

$$\forall i \in \{1,2,3\} : \forall \phi \in \mathcal{P}^k(\partial K_i) : \quad \langle (\boldsymbol{\Pi}_h^k \mathbf{q})^{\mathrm{T}} \nu, \phi \rangle_{\partial K_i} = \langle \mathbf{q}^{\mathrm{T}} \nu, \phi \rangle_{\partial K_i}, \tag{30}$$

$$\forall \mathbf{r} \in [\mathcal{P}^{k-1}(K)]^2 : \quad (\boldsymbol{\Pi}_h^k \mathbf{q}, \mathbf{r})_K = (\mathbf{q}, \mathbf{r})_K. \tag{31}$$

Restricted to $\mathbf{H}_0(\mathrm{div};\Omega)$ it maps into $\boldsymbol{\Gamma}_{0h}^k$. Its approximation properties are optimal in order of h:

PROPOSITION 3.4   ([**22**, Th. 3, p. 303]) Suppose the family $(\mathcal{T}_h)_h$ of triangulations of Ω is regular. Then there exists a constant C_F independent of h such that:

$$\forall T \in \mathcal{T}_h : \forall \mathbf{q} \in [H^{k+1}(\Omega)]^2 : |\mathbf{q} - \mathbf{\Pi}_h^k \mathbf{q}|_{0,T} \leq C_F h^{k+1} |\mathbf{q}|_{k+1,T}. \tag{32}$$

Now, by the commuting diagram property (29) we have that for all $v \in H^2(\Omega) \cap H_0^1(\Omega)$:

$$\operatorname{div} \mathbf{\Pi}_h^k \operatorname{\mathbf{curl}} v = \mathbf{\Pi}_h^k \operatorname{div} \operatorname{\mathbf{curl}} v = 0 \tag{33}$$

which by Remark 3.2 means that $\mathbf{\Pi}_h^k \operatorname{\mathbf{curl}} v \in \operatorname{\mathbf{curl}} V_{0h}^{k+1}$. This gives rise to the following definition.

DEFINITION 3.5   ([**4**, Def. 2.7], Fortin interpolation.) We define "Fortin interpolation" (the lower case f to distinguish it from the upper case interpolant above) as

$$\pi_{0h}^{k+1} : H^2(\Omega) \cap H_0^1(\Omega) \to V_{0h}^{k+1} : v \mapsto \pi_{0h}^{k+1} v \tag{34}$$

where $\pi_{0h}^{k+1} u$ is the unique element from V_{0h}^{k+1} satisfying

$$\mathbf{\Pi}_h^k \operatorname{\mathbf{curl}} v = \operatorname{\mathbf{curl}} \pi_{0h}^{k+1} v. \tag{35}$$

For $k = 1$, this *local* interpolation reduces to linear Lagrange interpolation on the nodes. For $k = 2$, it is a higher order perturbation of quadratic Lagrange interpolation on the nodes and the midpoints of the edges. For $k \geq 3$, it is essentially different from Lagrange interpolation. Its approximation properties in the $|\cdot|_{1,\Omega}$ semi-norm are of course those of Fortin interpolation (cf. Prop. 3.4). In the L^2-norm, they are as follows:

LEMMA 3.6   ([**4**, Lem. 2.9.] Let $k \geq 1$. Then there exist constants C_f independent of u and h such that for each $u \in H^{k+1}(\Omega) \cap H_0^1(\Omega)$:

$$|u - \pi_{0h}^k u|_{0,\Omega} \leq C_f h^{k+1} |u|_{k+1,\Omega}. \tag{36}$$

3.4   Standard and mixed method for Poisson's equation

Denote by $U_h \in V_{0h}^{k+1}$ the standard finite element approximation of the exact solution U of the Poisson equation with homogeneous Dirichlet boundary conditions and right-hand side f. Let $\mathbf{p}_h \in \mathbf{\Gamma}_{0h}^k$ be the mixed finite element approximation of the gradient $\operatorname{grad} u$ of the exact solution u of the Poisson equation with homogeneous Neumann boundary conditions and right-hand side f. We will not write down the discrete equations that are satisfied by those approximations, since they are quite similar to the equations in the one-dimensional case in the previous section, apart from the fact that now we deal with gradients and divergences in the formulation instead of mere (line-)derivatives. We refer to the standard literature.

3.5   Unified approach to superconvergence

Consider the standard finite method for Poisson's equation. Write $\chi_h = U_h - \pi_{0h}^{k+1} U$ and note that the substitution of **curl** for **grad** below is an isometry. Using the orthogonality of the error to the finite element space it follows that:

$$
\begin{aligned}
|\operatorname{\mathbf{grad}} \chi_h|_{0,\Omega}^2 = |\operatorname{\mathbf{curl}} \chi_h|_{0,\Omega}^2 &= (\operatorname{\mathbf{curl}} U_h - \operatorname{\mathbf{curl}} \pi_{0h}^{k+1} U, \operatorname{\mathbf{curl}} \chi_h)_\Omega \\
&= (\operatorname{\mathbf{curl}} U - \operatorname{\mathbf{curl}} \pi_{0h}^{k+1} U, \operatorname{\mathbf{curl}} \chi_h)_\Omega \\
&= ((\mathbf{I} - \mathbf{\Pi}_h^k) \operatorname{\mathbf{curl}} U, \operatorname{\mathbf{curl}} \chi_h)_\Omega.
\end{aligned} \tag{37}
$$

For the mixed finite element method, by manipulation of the second error equation find out that $\operatorname{div} \mathbf{p}_h = \mathrm{P}_h^k \operatorname{div} \mathbf{p}$. Therefore, by the commuting diagram property Eq. (29) we find that $\mathbf{p}_h - \mathbf{\Pi}_h^k \mathbf{p}$ is divergence-free and hence, by Remark 3.2 that $\mathbf{p}_h - \mathbf{\Pi}_h^k \mathbf{p} = \operatorname{\mathbf{curl}} \omega_h$ for some $\omega_h \in V_{0h}^{k+1}$. Manipulation of the first error equation gives that $\mathbf{p} - \mathbf{p}_h$ is orthogonal to the divergence-free elements from $\mathbf{\Gamma}_{0h}^k$, which gives:

$$
\begin{aligned}
|\operatorname{\mathbf{curl}} \omega_h|_{0,\Omega}^2 = (\mathbf{p}_h - \mathbf{\Pi}_h^k \mathbf{p}, \operatorname{\mathbf{curl}} \omega_h)_\Omega \\
= (\mathbf{p} - \mathbf{\Pi}_h^k \mathbf{p}, \operatorname{\mathbf{curl}} \omega_h)_\Omega = ((\mathbf{I} - \mathbf{\Pi}_h^k) \operatorname{\mathbf{grad}} u, \operatorname{\mathbf{curl}} \omega_h)_\Omega.
\end{aligned} \tag{38}
$$

Now, both expressions (37) and (38) can be analyzed simultaneously and moreover independent from the method they originate from by studying the linear functional $\mathcal{F}(\mathbf{q}) \in (V_h^{k+1})^*$ (below) for some fixed but arbitrary vector field $\mathbf{q} \in [H^1(\Omega)]^2$. Here, $(V_h^{k+1})^*$ is the (topological) dual of V_h^{k+1}.

$$
\mathcal{F}(\mathbf{q}) : V_h^{k+1} \to \mathbb{R} : v_h \mapsto \mathcal{F}(\mathbf{q})(v_h) = ((\mathbf{I} - \mathbf{\Pi}_h^k)(\mathbf{q}), \operatorname{\mathbf{curl}} v_h)_\Omega. \tag{39}
$$

Of course this also defines an operator $\mathcal{F}$ in the obvious way:

$$
\mathcal{F} : [H^1(\Omega)]^2 \to (V_h^{k+1})^* : \mathbf{q} \mapsto \mathcal{F}(\mathbf{q}). \tag{40}
$$

The following theorem is the summary of our analysis so far. Its proof is to combine Eqs. (37), (38) and (39).

THEOREM 3.7   ([**4**, Th. 3.2.])   Suppose that on a family of triangulations $(\mathcal{T}_h)_h$, for all $\mathbf{q} \in [H^{k+1}(\Omega)]^2$ there exists a constant $C(\mathbf{q})$ independent of h and a function $r(h)$ with $\lim_{h\to\infty} r(h) = 0$ such that for all $v_h \in V_{0h}^{k+1} \subset V_h^{k+1}$:

$$
|\mathcal{F}(\mathbf{q})(v_h)| \le C(\mathbf{q}) r(h) |\mathbf{q} - \mathbf{\Pi}_h^k \mathbf{q}|_{0,\Omega} |v_h|_{1,\Omega}. \tag{41}
$$

Then both the standard finite element method for the Poisson equation with homogeneous Dirichlet boundary conditions as well as the mixed finite element method for the Poisson equation with homogeneous Neumann boundary conditions are superconvergent.   □

REMARK 3.8   If similar conditions are satisfied on V_h^{k+1} (as opposed to V_{0h}^{k+1}), superconvergence is also obtained for the other boundary conditions in each case (cf. [**4**]).

3.6   Summary of the two dimensional preliminaries

Commuting diagram

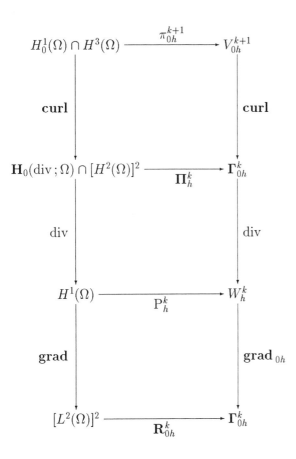

In the diagram above the essentials of this section are summarized. It is commuting throughout. The standard finite element operator maps from left to right on first level of this diagram (at the same position as π_{0h}^{k+1}), giving an approximation $U_h \in V_{0h}^{k+1}$ of a function u, and implicitly an approximation $\mathbf{curl}\, U_h \in \Gamma_{0h}^k$ of $\mathbf{curl}\, u$. The mixed finite element method can be considered as mapping from left to right on the second level, at the same position as $\mathbf{\Pi}_h^k$, giving an approximation $\mathbf{p}_h \in \mathbf{\Pi}_h^k$. On the lowest level, we have put the orthogonal projection $\mathbf{R}_{0h}^k$ on Γ_{0h}^k for completion.

We stess once again, that the operator π_{0h}^{k+1} is a local operator: it interpolates u at the nodes of each triangle, and can therefore be constructed (integrated) from the Fortin-interpolated curl of u.

4   OVERVIEW OF RESULTS IN THIS SETTING

As we saw in the previous section, studying the functional $\mathcal{F}(\mathbf{q})$ for some arbitrary vector field $\mathbf{q}$ can result in superconvergence with respect to Fortin interpolation for both standard and mixed finite element methods for Poisson's equation. Indeed, many of the known results for the one method can relatively easily be classified in this framework, and remarks can be made to the applicability of the result to the other method. We will give an overview of the state of affairs and give references to the literature, but before that we outline some main ideas and give some additional aspects of our approach that have not yet been mentioned.

4.1   Using the orthogonality properties

The basis of the success of our approach with Fortin interpolation, as opposed to Lagrange interpolation is the following simple proposition:

PROPOSITION 4.1   By the properties of Fortin's interpolation (cf. Def. 3.3) we have:

$$\forall \mathbf{q} \in [H^1(\Omega)]^2 : V_h^k \subset \operatorname{Ker} \mathcal{F}(\mathbf{q}) \tag{42}$$

$$[V_h^k]^2 \subset \operatorname{Ker} \mathcal{F}. \tag{43}$$

This proposition is an easy consequence of the definition of Fortin interpolation, that does not have its counterpart in the classical Lagrange interpolation approach. It gives us some – additional – space to play around and manipulate the inner product equations involved. For instance, the equality

$$\forall v_h \in V_h^{k+1} : \quad \mathcal{F}(\mathbf{q})(v_h) = \mathcal{F}(\mathbf{q})(v_h - \pi_h^k(v_h)) \tag{44}$$

makes that we can restrict ourselves in the analysis to considering the action of $\mathcal{F}(\mathbf{q})$ on hierarchical bases for the finite element spaces.

In order to be able to apply the Bramble Hilbert lemma (cf. [1]) at a certain stage, we need the polynomial fields of degree $k + 1$ to be in the kernel of $\mathcal{F}(\cdot)$, while we already know that they are for degree k. A certain amount of point-symmetry in the grid seems necessary to establish this: if the supports of the hierarchical basis functions are point-symmetric and restricted to a fixed number of elements, things work out fine again. For rectangular elements, this point-symmetry is always there, as well as for linear and quadratic triangular elements. For cubics on triangles it is not: the basis function that is zero on each of the element edges (and non-zero in the interior) does not have a pointsymmetric support, and the analysis fails. Not only the analysis fails, but also numerical experiments: the Fortin interpolant approach does not (immediately) seem to give superconvergence for the cubics, as we will see in Sect. 4.7.

4.2   Additional features of the approach

The additional features of our approach indicated above are that because of the orthogonality properties of Fortin interpolation, it is very easy to make duality estimates and obtain results for the scalar function too. That is, apart from the lowest order method (just as was the case in the one-dimensional setting). Under some regularity conditions, it can easily be proved that for the functions ω_h and χ_h it holds that:

THEOREM 4.2   ([4], Duality estimates.)

$$|\omega_h|_{0,\Omega} \leq Ch|\omega_h|_{1,\Omega} \quad \text{and} \quad |\chi_h|_{0,\Omega} \leq Ch|\chi_h|_{1,\Omega}. \tag{45}$$

As a consequence, the standard finite element approximation can also be seen to be L^∞ superclose to $\pi_h^{k+1}(u)$, by using the discrete inequality $|v_h|_{\infty,\Omega} \leq Ch^{-1}|v_h|_{0,\Omega}$. Moreover, since $\pi_h^{k+1}(u)$ is superclose to u at the nodes, and their respective tangential derivatives at the Gauss points on the edges of the elements, we can also obtain pointwise superconvergence results too. Those results will, however, not always be the strongest possible.

We will now proceed to summarize the results that can be derived using our approach. In the experiments we will mainly concentrate on the results for the mixed method, since the majority of the results for the standard method are already well-known.

4.3   About the numerical experiments

In the sequel, we will present numerical experiments that support the claims – as far as those claims are ours. Those experiments were carried out using MatLab 4.2c. This means that the pictures originate from the graphics toolbox from MatLab, and that the representation of piecewise linear quadratic and cubic functions is done by piecewise linear interpolation on the degrees of freedom. This makes the pictures slightly inaccurate, although the main features stay in tact.

4.4   Rectangular elements, all polynomial orders

Implicitly, in [12] the functional $\mathcal{F}(\mathbf{q})$ is studied, and superconvergence is derived in the mixed finite element setting. The results (and also the post-processing techniques given in the same paper) can however also be applied to the standard method, rediscovering some of the results of [11]. The duality estimates in the standard setting with $k \geq 1$ will result in L^2-superconvergence not only for the gradient but also for the scalar function.

4.5   Linear triangular elements $(k = 0)$

In [2], the functional $\mathcal{F}(\mathbf{q})$ was estimated supersmall if the grids considered form a regular and uniform family. Although attention was paid to the mixed method only,

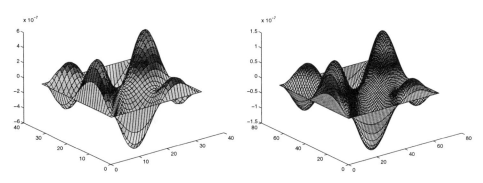

Figure 2   The function $\omega_h \in V_{0h}^1$ for 2048 resp. 8192 triangles.

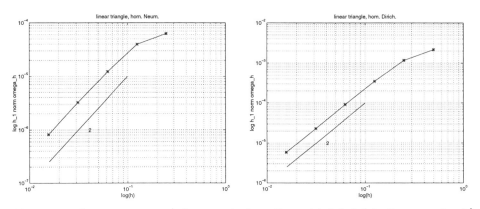

Figure 3   Superconvergence for ω_h. A short line with label k indicates order h^k behaviour.

the proofs are directly applicable to the standard method, rediscovering the well-known superconvergence for this case. In the analysis, considering the functional $\mathcal{F}(\mathbf{q})$ on the extended domain V_h^{k+1} instead of V_{0h}^{k+1} led to superconvergence for the standard Neumann problem and the mixed Dirichlet problem at the cost of an order $\sqrt{h}$. In numerical experiments however (see further on), this (partial) loss of superconvergence does not really show.

Homogeneous Neumann boundary conditions

In Figure 2, we see the function ω_h of which the curl is equal to $\mathbf{p}_h - \mathbf{\Pi}_h^k \mathbf{p}$, where we partitioned the unit square into squares and then put in the slope -1 diagonals. In Figure 3, a log-log plot of the L^2 norm of $\mathbf{curl}\,\omega_h$ is shown on the left. The exact solution was the polynomial $[x(1-x)y(1-y)]^2$.

In Figure 4, we see ω_h for a homogeneous Dirichlet problem. The exact solution was $[x(1-x)y(1-y)]^2$. Although this is the same exact solution as for the Neumann problem above, the mixed finite element formulation is different in the sense that $\mathbf{\Gamma}_{0h}^k$ has to be instead of $\mathbf{\Gamma}_h^k$. The theory in [2] gives only superconvergence of order $h\sqrt{h}$, but as we can see in the right picture of Figure 2, this experiment still shows order h^2 behaviour, just as for the experiment above. We did more experiments,

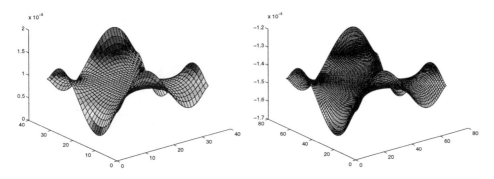

Figure 4   The function $\omega_h \in V_h^1$ for 2048 resp. 8192 triangles.

but neither one of them showed the $h\sqrt{h}$ behaviour. This is quite remarkable since the corresponding loss of $\sqrt{h}$ is indeed visible in case $k = 1$ (see below).

In both experiments, the preservation of the shape of the function ω_h is quite remarkable. It suggests that $h^{-2}\omega_h$ converges to some (fixed) function. It would be interesting to know which function, and whether in the standard case something similar is happening. Further research could clarify this, moreover, it might give an answer to the question:

Given a Poisson problem and the sequence of functions χ_h arising from standard finite element discretizations (note: the χ_h are not the approximations, but the "superconvergent difference" with the Fortin interpolant). Does there exist a Poisson problem such that the similar sequence ω_h that arises from the mixed discretization is *equal to* the sequence χ_h? For the homogeneous Poisson problem (i.e., the Dirichlet problem), this is true (cf. [4]).

4.6   Quadratic triangular elements ($k = 1$)

In [7], the functional $\mathcal{F}(\mathbf{q})$ was estimated supersmall if the grids considered form a regular and uniform family. The analysis was done explicitly for standard and mixed elements simultaneously. In contrast to the lowest order case, in numerical experiments it showed clearly the loss of a factor $\sqrt{h}$ when the boundary conditions were relaxed. We note that the post-processing mechanism for the standard method developed in [15] is applicable to the mixed setting too by replacing "tangential" components by "normal" components throughout. We plan to test the resulting a posteriori error estimator in the near future.

Homogeneous Neumann boundary conditions

In Figure 5, the function ω_h is shown, with the same grids and solution as before.

This time, since $k \geq 1$, we did not only calculate the semi-norm $|\omega_h|_{1,\Omega}$ but also $|\omega_h|_{0,\Omega}$. These are both shown on the left in Figure 6, and indeed, the superconvergence of order h^3 and h^4 respectively are both clearly visible.

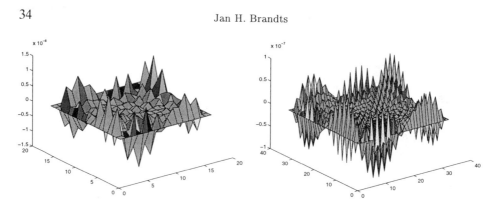

Figure 5   The function $\omega_h \in V_{0h}^2$ for 128 resp. 512 triangles.

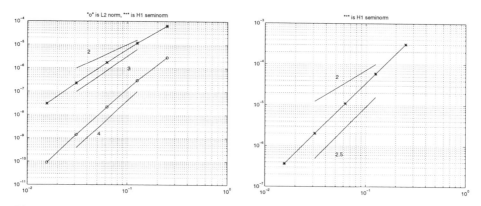

Figure 6   Superconvergence for ω_h in L^2 norm ('o') and H^1 seminorm ('*').

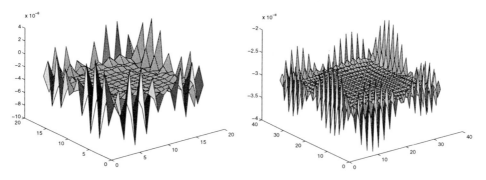

Figure 7   The function $\omega_h \in V_h^2$ for 128 resp. 512 triangles.

Homogeneous Dirichlet boundary conditions

The Dirichlet problem, with exact solution $[x(1-x)y(1-y)]^2$ gives the function ω_h as depicted in Figure 7.

According to the theoretical expectations (but contrary to our numerical experi-

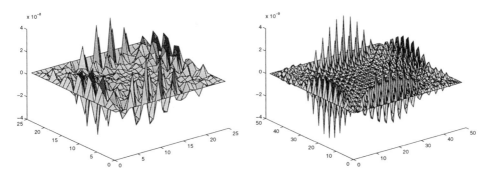

Figure 8   The function $\omega_h \in V_{0h}^3$ for 128 resp. 512 triangles.

ences in the lowest order case $k = 0$), we observed only superconvergence of order $h^2\sqrt{h}$, as is shown on the right of Figure 6.

4.7   Cubic triangular elements ($k = 2$)

As mentioned before, the method of proving superconvergence with respect to Fortin interpolation, that worked out quite well for linear and quadratic triangular elements, does not immediately apply to the cubic case. The reason was, that there does not (seem to) exist (additional) hierarchical basis functions consisting of odd functions with point-symmetric supports. Indeed, the cubic bubble function (zero on the edges of the triangle and non-zero in the interior) has a triangle as support, and a straightforward generalization of the principles in the proof comes to a halt. Numerical experiments suggest that superconvergence in the sense that $|\omega_h|_{1,\Omega}$ is supersmall, does not hold (which is a well-established fact for the standard method). We do not exclude the possibility that post-processing can be successful, though. There might still be points in which the rate of convergence is superhigh and those could be used in a suitable post-processing scheme.

Homogeneous Neumann boundary conditions

In Figure 8, we see again the function ω_h for the exact solution $[x(1-x)y(1-y)]^3$ where homogeneous Neumann boundary conditions are imposed, and in Figure 9 on the left, the corresponding (non)superconvergence plot. Indeed, the results look like order h^3 behaviour only, which is exactly the a priori convergence rate.

Homogeneous Dirichlet boundary conditions

Finally, in Figure 10 we see plots of the function ω_h where $[x(1-x)y(1-y)]^3$ is the exact solution and homogeneous Dirichlet boundary conditions are imposed. The corresponding (non)superconvergence plot is visible in Figure 9 on the right. Here too, the result does not seem to exceed the order order h^3 behaviour.

   In both cases, it seems from the picture that perhaps in the interior of the domain, things are less worse than they seem: by far the largest contributions to the semi-

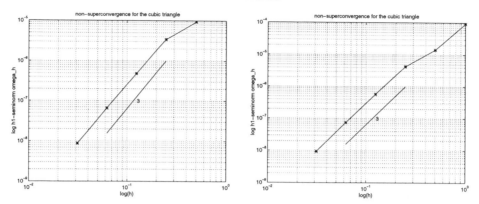

Figure 9   Non-superconvergence for ω_h in the H^1 seminorm ('*'). A short line with label k indicates order h^k behaviour.

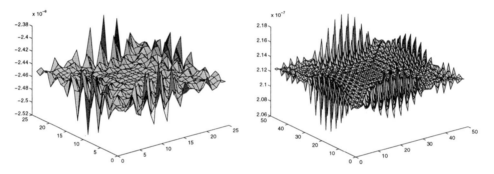

Figure 10   The function $\omega_h \in V_h^2$ for 128 resp. 512 triangles.

norm $|\omega_h|_{1,\Omega}$ come from the peaks near and at the boundary of the domain.

4.8   Incomplete cubic triangular elements. The triangle of type 3'

In his famous book on finite element method, Ciarlet [8] mentions the so-called *triangle of type 3'*. By this, he means the nine-degree of freedom finite element that consists of all the cubics – except the cubic bubble function! The resulting space $V_{0h}^3 \setminus \mathbf{curl}\, B_h$, with B_h the space of cubic bubble functions, seems suitable for application of our approach to prove superconvergence. The corresponding mixed finite element space could be the space $\Gamma_{0h}^2 \setminus \mathbf{curl}\, B_h$. Of course, details need to be worked out carefully. We will investigate this in a forthcoming paper.

4.9   Conclusions and remarks for the two-dimensional analysis

Here we relate some of the results for the two-dimensional analysis back to the one-dimensional analysis (cf. Sect. 2.4).

- In MFE, it were again the homogeneous Neumann conditions which were the easiest to analyze, and in SFE the homogeneous Dirichlet conditions. Going to the other conditions for each case theoretically gives a loss of order $\sqrt{h}$. This shows in experiments for the quadratic triangle, but not for the linear triangle.

- The duality estimate (cf. Th. 4.2) gives superconvergence results for the scalar function too, but only if $k \geq 1$ and if the gradient is superconvergent. The lowest order case does not have such a result (and numerical experiments – that we did not administrate – indicate there is no superconvergence in this case).

- Post-processing techniques developed in the one setting (standard or mixed) are in general applicable to the other setting by simple "pointwise rotation of the vector field", i.e., by interchanging normal and tangential components of the vector fields.

- Semi-discrete parabolic and hyperbolic systems inherit some of the superconvergence properties from the elliptic space-part, although not as convincing as in the one-dimensional case. The analysis can be done using (mixed) elliptic projection and the inevitable Gronwall inequality (cf. [5]).

- The operator π_h^k, of which the curl is equal to Fortin interpolation applied to the curl, is a local operator with nice properties. For instance, it interpolates at the corner points of the elements (which is exactly the key property that makes it local) and the tangential derivatives at the boundary have super-closeness properties at the Gauss points on the boundary.

- More general elliptic problems, also with smooth non-constant (matrix-)coefficients, can also be tackled for superconvergence, for example by locally estimating the coefficients by piecewise constants.

5   FINAL CONCLUSIONS AND REMARKS

We have seen connections between, one the one hand, standard and mixed finite elements in one dimension, and, on the other hand, mixed and standard methods in two dimensions. Properties in the two-dimensional case were often already present, in simplified form, in the one-dimensional case. For example, without having stated it explicitly, we already used a discrete Helmholtz decomposition of the mixed finite element space in dimension one, considering $L^2(I) = \frac{\partial}{\partial x} H_0^1(I) \oplus \frac{\partial}{\partial x} \mathcal{H}(I)$. Also, we made use of operators satisfying commuting diagram properties; diagrams in which the differential operators played a role. Even though the simplifications one encounters by considering the one-dimensional model first, we believe that it is always worthwhile to do so, moreover, since mathematically it is often not very difficult, and sometimes gives surprising insights.

We have also considered connections between standard finite elements on the one hand, and mixed finite elements on the other. The superconvergence properties of both can be seen as coming from a matter in approximation theory: approximating vector fields using Fortin interpolation. Apart from our global-norm approach, we suspect that pointwise properties too can be treated in one unifying theory.

After all, both methods can still be seen as taking place in Raviart–Thomas space, although for the standard method this is not a conventional way of looking at things. In fact, the methods can be characterized as follows:

- The (reduced integration–a subtlety) standard finite element method projects the Fortin interpolant $\mathbf{\Pi}_h^k \mathbf{grad}\, u$ of the gradient of the exact solution on $\mathbf{grad}\, V_{0h}^{k+1}$.

- The mixed finite element projects the Fortin interpolant $\mathbf{\Pi}_h^k \mathbf{grad}\, u$ of the gradient of the exact solution on $\mathbf{curl}\, V_{0h}^{k+1}$ – and gets rid of the result since it is a curl that cannot possibly approximate a gradient – quite clever, but we do lose the locality of Fortin interpolant this way. However, if this curl part is supersmall . . .

This curl is indeed supersmall for rectangular elements and linear and quadratic triangular elements. So in this case, both methods are higher order perturbations of a (the same) local interpolation scheme.

REFERENCES

1. J.H. Bramble and S.R. Hilbert, *Estimation of linear functionals on Sobolev spaces with applications to Fourier transforms and spline interpolation*, SIAM J. Numer. Anal. **7** (1970).
2. J.H. Brandts, *Superconvergence and a posteriori error estimation in triangular mixed finite elements.*, Numer. Math. **68(3)** (1994), 311–324.
3. J.H. Brandts, "Superconvergence phenomena in finite element methods," Ph.D. thesis, Utrecht University, 1995.
4. J.H. Brandts, *A unified treatment for mixed and standard finite element methods for the Poisson equation*, Revised version submitted to Numer. Math.
5. J.H. Brandts, *A short note on time integration of spatial superconvergence for mixed finite element discretizations using mixed elliptic projection techniques*, Revised version submitted to Numer. Math.
6. J.H. Brandts, "A posteriori foutschatting bij de eindige elementen methode en de gemengde eindige elementen methode," Master's Thesis, Utrecht University, 1990.
7. J.H. Brandts, *Superconvergence for second order triangular mixed and standard finite elements*, Report of Laboratory of Scientific Computing, University of Jyväskylä, Finland **9** (1996).
8. P. Ciarlet, "The finite element method for elliptic problems," North Holland, Amsterdam, 1978.
9. J. Douglas and J.E. Roberts, *Global estimates for mixed methods for second order elliptic problems*, Math. Comp. **44 (169)** (1985), 39–52.
10. J. Douglas and J. Wang, *Superconvergence of mixed finite element methods on rectangular domains*, Calcolo **26** (1989), 121–134.
11. J.J. Douglas, T. Dupont, and M.F. Wheeler, *An l^∞ estimate and a superconvergence result for a Galerkin method for elliptical equations based on tensor products of piecewise polynomials*, RAIRO Anal, Numér. **8** (1974), 61–66.
12. R. Durán, *Superconvergence for rectangular mixed finite elements*, Numer. Math. **58** (1990), 2–15.

13. R.E. Ewing, R.D. Lazarov, and J. Wang, *Superconvergence of the velocity along the Gauss lines in mixed finite element methods*, SIAM J. Numer. Anal. **28(4)** (1991), 1015–1029.

14. G. Goodsell and J.R. Whiteman, *A unified treatment of superconvergent recovered gradient functions for piecewise linear finite element approximations*, Internat. J. Numer. Methods Engrg. **27** (1989), 469–481.

15. G. Goodsell and J.R. Whiteman, *Superconvergence of recovered gradients for piecewise quadratic finite element approximations. Part 1: l_2-error estimates*, Numer. Methods Partial Differential Equations **1** (1991), 61–83.

16. G. Goodsell and J.R. Whiteman, *Superconvergence of recovered gradients for piecewise quadratic finite element approximations. Part 2: l_∞-error estimates*, Numer. Methods Partial Differential Equations **1** (1991), 85–99.

17. M. Křížek and P. Neittaanmäki, *On superconvergence techniques*, Acta Appl. Math. **9** (1987), 175–233.

18. M. Křížek and P. Neittaanmäki, *On a global superconvergence of the gradient of linear triangular elements*, J. of Comput. Appl. Math. **18** (1987), 221–233.

19. N. Levine, *Superconvergent recovery of the gradient from piecewise linear finite element approximations*, IMA J. Numer. Anal. **5** (1985), 407–427.

20. L.A. Oganesjan and L.A. Ruhovets, *Study of the rate of convergence of variational difference schemes for second-order elliptic equations in a two dimensional field with a smooth boundary*, Z. Vychisl. Mat. Mat. Fiz. **9** (1969), 1102–1120.

21. P. Peisker and D. Braess, *Uniform convergence of mixed interpolated elements for Reissner–Mindlin plates*, Technical report, Fakultät und Institut für Mathematik der Ruhr-Universität Bochum, Preprint nr. **142** (1990).

22. P.A. Raviart and J.M. Thomas, *A mixed finite element method for second order elliptic problems*, Lecture Notes in Math. **606** (1977), 292–315.

23. A.H. Schatz and I.H. Sloan and L.B. Wahlbin, *Superconvergence in finite element methods and meshes that are symmetric with respect to a point*, SIAM J. Numer. Anal. **33** (1996), 505–521.

24. J. Wang, *Superconvergence and extrapolation for mixed finite element methods on rectangular domains*, Math. Comp. **56(194)** (1991), 477–503.

25. M.F. Wheeler and J.R. Whiteman, *Superconvergent recovery of gradients on subdomains from piecewise linear finite element approximations*, Numer. Methods Partial Differential Equations **3** (1987), 357–374.

On the FEM for the Navier–Stokes equations in the domains with corner singularities

P. BURDA*   Department of Mathematics, Faculty of Mechanical Engineering,
Czech Technical University, Karlovo náměstí 13, 121 35 Prague 2, Czech Republic,
e-mail: burda@fsik.cvut.cz

Abstract   We study the axisymmetric flow governed by the Navier–Stokes equations, in tubes with abrupt changes of diameter. Our first concern is the asymptotic behaviour of the solution near the corners.

The rotational symmetry allows us to restrict the problem to two dimensions using the cylindrical coordinates r, z. In the case of the Stokes flow the stream function ψ then satisfies the equation

$$\frac{1}{r}\left(\frac{\partial^4\psi}{\partial z^4} + 2\frac{\partial^4\psi}{\partial z^2\partial r^2} + \frac{\partial^4\psi}{\partial r^4}\right) - \frac{1}{r^2}\left(\frac{\partial^3\psi}{\partial z^2\partial r} + \frac{\partial^3\psi}{\partial r^3}\right) + \frac{1}{r^3}\frac{\partial^2\psi}{\partial r^2} - \frac{3}{r^4}\frac{\partial\psi}{\partial r} = 0.$$

This equation is different from that for the stream function in desk geometry, where the asymptotic behaviour of the solution near the corners is known (cf., e.g., Kondratiev and Olejnik [15]). In our paper we transform the equation into the polar coordinates ρ, ϑ. Using the Fourier transform technique and some of the results of Kondratiev and Olejnik [15], we obtain the asymptotic behaviour of the stream function ψ near the corner. E.g., for the internal angle of $\frac{3}{2}\pi$ there exists a function φ, independent of the radius ρ, such that

$$\psi(\rho, \vartheta) = \rho^{1.54448374}\varphi(\vartheta) + \dots,$$

which means that the velocity components v_1, v_2 behave near the corner like

$$v_l(\rho, \theta) = \rho^{0.54448374}\varphi_l(\theta) + \dots, \quad l = 1, 2.$$

These asymptotics agree with those obtained for planar Stokes flow.

In the second part of the paper we deal with the numerical solution of flow of the incompressible fluid in a tube with sharp changes of diameter. Our aim is to

*This research has been supported partly by grant No. 101/94/0280 and partly by grant No. 101/96/1051, both from the GACR Agency.

41

make use of the information on the local behaviour of the solution near the corner point, in order to suggest local meshing in correspondence with the asymptotics. We present a cheap strategy for various families of triangular elements. For linear and quadratic elements we show tables of adequate meshings.

1   INTRODUCTION

We are concerned with the numerical solution of flow of incompressible fluid in tubes with abrupt changes of diameter. We study the axisymmetric flow governed by the Navier–Stokes equations.

Our first concern is the asymptotic behaviour of the solution near the corners. We are not aware of any published results on the asymptotics for axisymmetric flow. So in the first part of the paper we study the asymptotics for the axisymmetric Stokes flow.

In the second part of the paper we deal with the numerical solution of the flow in a tube with sharp changes of diameter.

In plane flow, the finite element method has been successfully used, cf., e.g., Skalák [20]. The singularity at the corner is not so strong, and appropriate local refinement of the mesh near the corner gives quite successful results.

In the case of axisymmetric flow, we used for the space discretization the MAC method which may be considered as a variant of the finite element method, cf. Girault, Lopez [11]. Some numerical results for pulsatile axisymmetric flow in tubes with step changes of diameter were published in [2], [3], [6], [7].

In this paper our aim is to make use of the information on the local behaviour of the solution near the corner point, in order to suggest local meshing subordinate to the asymptotics. We present a cheap strategy to be applied to families of triangular elements. For linear and quadratic elements we show examples of adequate meshings.

2   ON THE NAVIER–STOKES EQUATIONS NEAR THE CORNER, IN THE AXISYMMETRIC CASE

The asymptotic behaviour of *plane flow* with corner singularities has been studied, e.g., by Lugt and Schwiderski [18], Kondratiev [13], Ladevéze and Peyret [17], Kufner and Sändig [16]. The asymptotics of the biharmonic equation for the stream function ψ are basic.

In our paper we concentrate on *pipe flow* (axially symmetric). To study the asymptotic behaviour of the solution of the Navier–Stokes equations for an incompressible fluid, we utilize the stream function - vorticity formulation, which in

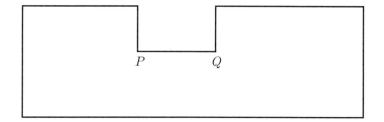

Figure 2.1 The solution domain Ω.

cylindrical geometry reads

$$\frac{\partial \omega}{\partial t} + u\frac{\partial \omega}{\partial z} + v\frac{\partial \omega}{\partial r} + v\frac{\omega}{r} = \nu\left(\frac{\partial^2 \omega}{\partial z^2} + \frac{\partial^2 \omega}{\partial r^2} + \frac{1}{r}\frac{\partial \omega}{\partial r} - \frac{\omega}{r^2}\right), \tag{2.1}$$

$$\frac{\partial^2 \psi}{\partial z^2} + \frac{\partial^2 \psi}{\partial r^2} - \frac{1}{r}\frac{\partial \psi}{\partial r} = -r\omega, \tag{2.2}$$

$$v_1 = \frac{1}{r}\frac{\partial \psi}{\partial r}, \tag{2.3}$$

$$v_2 = -\frac{1}{r}\frac{\partial \psi}{\partial z}, \tag{2.4}$$

where r, z are cylindrical coordinates, $v_1 = V_z, v_2 = V_r$ are velocity components, ω is the vorticity, ψ is the stream function, and ν is the viscosity. We assume that all derivatives exist here at least in the generalized sense.

First, let us study the *stationary flow*. Then, substituting ω, u, v from (2.2)–(2.4) into (2.1) we get

$$\frac{1}{r}\frac{\partial \psi}{\partial r}\frac{\partial}{\partial z}\left(-\frac{1}{r}\frac{\partial^2 \psi}{\partial z^2} - \frac{1}{r}\frac{\partial^2 \psi}{\partial r^2} + \frac{1}{r^2}\frac{\partial \psi}{\partial r}\right) - \frac{1}{r}\frac{\partial \psi}{\partial z}\frac{\partial}{\partial r}\left(-\frac{1}{r}\frac{\partial^2 \psi}{\partial z^2} - \frac{1}{r}\frac{\partial^2 \psi}{\partial r^2} + \frac{1}{r^2}\frac{\partial \psi}{\partial r}\right)$$

$$-\frac{1}{r^2}\frac{\partial \psi}{\partial z}\left(-\frac{1}{r}\frac{\partial^2 \psi}{\partial z^2} - \frac{1}{r}\frac{\partial^2 \psi}{\partial r^2} + \frac{1}{r^2}\frac{\partial \psi}{\partial r}\right)$$

$$= \nu\Bigg\{\frac{\partial^2}{\partial z^2}\left(-\frac{1}{r}\frac{\partial^2 \psi}{\partial z^2} - \frac{1}{r}\frac{\partial^2 \psi}{\partial r^2} + \frac{1}{r^2}\frac{\partial \psi}{\partial r}\right) + \frac{\partial^2}{\partial r^2}\left(-\frac{1}{r}\frac{\partial^2 \psi}{\partial z^2} - \frac{1}{r}\frac{\partial^2 \psi}{\partial r^2} + \frac{1}{r^2}\frac{\partial \psi}{\partial r}\right)$$

$$+ \frac{1}{r}\frac{\partial}{\partial r}\left(-\frac{1}{r}\frac{\partial^2 \psi}{\partial z^2} - \frac{1}{r}\frac{\partial^2 \psi}{\partial r^2} + \frac{1}{r^2}\frac{\partial \psi}{\partial r}\right) + \frac{1}{r^3}\frac{\partial^2 \psi}{\partial z^2} + \frac{1}{r^3}\frac{\partial^2 \psi}{\partial r^2} - \frac{1}{r^4}\frac{\partial \psi}{\partial r}\Bigg\}. \tag{2.5}$$

We are interested in the asymptotic behaviour of the solution near the corners. One example of our solution domain is shown in Figure 2.1, where the corners are the points P, Q.

To study the asymptotics near the corner $P = [z_0, r_0]$, we perform the transformation into the polar coordinates ρ, ϑ:

$$z - z_0 = \rho\cos\vartheta,$$
$$r - r_0 = \rho\sin\vartheta, \tag{2.6}$$

Then equation (2.5) transforms into

$$
\frac{1}{r}\frac{\partial\psi}{\partial\rho}\left[-\frac{1}{\rho}\frac{\partial^3\psi}{\partial\rho^2\partial\vartheta}-\frac{1}{\rho^3}\frac{\partial^3\psi}{\partial\vartheta^3}\right]+\frac{1}{r}\frac{\partial\psi}{\partial\vartheta}\left[\frac{1}{\rho}\frac{\partial^3\psi}{\partial\rho^3}+\frac{1}{\rho^3}\frac{\partial^3\psi}{\partial\rho\partial\vartheta^2}\right]
$$

$$
+\frac{1}{r^2}\frac{\partial\psi}{\partial\rho}\left[\frac{1}{\rho}\frac{\partial^2\psi}{\partial\rho\partial\vartheta}\sin^3\vartheta+\frac{1}{\rho^2}\frac{\partial^2\psi}{\partial\vartheta^2}\cos\vartheta\right]+\frac{1}{r^2}\frac{\partial\psi}{\partial\vartheta}\left[-\frac{1}{\rho}\frac{\partial^2\psi}{\partial\rho^2}\sin\vartheta-\frac{1}{\rho^2}\frac{\partial^2\psi}{\partial\rho\partial\vartheta}\cos^3\vartheta\right]
$$

$$
+\frac{1}{r^3}\left[-\left(\frac{\partial\psi}{\partial\rho}\right)^2\sin\vartheta\cos\vartheta+\frac{1}{\rho}\frac{\partial\psi}{\partial\rho}\frac{\partial\psi}{\partial\vartheta}\left(1-2\cos^2\vartheta\right)+\frac{1}{\rho^2}\left(\frac{\partial\psi}{\partial\vartheta}\right)^2\sin\vartheta\cos\vartheta\right]
$$

$$
=\nu\Bigg\{\left[\frac{\partial^4\psi}{\partial\rho^4}+\frac{2}{\rho^2}\frac{\partial^4\psi}{\partial\rho^2\partial\vartheta^2}+\frac{1}{\rho^4}\frac{\partial^4\psi}{\partial\vartheta^4}\right]
$$

$$
-\frac{2}{r}\left[\sin\vartheta\frac{\partial}{\partial\rho}\left(\frac{\partial^2\psi}{\partial\rho^2}+\frac{1}{\rho^2}\frac{\partial^2\psi}{\partial\vartheta^2}\right)+\frac{1}{\rho}\cos\vartheta\frac{\partial}{\partial\vartheta}\left(\frac{\partial^2\psi}{\partial\rho^2}+\frac{1}{\rho^2}\frac{\partial^2\psi}{\partial\vartheta^2}\right)\right]
$$

$$
+\frac{3}{r^2}\left[\frac{\partial^2\psi}{\partial\rho^2}\sin^2\vartheta+\frac{2}{\rho}\frac{\partial^2\psi}{\partial\rho\partial\vartheta}\sin\vartheta\cos\vartheta+\frac{1}{\rho^2}\frac{\partial^2\psi}{\partial\vartheta^2}\cos^2\vartheta\right]
$$

$$
-\frac{3}{r^3}\left[\frac{\partial\psi}{\partial\rho}\sin\vartheta+\frac{1}{\rho}\frac{\partial\psi}{\partial\vartheta}\cos\vartheta\right]\Bigg\}.
$$

$$(2.7)$$

This equation seems to be rather complicated for the study of the asymptotics. In the next paragraph we restrict ourselves to the Stokes flow.

3 ON THE ASYMPTOTIC BEHAVIOUR OF THE STATIONARY STOKES FLOW NEAR THE CORNER

In the case of the Stokes flow, equation (2.5) reduces to

$$
\nu\Bigg\{\frac{1}{r}\left(\frac{\partial^4\psi}{\partial z^4}+2\frac{\partial^4\psi}{\partial z^2\partial r^2}+\frac{\partial^4\psi}{\partial r^4}\right)
$$

$$
-\frac{1}{r^2}\left(\frac{\partial^3\psi}{\partial z^3}+\frac{\partial^3\psi}{\partial z^2\partial r}+\frac{\partial^3\psi}{\partial z\partial r^2}+\frac{\partial^3\psi}{\partial r^3}\right)+\frac{3}{r^3}\frac{\partial^2\psi}{\partial z^2}-\frac{3}{r^4}\frac{\partial\psi}{\partial z}\Bigg\}=0.
$$

$$(3.1)$$

Substituting

$$
z-z_0=x,
$$
$$
r-r_0=y,
$$

$$(3.2)$$

we come to the equation

$$
\nu\Bigg\{\frac{1}{y+r_0}\left(\frac{\partial^4\psi}{\partial x^4}+2\frac{\partial^4\psi}{\partial x^2\partial y^2}+\frac{\partial^4\psi}{\partial y^4}\right)
$$

$$
-\frac{1}{(y+r_0)^2}\left(\frac{\partial^3\psi}{\partial x^3}+\frac{\partial^3\psi}{\partial x^2\partial y}+\frac{\partial^3\psi}{\partial x\partial y^2}+\frac{\partial^3\psi}{\partial y^3}\right)
$$

$$
+\frac{3}{(y+r_0)^3}\frac{\partial^2\psi}{\partial x^2}-\frac{3}{(y+r_0)^4}\frac{\partial\psi}{\partial x}\Bigg\}=0,
$$

$$(3.3)$$

which, to be a bit more general, we consider on the domain Ω_0 shown in Figure 3.1, the corner being in the origin of the coordinates, with the internal angle ω, $0<\omega\leq 2\pi$.

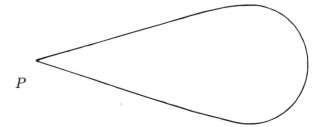

Figure 3.1   The auxiliary domain Ω_0.

The coefficients in (3.3) are infinitely differentiable in Ω_0. To study the asymptotic behaviour of the solution of (3.3) near the corner P, we first restrict ourselves (cf. Kondratiev [**13**]) to the principal part of equation (3.3), namely

$$\frac{\partial^4 \psi}{\partial x^4} + 2\frac{\partial^4 \psi}{\partial x^2 \partial y^2} + \frac{\partial^4 \psi}{\partial y^4} = f, \tag{3.4}$$

where $f = 0$.

Now we perform the transformation into the polar coordinates ρ, ϑ,

$$\begin{aligned} x &= \rho\cos\vartheta, \\ y &= \rho\sin\vartheta. \end{aligned} \tag{3.5}$$

Equation (3.4) in polar coordinates reads

$$\begin{aligned} S_0(\psi) \equiv \Bigg[&\frac{1}{\rho}\frac{\partial}{\partial\rho}\rho\frac{\partial}{\partial\rho}\frac{1}{\rho}\frac{\partial}{\partial\rho}\rho\frac{\partial\psi}{\partial\rho} + \frac{1}{\rho}\frac{\partial}{\partial\rho}\rho\frac{\partial}{\partial\rho}\frac{1}{\rho^2}\frac{\partial^2\psi}{\partial\vartheta^2} \\ &+ \frac{1}{\rho^2}\frac{\partial^2}{\partial\vartheta^2}\frac{1}{\rho}\frac{\partial}{\partial\rho}\rho\frac{\partial\psi}{\partial\rho} + \frac{1}{\rho^2}\frac{\partial^2}{\partial\vartheta^2}\frac{1}{\rho^2}\frac{\partial^2\psi}{\partial\vartheta^2} \Bigg] = 0. \end{aligned} \tag{3.6}$$

The boundary conditions are

$$\begin{aligned} \psi\big|_{\partial\Omega_0} &= 0, \\ \frac{\partial\psi}{\partial n}\bigg|_{\partial\Omega_0} &= 0, \end{aligned} \tag{3.7}$$

where n is the outgoing normal to the boundary $\partial\Omega$.

The problem (3.6), (3.7) is the same as in Kondratiev [**14**], and we follow his procedure. To study the asymptotic behaviour in Ω_0 near the corner P, we first consider the infinite cone $\widetilde{\Omega}_0$,

$$\widetilde{\Omega}_0 = \{(\rho,\vartheta),\ 0 < \rho < \infty,\ \alpha < \vartheta < \beta\}, \tag{3.8}$$

where $\alpha \in [0, 2\pi)$, $\beta \in (0, 2\pi]$, $\alpha < \beta$, are given angles, $\beta - \alpha = \omega$. Substituting

$$\tau = \ln\frac{1}{\rho} \tag{3.9}$$

into (3.6), we get

$$[(\psi_{\tau\tau\tau\tau} + 4\psi_{\tau\tau\tau} + 4\psi_{\tau\tau}) + 4\psi_{\tau\vartheta\vartheta} + 2\psi_{\tau\tau\vartheta\vartheta} + \psi_{\vartheta\vartheta\vartheta\vartheta} + 4\psi_{\vartheta\vartheta}] = 0 \qquad (3.10)$$

on the infinite strip

$$\tau \in (-\infty, +\infty), \quad \vartheta \in (\alpha, \beta). \qquad (3.11)$$

Now we can perform the Fourier transform with respect to τ, (i being the imaginary unit)

$$\widehat{\psi}(\lambda, \vartheta) = (2\pi)^{-1/2} \int_{-\infty}^{+\infty} e^{-i\lambda\tau} \psi(\tau, \vartheta) d\tau, \qquad (3.12)$$

and then equation (3.10) transforms to the ordinary differential equation

$$\widehat{L}(\vartheta, i\lambda)\widehat{\psi} \equiv \widehat{\psi}_{\vartheta\vartheta\vartheta\vartheta} + (-2\lambda^2 + 4i\lambda + 4)\widehat{\psi}_{\vartheta\vartheta} + (\lambda^4 - 4i\lambda^3 - 4\lambda^2)\widehat{\psi} = 0, \qquad (3.13)$$

where $\vartheta \in (\alpha, \beta)$. The operator $\widehat{L}$ depends analytically (in fact polynomially) on λ. Thus the inverse operator $R(\lambda)$ (if it exists at least for one point $\lambda \in \mathbb{C}$) is a meromorphic operator-valued function of λ, each pole of $R(\lambda)$ having finite multiplicity.

Equation (3.13) is a fourth order equation with constant coefficients. We denote them

$$\begin{aligned} A &= -2\lambda^2 + 4i\lambda + 4, \\ B &= \lambda^4 - 4i\lambda^3 - 4\lambda^2. \end{aligned} \qquad (3.14)$$

To find the general solution of (3.13), we have to solve the biquadratic equation

$$\mu^4 + A\mu^2 + B = 0. \qquad (3.15)$$

The squares of the solutions of equation (3.15) are

$$\begin{aligned} (\mu^2)_1 &= \lambda^2, \\ (\mu^2)_2 &= (\lambda - 2i)^2. \end{aligned} \qquad (3.16)$$

So the solutions of (3.15) are

$$\mu_{1,2} = \pm\lambda, \qquad (3.17)$$

$$\mu_{3,4} = \pm(\lambda - 2i). \qquad (3.18)$$

We can prove the following Lemma.

LEMMA 3.1   There are no poles of the resolvent $R(\lambda)$ on the line $\operatorname{Im}\lambda = 2$.

Now we can use the following theorem proved by Kondratiev [14]. Here we use the Sobolev spaces $\overset{0}{W}{}_{\delta}^{k}(\Omega)$ supplied with the norm

$$\|u\|_{\overset{0}{W}{}_{\delta}^{k}(\Omega)}^{2} = \sum_{m=0}^{k} \iint_{\Omega} \rho^{\delta - 2(k-m)} |D^m u|^2 \, dx, \quad D^m = \frac{D^{|m|}}{\partial x_1^{m_1} \partial x_2^{m_2}}, \quad |m| = m_1 + m_2.$$

$$(3.19)$$

Let us return to the full Stokes equation (3.3). We rewrite equation (3.3) in the form

$$\frac{\partial^4 \psi}{\partial x^4} + 2\frac{\partial^4 \psi}{\partial x^2 \partial y^2} + \frac{\partial^4 \psi}{\partial y^4} = f, \tag{3.20}$$

where

$$f = \frac{1}{\nu}\left\{ \frac{1}{(y+r_0)}\left(\frac{\partial^3 \psi}{\partial x^3} + \frac{\partial^3 \psi}{\partial x^2 \partial y} + \frac{\partial^3 \psi}{\partial x \partial y^2} + \frac{\partial^3 \psi}{\partial y^3} \right) + \frac{3}{(y+r_0)^2}\frac{\partial^2 \psi}{\partial x^2} - \frac{3}{(y+r_0)^3}\frac{\partial \psi}{\partial x} \right\}. \tag{3.21}$$

It is well known that $\psi \in W^4(\widetilde{\Omega}_0 \cap \{(x,y), x^2 + y^2 > R\})$ for all $R > 0$. It can then be proved that $\psi \in \overset{0}{W}{}^4_4(\widetilde{\Omega}_0)$. We can also prove that function f in (3.21) satisfies $f \in \overset{0}{W}{}^0_2(\widetilde{\Omega}_0)$. Now we are in the position to use the following theorem by Kondratiev and Olejnik [15].

THEOREM 3.1   Let $f \in \overset{0}{W}{}^{k_1}_{\delta_1}(\widetilde{\Omega}_0)$ and let $\psi \in \overset{0}{W}{}^{k+4}_{\delta}(\widetilde{\Omega}_0)$ be the solution of (3.4) satisfying the boundary conditions (3.7) on $\partial\widetilde{\Omega}$. Let

$$h_1 \equiv \frac{-\delta_1 + 2k_1 + 6}{2} > \frac{-\delta + 2k + 6}{2} \equiv h, \quad k_1 \geq k. \tag{3.22}$$

Suppose that the resolvent function $R(\lambda)$ has no poles on the line Im $\lambda = h_1$. Then the solution ψ has the form

$$\psi(x,y) = \sum_j \sum_{s=0}^{p_j-1} a_{js}\rho^{-i\lambda_j} \ln^s \rho \cdot \psi_{sj}(\vartheta) + w(x,y), \tag{3.23}$$

where w satisfies (3.7),

$$w \in \overset{0}{W}{}^{k_1+4}_{\delta_1}(\widetilde{\Omega}_0), \quad \psi_{sj} \in C^\infty(\widetilde{\Omega}_0), \quad a_{js} = \text{const},$$

and λ_j are the poles of multiplicity p_j of the function $R(\lambda)$, satisfying

$$h < \text{Im}\lambda_j < h_1.$$

Now we can apply Theorem 3.1 to equation (3.20), with $k = 0$, $\delta = 4$. We put $h_1 = 2$ according to Lemma 3.1, $k_1 = 0$, $\delta_1 = 2$.

Theorem 3.1 deals with the infinite cone $\widetilde{\Omega}_0$. The situation is a bit more complicated in the conical domain Ω_0, as can be seen from the following theorem proved by Kondratiev and Olejnik [15].

THEOREM 3.2   Let the assumptions of Theorem 3.1 be satisfied on the domain Ω_0. Then the solution ψ of problem (3.4), (3.7) on Ω_0 has the form

$$\psi(x,y) = \sum_j \sum_{s=0}^{p_j-1} \sum_{q=0}^{Q_j} a_{sj}\rho^{-i\lambda_j}(\ln\rho)^s P_{sjq}(\rho(\ln\rho)^q) + w(x,y), \tag{3.24}$$

where w satisfies (3.7),

$$w \in \overset{0}{W}{}^{k_1+4}_{\delta_1}(\Omega_0), \quad \psi_{sj} \in C^\infty(\Omega_0), \quad a_{js} = \text{const}, \tag{3.25}$$

λ_j are the poles of multiplicity p_j of the function $R(\lambda)$, satisfying

$$h < \text{Im}\lambda_j < h_1, \tag{3.26}$$

Q_j are integer numbers, P_{sjq} is a polynomial of the degree $[h_1 - \text{Im}\lambda_j]$ with respect to $\rho(\ln \rho)^q$, the coefficients of which are infinite differentiable functions of ϑ.

Now let us try to find the poles of $R(\lambda)$. According to (3.17), (3.18), the general solution of (3.13) is

$$\widehat{\psi} = c_1 \exp(\lambda\vartheta) + c_2 \exp(-\lambda\vartheta) + c_3 \sin(2\vartheta) + c_4 \cos(2\vartheta). \tag{3.27}$$

The boundary conditions are

$$\widehat{\psi}\Big|_{\vartheta=\alpha} = 0, \qquad \widehat{\psi}\Big|_{\vartheta=\beta} = 0,$$
$$\frac{\partial\widehat{\psi}}{\partial\vartheta}\Big|_{\vartheta=\alpha} = 0, \qquad \frac{\partial\widehat{\psi}}{\partial\vartheta}\Big|_{\vartheta=\beta} = 0. \tag{3.28}$$

To obtain a nontrivial solution of the problem (3.27), (3.28), the necessary and sufficient condition is that the following determinant be zero,

$$R(\lambda) \equiv \det \begin{pmatrix} \exp(\lambda\alpha) & \exp(\lambda\beta) & \lambda\exp(\lambda\alpha) & \lambda\exp(\lambda\beta) \\ \exp(-\lambda\alpha) & \exp(-\lambda\beta) & -\lambda\exp(-\lambda\alpha) & -\lambda\exp(-\lambda\beta) \\ \sin(2\alpha) & \sin(2\beta) & 2\cos(2\alpha) & 2\cos(2\beta) \\ \cos(2\alpha) & \cos(2\beta) & -2\sin(2\alpha) & -2\sin(2\beta) \end{pmatrix} = 0. \tag{3.29}$$

Let us for example take the angle $\omega = \frac{3}{2}\pi$, i.e.,

$$\alpha = 0, \quad \beta = \frac{3}{2}\pi. \tag{3.30}$$

We come to the first root of (3.29)

$$i\lambda_1 = -1.54448374, \tag{3.31}$$

which is simple, so that, according to Theorem 3.2, the first term of the asymptotic expansion is $\rho^{1.54448374}$, i.e.

$$\psi(\rho,\vartheta) = \rho^{1.54448374}\,\phi(\vartheta) + \dots. \tag{3.32}$$

This result is the same as that obtained in desk geometry by Kondratiev [14], Ladevéze and Peyret [17], M. Dauge [8], where

$$\psi^{desk}(\rho,\vartheta) = \rho^{1.5445}\,\phi^d(\vartheta) + \dots. \tag{3.33}$$

Now according to (2.3), (2.4), we get for the velocities, the expansion

$$v_l(\rho, \vartheta) = \rho^{0.54448374} \varphi_l(\vartheta) + \ldots, \quad l = 1, 2, \tag{3.34}$$

where the functions φ_l do not depend on ρ.

4   STRATEGY FOR ADEQUATE MESHING NEAR THE CORNER

Now we are concerned with the finite element solution of the stationary Stokes flow. Especially we would like to use the information on the asymptotic behaviour of the flow near the singular point, in order to choose adequate local meshing.

In Section 3 we have shown that the asymptotic behaviour near the singular point, for the axisymmetric flow and for the plane flow, is the same. So in what follows, for simplicity we deal with the plane flow, in the domain Ω which has the same shape as in Figure 2.1.

We consider the stationary Stokes equation in primitive variables V (velocity vector) and p (pressure)

$$-\nu \Delta V + \nabla p = 0 \quad \text{in } \Omega, \tag{4.1}$$

$$\operatorname{div} V = 0 \quad \text{in } \Omega, \tag{4.2}$$

and the boundary condition

$$V(x) = g(x) \quad \text{on } \partial\Omega, \tag{4.3}$$

where the function $g(x)$ is given on $\partial\Omega$ and satisfies the compatibility condition

$$\int_{\partial\Omega} g \cdot n \, ds = 0. \tag{4.4}$$

As the numerical approximation we take the stabilized finite element approximation V_h, p_h by Douglas and Wang [9], where $V_h \in X_h$, $p_h \in M_h$, X_h being the subspace of $\mathbf{H}_0^1(\Omega) = (H_0^1(\Omega))^2$ of vector-functions whose components are polynomials of degree k on triangles T of a triangulation $\mathcal{T}_h$; similarly M_h is the subspace of $L_0^2(\Omega)$, of piecewise polynomials of degree $k - 1$ on triangles. Here $H_0^1(\Omega)$ is the standard Sobolev space with first generalized derivatives square integrable and with zero traces on the boundary. $L_0^2(\Omega)$ is the subspace of $L^2(\Omega)$ containing functions with mean value zero.

Let us denote by h_T the diameter of the triangle $T \in \mathcal{T}_h$.

The following theorem is a variant of Theorems 3.1 and 3.2 of Douglas and Wang [9]. As the proof is also only a modification of these Theorems, we omit it.

THEOREM 4.1   Let the triangulation $\mathcal{T}_h$ be shape regular (i.e. the ratio of the diameter of the circumscribed ball for $T \in \mathcal{T}_h$ to that of the inscribed ball is bounded, independently of $T \in \mathcal{T}_h$). Then

$$\|\nabla(V - V_h)\|_0 \leq C \left[\left(\sum_T h_T^{2k} \mid V \mid_{\mathbf{H}^{k+1}(T)}^2 \right)^{1/2} + \left(\sum_T h_T^{2k} \mid p \mid_{H^k(T)}^2 \right)^{1/2} \right], \tag{4.5}$$

$$\|p - p_h\|_0 \leq C \left[\left(\sum_T h_T^{2k} \mid V \mid_{\mathbf{H}^{k+1}(T)}^2 \right)^{1/2} + \left(\sum_T h_T^{2k} \mid p \mid_{H^k(T)}^2 \right)^{1/2} \right], \tag{4.6}$$

provided that $(V, p) \in \mathbf{H}^{k+1}(T) \times H^k(T)$, for all $T \in \mathcal{T}_h$.

For our problem with the corner singularity it seems to be advantageous to make use of the information on the local behaviour of the solution. Several techniques have already been used. E.g., Strang and Fix [21], Olson et al [19] use special singular functions in addition to standard finite element polynomials. Fried and Yang [11] suggest particular refinement near the corner. Johnson's idea [12] is to balance the meshsize according to the norm on the element. In [4], [5] partial problem was solved by cubic elements. In this paper our approach is close to the idea of Johnson [12].

In Section 3 we have proved that near the corner, the velocity has the form (cf. (3.34))

$$v_l(\rho, \vartheta) = \rho^\gamma \varphi_l(\vartheta) + \ldots, \quad l = 1, 2, \tag{4.7}$$

where γ is determined by the angle of the corner. So in the formulae (4.5), (4.6) we may estimate the seminorm

$$|V|^2_{\mathbf{H}^{k+1}(T)} \approx C \int_0^{r_T} \rho^{2(\gamma-k-1)} \rho \, d\rho \approx C r_T^{2(\gamma-k)}.$$

We would like to balance the contribution of different elements to the error in (4.5), (4.6). This means we have to guarantee

$$h_T^{2k} \, r_T^{2(\gamma-k)} \approx h^{2k}, \tag{4.8}$$

h being some auxiliary stepsize. By (4.8) we come to the formula for the local refinement

$$h_T = h \cdot r_T^{1-\frac{\gamma}{k}}, \tag{4.9}$$

where r_T is the distance of the element T from the corner.

The formula (4.9) suggests the following

Spacing algorithm Let r_1 be the distance of the large element from the corner. For given auxiliary stepsize h we compute recurrently:

$$\left.\begin{array}{l} h_i = h \cdot (r_i)^{1-\frac{\gamma}{k}}, \\ r_{i+1} = r_i - h_i, \end{array}\right\} \quad \text{for } i = 1, 2, \ldots, N \tag{4.10}$$

EXAMPLE 4.1 Let us consider linear elements on triangles, the corner $\omega = \frac{3}{2}\pi$ (as on Figure 2.1). Then by (3.34), $\gamma = 0.5444837$. Let us choose $r_1 = 10^{-4}$, $h = 10^{-3}$. So by (4.10), we come to Table 4.1.

EXAMPLE 4.2 Let us now consider quadratic elements on triangles, the corner $\omega = \frac{3}{2}\pi$ (as on Figure 2.1). Again, by (3.34), $\gamma = 0.5444837$. Let us choose $r_1 = 10^{-4}$, $h = 0.02$. So by (4.10), we come to Table 4.2.

Table 4.1 Meshing for linear elements

i	r_i	h_i
1	1.E-4	1.507E-5
2	8.493E-5	1.399E-5
3	7.095E-5	1.288E-5
4	5.806E-5	1.176E-5
5	4.630E-5	1.061E-5
6	3.569E-5	9.42E-6
7	2.627E-5	8.195E-6
8	1.807E-5	6.910E-6
9	1.116E-5	5.549E-6
10	5.61E-6	4.060E-6
11	1.555E-6	1.555E-6

Table 4.2 Meshing for quadratic elements

i	r_i	h_i
1	1.E-4	2.455E-5
2	5.545E-5	2.000E-5
3	5.546E-5	1.598E-5
4	3.947E-5	1.248E-5
5	2.699E-5	9.464E-6
6	1.753E-5	6.913E-6
7	1.062E-5	4.800E-6
8	5.818E-6	3.098E-6
9	2.720E-6	1.781E-6
10	9.390E-7	8.214E-7
11	1.175E-7	1.175E-7

REFERENCES

1. Ainsworth, M., Oden, J. T., *A procedure for a posteriori error estimation for $h - p$ finite element methods*, Comput. Methods Appl. Mech. Engrg. **101** (1992), 73–96.

2. Burda, P., Kořenář, J., *Numerical solution of pulsatile flow in a round tube with axisymmetric constrictions*, Comm. Inst. Hydrodyn. **19** (1993), 87–110.

3. Burda, P., Kořenář, J., *Numerical solution of Navier–Stokes equations for constricted pulsations*, ZAMM **74** (1994), T649–T651.

4. Burda, P., *Cubic finite elements in elliptic problems with interfaces and singularities*, in "Abstract in the Mathematics of Finite Elements and Applications," J. R. Whiteman (ed.), J. Wiley, Chichester, 1994, p. 363.

5. Burda, P., *Cubic version of FEM in elliptic problems with interfaces and singularities*, in "Finite Element Methods, Fifty years of the Courant element," M. Křížek, P. Neittaanmäki and R. Stenberg (eds.), Marcel Dekker, New York, 1994, pp. 95–104.

6. Burda, P., Kořenář, J., *Second order discretization of pulsatile flow in a constricted tube*, in "Proc. Conf. Numerical Modelling Prague 1994," M. Feistauer, R. Rannacher, K. Kozel (eds.), FMF Prague 1995, pp. 53–58.

7. Burda, P., Kořenář, J., *MAC scheme in the numerical solution of pulsatile flow*, ICIAM Conf. Hamburg 1995, ZAMM **76**, S1 (1996), 365–366.

8. Dauge, M., *Stationary Stokes system in two or three dimensional domains with corners*, in "Séminaire Equations aux Dérivées Partielles 1987," exposé no. 10, Univ. de Nantes, pp. 315–357.

9. Douglas, J., Jr., Wang, J., *An absolutely stabilized finite element method for the Stokes problem*, Math. Comp. **52** (1989), 495–508.

10. Fried, I., Yang, S. K., *Best finite element distribution around a singularity*, AIAA Journal **10** (1972), 1244–1246.

11. Girault, V., Lopez, H., *Finite element error estimates for the MAC scheme*,

Preprint 1994.

12. Johnson, C., "Numerical Solution of Partial Differential Equations by the Finite Element Method," Cambridge University Press, 1994.

13. Kondratiev, V. A., *Asimptotika rešenija uravněnija Nav'je-Stoksa v okrestnosti uglovoj točki granicy*, Prikl. Mat. i Mech. **1** (1967), 119–123.

14. Kondratiev, V. A., *Krajevyje zadači dlja elliptičeskich uravněnij v oblastach s koničeskimi i uglovzmi točkami*, Trudy Moskov. Mat. Obshch. **16** (1967), 209–292.

15. Kondratiev, V. A. and Olejnik, O. A., *Krajevyje zadači dlja uravněnij s častnymi proizvodnzmi v negladkich oblastach*, Uspechi Mat. Nauk **38** (1983), 3–76.

16. Kufner, A., Sändig, A.-M., "Some Applications of Weighted Sobolev Spaces," Teubner, Leipzig, 1987.

17. Ladevéze J., Peyret, R., *Calcul numérique d'une solution avec singularité des équations de Navier–Stokes: écoulement dans un canal avec variation brusque de section*, Journal de Mécanique **13** (1974), 367–396.

18. Lugt, H. J., Schwiderski, E. N., *Flow around dihedral angles II. Analysis of regular and singular motions*, Proc. Royal Soc. A **285** (1965), 400–412.

19. Olson, L. G., Georgiou, G. C., Schultz, W. W., *An efficient finite element method for treating singularities in Laplace's equation*, J. Comp. Phys. **96** (1991), 391–410.

20. Skalák, Z., *Numerical analysis of the flow in corrugated channels*, in "Proc. Conf. Numerical Modelling Prague 1994," M. Feistauer, R. Rannacher, K. Kozel (eds.), FMF Prague, 1995, pp. 256–263.

21. Strang, G., Fix, G. J., "An Analysis of the Finite Element Method," Prentice-Hall, Englewood Cliffs, NJ, 1973.

22. Szabó, B., Babuška, I., "Finite Element Analysis," J. Wiley, New York, 1991.

Projections in finite element analysis and application

GRAHAM F. CAREY   ASE/EM Dept. WRW 301, The University of Texas at Austin, Austin, TX 78712, USA, e-mail: carey@cfdlab.ae.utexas.edu

Abstract   Projection between finite element grids and bases arise naturally in the formulation of the finite element method. These projections also form the basis of many important post-processing techniques for superconvergence extractions, *a posteriori* error estimators and multilevel strategies. In this study we consider the interrelation of these projections for both standard and mixed type finite element methods. Representative examples of projection strategies and their use in superconvergence and error post processing are considered together with supporting numerical experiments. The work also includes results from ongoing investigations of divergence-free projections, the use of least squares mixed methods for adaptive refinement and projections for multigrid and hierarchic multilevel finite element schemes.

1   INTRODUCTION

At a fundamental level the notion of projections is central to the finite element method. Perhaps the simplest finite element problem one can pose is to interpolate a give function $u \in H$ in some finite element subspace H^h where h denotes the mesh parameter for the finite element discretization of the domain. Since the approximation $u_h^I \in H^h$ is interpolatory it satisfies $u_h^I(x_i) = u(x_i)$ for each node in the discretization. Here we have assumed, for convenience, that only function values have been interpolated. This interpolation condition is equivalent to requiring that the interpolation error $e_I = u - u_h^I$ satisfy $e_I(x_i) = 0$ at the nodes and this can be interpreted as a collocation projection or duality pairing of the error with the dirac distribution at $x - x_i$. A more interesting case is the least-squares approximation problem on the finite element space. For $u \in L^2(\Omega)$, the approximation $u_h \in H^h$ minimizes the least squares error functional $I = \int_\Omega e^2 \, dx$, where $e = u - u_h$. It is a trivial conclusion that $\|e\|_0 \leq \|e_I\|_0$ and the error estimate of interpolation theory immediately provides a global bound and estimate for $\|e\|_0$. This is, in fact, the essence of *a priori* error estimation in finite element analysis.

As an instructive example to motivate superconvergence extraction, consider a continuous function u on $[-1, 1]$ approximated by a polynomial of degree p. Since $u(x)$ is continuous, by Weierstrass Theorem it can be approximated uniformly by polynomials. Now the error orthogonality condition for the least squares minimizer implies that the components of error e in the polynomial basis ψ_i, $i = 1, 2, \ldots, N = p + 1$ are zero and hence the error is uniformly approximated by a polynomial with first non-zero term of degree $p + 1$. If this high degree polynomial representation of the error is expressed using the Legendre polynomials then the leading non-zero term will vanish at the zeroes of $P_{p+1}(x)$, i.e., at the Gauss–Legendre points. These are then the superconvergent recovery points for the approximation.

In practice we are usually more interested in approximating the solution to a boundary-value problem. Let us consider the linear elliptic PDE $Lu = f$ in Ω. The finite element approach may be interpreted using the idea of weighted residual projections as follows: for admissible trial function v define the residual $r = f - Lv$ and take its duality pairing (projection) to obtain a weighted residual condition $< r, w > = 0$ for all admissible w. Various finite element schemes can be constructed depending on the choices of trial and test spaces and regularity assumptions. For example, if v is sufficiently smooth then we can collocate at discrete points to write $r(x_i) = 0$, $i = 1, 2, \ldots, N$. As in the previous interpolation problem, this can be interpreted as projecting $r(x)$ with the dirac distribution $\delta(x - x_i)$. It is more standard practice to use integration by parts to obtain a weak variational form so that the projection involves the same type of trial and test functions. Finally, if the finite element approximation is obtained by minimizing the residual in L^2 then this corresponds to a projection involving the adjoint operator applied to the test functions. Different forms of local superconvergence can be demonstrated for the approximations obtained using these methods and superconvergent post-processing schemes can also be developed. Of particular interest are post-processing strategies that utilize the differential equation as an auxiliary relation. We elaborate on some of these points later.

Projections also arise in several other avenues of finite element analysis and application: e.g., the extraction of viable *a posteriori* error indicators is a post-processing step that involves projection of the approximation. Likewise, operations between successively refined or coarsened (unrefined) grids in an adaptive mesh strategy involve local projections. Finally, in the solution algorithm for multilevel and domain decomposition strategies various projection operations are employed. In the present work we present some of the key ideas associated with finite element projections in each of the above categories. Elementary examples are included for clarity of exposition and references are provided for more detailed results and analysis.

2   SUPERCONVERGENT EXTRACTION/PROJECTION

In certain special situations the discrete solution to an approximation problem or to a boundary-value problem is exact at the nodes or the flux is exact at other special points such as the zeros of an associated set of orthogonal polynomials. More generally it has been observed in numerical experiments and confirmed in theoretical studies that local values at special points, while not exact, may exhibit exceptional accuracy and higher rates of convergence than anticipated. Similarly, it is well known that various post-processing strategies applied to the approximation can yield more accurate results with improved asymptotic rates. This behavior is now generally referred to as superconvergence.

Numerical observations of superconvergence were described in a number of early case studies, especially in finite element applications (e.g., see Hinton and Campbell [16], Barlow [4], Carey and Finlayson [10], and Carey and Wheeler [12]). Among the first theoretical studies providing superconvergent error estimates are those of Douglas and Dupont [14], Dupont [15], and Zlámal [31]. More recently, attention has focused on different forms of post-processing formulas. For example, averaging derivatives from adjacent elements may lead to local error cancellation (e.g., Schatz and Wahlbin [24], Wheeler and Whiteman [28], Pehlivanov [23]). Using auxiliary data such as the differential equation to post-process the approximate solution to a boundary-value problem and develop superconvergent projection formulas is a very effective strategy (e.g., Carey [8], Babuška and Miller [2], Carey, Chow and Seager [9], Lazarov *et al* [19], MacKinnon and Carey [20]. A more extensive treatment of superconvergence is provided in the recent monograph of Wahlbin [26] and related work by Chen and Huang [13] and Zhu and Lin [29].

Let us first consider the least-squares approximation problem with discontinuous piecewise-linear basis. The linear least-squares fit u_h on Ω_e must intersect the given function u so there are points in the element interior where u_h is exact. If u is an arbitrary quadratic polynomial then using the orthogonality property $\int_{\Omega_e} E\psi_i \, dx = 0$ and the properties of the Legendre polynomials it is easy to show that u_h is exact at the two interior Gauss points $(x_1 + x_2)/2 + \xi(x_1 - x_2)/2$ where $\xi = \pm 1/\sqrt{3}$.

An analogous situation arises for numerical solution of the elementary two-point boundary-value problem $-u'' = 1$, on $[0, 1]$ with $u(0) = u(1) = 0$. The exact solution is $u(x) = \frac{1}{2}x(1-x)$. The finite element Galerkin approximation with linear elements (or equivalently the central difference approximation) yields the discrete equation $-(u_{i+1} - 2u_i + u_{i-1})/h^2 = 1$, at interior nodes $i = 1, 2, \ldots, N - 1$. For $N = 2$, $h = \frac{1}{2}$ and the approximate solution $u_1 = 1/8$ is exact. This interpolatory property holds for all $N \geq 2$ and even for the corresponding difference scheme on nonuniform grids. This result is to be expected since the exact solution is quadratic and the central difference approximation is an exact representation of the curvature u''. All higher-order derivatives in the Taylor series truncation error vanish.

Even when the approximate solution is not interpolatory, there are instances when exceptional accuracy as well as higher-order asymptotic rates of convergence occur at certain points. This behavior can be demonstrated more formally using the Green's function as shown in the standard argument below: Consider the variational problem find $u \in H_0^1(0, 1)$ such that

$$\int_0^1 (au'v' + buv) \, dx = \int_0^1 fv \, dx \quad \text{for all } v \in H_0^1(0, 1). \tag{1}$$

The error $e = u - u_h$ for the Galerkin approximation satisfies the orthogonality property

$$\int_0^1 (ae'v_h' + bev_h) \, dx = 0 \quad \text{for all } v_h \in H^h. \tag{2}$$

The Green's function satisfies

$$\int_0^1 (aG'v' + bGv)\,dx = \int_0^1 \delta(x - \xi)v\,dx \quad \text{for all } v \in H_0^1(0,1). \tag{3}$$

Setting $v = e$ in (3), we obtain the extraction formula

$$e(\xi) = \int_0^1 (aG'e' + bGe)\,dx. \tag{4}$$

Since e also satisfies the orthogonality property (2), it follows that for any $w_h \in H^h$

$$e(\xi) = \int_0^1 [a(G' - w_h')e' + b(G - w_h)e]\,dx \tag{5}$$

and hence

$$|e(\xi)| \leq \|G - w_h\|_1 \|e\|_1 . \tag{6}$$

In particular, if $\xi = \hat{x}_t$ is an interface knot, then G is smooth on the open intervals excluding ξ and w_h can be constructed from the elliptic projection of G such that

$$\|G - w_h\|_1 \leq Ch^k. \tag{7}$$

Substituting in (6) and using the global estimate for $\|e\|_1$, the optimal supercon-vergence estimate for smooth u is

$$|e(\hat{x}_t)| \leq Ch^{2k} \|u\|_{k+1} . \tag{8}$$

If $b = 0$ and $a = 1$, then the Green's function $G(x;\xi)$ is piecewise linear and $G(x;\hat{x}_k) \in H^h$. Setting $w_h = G(x;\hat{x}_t)$ in (5) yields the interpolatory property $e(\hat{x}_t) = 0$ at interface nodes $\{\hat{x}_t\}$.

Exceptional accuracy has also been observed in derived or post-processed quantities. For example, the derivative of the approximation may be very accurate at the Gauss points interior to an element or cell. Here this behavior also suggests that such superconvergence properties might also be exploited to construct error estimators. For example, in the Galerkin finite element method the derivative is exceptionally accurate at the interior Gauss points of an element. Values of the derivative at these points can be compared with the average value of the approximate derivative (or a similar quantity) to get a local error estimate for the derivative.

As an illustrative example, consider the simple problem of piecewise linear interpolation of a smooth function u. Let u_h denote this interpolant of grid point values $\{u_i\}$ at $\{x_i\}$. The derivative of the linear interpolant on interval i is $u_h'(x) = (u_{i+1} - u_i)/h$, $h = x_{i+1} - x_i$ for $x \in (x_i, x_{i+1})$. Using Taylor series expansion for u_{i+1} and u_i about interior point x with $\delta_i = x - x_i$ and $\delta_{i+1} = x_{i+1} - x$

$$u'(x) = \frac{(u_{i+1} - u_i)}{h} + \frac{(\delta_{i+1} - \delta_i)}{2}u''(x) + \frac{(\delta_{i+1} + \delta_i)^3}{3!h}u'''(x) + \ldots \tag{9}$$

Clearly, if x is at the mid-point $\bar{x}$ of the interval (the central Gauss point or equivalently the zero of the corresponding linear Legendre polynomial on this interval), then $\delta_{i+1} = \delta_i = h/2$ and the chord slope in (9) is an $O(h^2)$ accurate approximation to $u'(\bar{x})$.

In a similar manner we can differentiate the quadratic Lagrange polynomial interpolant to obtain

$$u_h'(x) = \sum_{j=i-1}^{i+1} u_j \psi_j'(x), \tag{10}$$

where $\psi_j(x)$ are the local Lagrange quadratic basis functions on $[x_{i-1}, x_{i+1}]$. Expanding u_{i-1}, u_i, u_{i+1} in Taylor series about point $x \in [x_{i-1}, x_{i+1}]$, regrouping terms and simplifying,

$$u'(x) = u_h'(x) - \sum_{j=i-1}^{i+1} \psi_j'(x) \frac{\delta_j^3}{3!} u'''(x) + O(h^3), \tag{11}$$

where δ_j again is the distance from x to grid point x_j and now $h = x_{i+1} - x_{i-1}$. Straightforward calculation reveals that the leading error term on the right vanishes at the pair of interior Gauss points $\bar{x}_g$, $g = 1, 2$, corresponding to the zeros of the shifted quadratic Legendre polynomial on $[x_{i-1}, x_{i+1}]$.

This result can be directly extended to higher-degree elements and in general we find that the leading term in the Taylor series error vanishes at the corresponding interior Legendre–Gauss points. For the triangle or quadrilateral a similar analysis demonstrates that the tangential derivatives are superconvergent at the corresponding Gauss points on the sides. Further details, a constructive proof, and extensions, are given in MacKinnon and Carey [20].

Superconvergent formulas can be constructed by applying averaging or similar strategies to the approximation. As an example, consider a rectangle 1-2-3-4 of side lengths h, k partitioned to two right triangles 1-2-3 and 1-3-4 by diagonal 1-3. The partial derivative u_x at midpoint $(\bar{x}, \bar{y})$ of side 1-3 satisfies

$$u_x(\bar{x}, \bar{y}) = \frac{u_2 - u_1}{h} + \frac{k}{2} u_{xy}(\bar{x}, \bar{y}) + O(\delta^2), \tag{12}$$

$$u_x(\bar{x}, \bar{y}) = \frac{u_3 - u_4}{h} + \frac{k}{2} u_{xy}(\bar{x}, \bar{y}) + O(\delta^2), \tag{13}$$

in the respective triangles with a similar pair of expressions for $u_y(\bar{x}, \bar{y})$. Averaging yields an $O(\delta^2)$ approximation to u_x at the midpoint, where $\delta = \max(h, k)$.

In post-processing the approximate solution to a boundary-value problem, the differential equation can be viewed as "an auxiliary property" and used to construct superconvergent projections. As an example, let us assume that u in the previous interpolation example also satisfies the differential equation

$$pu'' + qu' + ru = f, \tag{14}$$

with p, q, r and f specified functions of position. Now (14) can be used to express u'' in terms of u and u'. Substituting for u'' in the leading Taylor series remainder term of (9)

$$u'(x) = \frac{u_{i+1} - u_i}{h} + \frac{\delta_{i+1} - \delta_i}{2p} (f - qu' - ru)|_x + O(h^2), \tag{15}$$

Table 1 Comparison of derivative errors e_{hx} and e^*_{hx} (superconvergent) at $x = \frac{3}{4}$.

	Linear		Quadratic	
h	e_{hx}	e^*_{hx}	e_{hx}	e^*_{hx}
1/4	1.62	0.35	0.29	0.07
1/8	0.64	0.07	0.06	0.008
1/16	0.29	0.02	0.01	0.0009
Rate	$\approx O(h)$	$\approx O(h^2)$	$\approx O(h^2)$	$\approx O(h^3)$

so that

$$\left[1 + \frac{q}{2p}(\delta_{i+1} - \delta_i)\right] u'(x) = \frac{u_{i+1} - u_i}{h} + \frac{\delta_{i+1} - \delta_i}{2p}(f - ru)|_x + O(h^2). \qquad (16)$$

This yields a post-processing formula for u' that is $O(h^2)$ accurate at any point x. Superconvergent formulas can be similarly constructed for quadratic and higher-degree elements by repeatedly differentiating the differential equation to develop expressions for $u'''(x), \ldots$ as needed. Note that, if the knot values are not inter-polatory but are still superconvergent then the above strategy can still be applied. Representative numerical results are given in Table 1 for the error in the deriva-tive at $x = 3/4$ computed using the local grid point values from the finite element solution to $-u'' + u' + 2u = -e^{-3x}$, $u(0) = u(1) = 0$. Both linear and quadratic elements are considered. In Table 1 e_{hx} is the error at $x = 3/4$ obtained by simply differentiating the approximation and e^*_{hx} is the result using the superconvergent formula.

The Green–Gauss formulas can be similarly employed to provide an auxiliary integral relation based on the differential equation. This then yields a superconver-gent projection for the flux. For example, using the Green–Gauss formula for u'' on (0,1) with test function satisfying $v(1) = 1$, $v(0) = 0$ yields

$$u'(1) = \int_0^1 (u'v' - uv + fv) \, dx. \qquad (17)$$

Setting u_h for u in the right side integrand we have a superconvergent post-processing formula for the derivative. This extraction formula can be generalized for any in-terior point. The difference between the standard formula u'_h and the superconver-gence result u'_* can then be used to construct a local error indicator.

Similarly, for the corresponding problem in two dimensions we get the post-processing projection formula

$$\int_{\partial\Omega} \sigma_h \phi_i \, ds = \int_\Omega (\nabla u_h \cdot \nabla \psi_i - u_h \psi_i + f \psi_i) \, dx \, dy, \qquad (18)$$

where σ_h is the superconvergent normal flux approximation. Application of reduced integration in the boundary integral generates an explicit point-extraction formula.

We have also used a similar idea in Navier-Stokes calculations to extract vorticity boundary data in the decoupled stream function-vorticity algorithm (MacKinnon,

Carey and Murray [21]). Rewriting (18) for the stream function equation $\Delta\psi = \omega$ and rearranging terms

$$\int_\Omega \omega\phi_i\, dx\, dy = \int_\Omega \nabla\psi_h\nabla\psi_i\, dx\, dy - \int_{\partial\Omega} \frac{\partial\psi}{\partial n}\phi_i\, ds \qquad (19)$$

for nodes i on the boundary. Applying Lobatto (nodal) quadrature, the strip integral on the left reduces to an explicit extraction formula for nodal vorticity values ω_i.

Recently, we have developed several results in the form of higher-order compact difference schemes which yield superconvergent approximations at the grid points (e.g., see Spotz and Carey [25]). The basic approach is to again use the differential equation as auxiliary data to approximate high-order derivatives in the Taylor series expansions and generate compact high-order discrete formulas.

To illustrate this approach, consider the standard central difference approximation to the model problem (14) on (0,1) with $p = 1$, $r = 1$, $q = q(x)$, $f = f(x)$, and $u(0) = u(1) = 0$. At interior node i, we have

$$\frac{(u_{i+1} - 2u_i + u_{i-1})}{h^2} + q\frac{(u_{i+1} - u_{i-1})}{2h} + u_i = f_i + \tau_i, \qquad (20)$$

where

$$\tau_i = \frac{h^2}{12}\left(2q_i u'''(x_i) - u^{(4)}(x_i)\right) + O(h^4) \qquad (21)$$

is the truncation error and the scheme is $O(h^2)$. Differentiating (14), substituting in the leading truncation error term of τ_i, differencing, and finally regrouping terms in (20)

$$\hat{p}_i\frac{(u_{i+1} - 2u_i + u_{i-1})}{h^2} + \hat{q}_i\frac{(u_{i+1} - u_{i-1})}{2h} + u_i = \hat{f}_i, \qquad (22)$$

where $\hat{p}_i = 1 + \frac{h^2}{12}(q_i^2 - 2\delta_x q_i)$, $\hat{q}_i = q_i + \frac{h^2}{12}(\delta_x^2 q_i - q_i\delta_x q_i)$ and $\hat{f}_i = f_i - \frac{h^2}{12}(-\delta_x^2 f_i - q_i\delta_x f_i)$ with δ_x, δ_x^2 representing the first and second difference operators. The scheme in (22) is $O(h^4)$ accurate but still involves the same tridiagonal matrix structure. This can be extended to nonuniform grids by mapping to a uniform grid in a reference domain $\xi \in (0,1)$ and then using (22) for the mapped equation. In the above example $u'' + qu' + u = f$ maps to $u_{\xi\xi} + \tilde{q}u_\xi + u = \tilde{f}$ where $\tilde{q}(\xi) = (q\xi_x - \xi_{xx})/\xi_x^2$, $\tilde{f}(\xi) = f/\xi_x^2$ and ξ_x, ξ_{xx} are the metric coefficients associated with the map. The higher-order compact scheme (22) is then applied to the transformed problem.

Higher-order compact schemes can also be developed in two and three dimensions. For example, a weighted-average of the standard 5-point stencil for Laplace's equation in two-dimensions and the 5-point stencil rotated through $45°$ can be taken to obtain a 9-point scheme that is $O(h^6)$. If the convective term is included, the higher-order scheme is $O(h^4)$, even if the convective coefficients are smooth functions of position. However, the derivatives in this case and in three-dimensions become algebraically tedious.

The idea can be used similarly in a defect correction strategy as follows: Let $\hat{u}_i$ denote the central difference approximation on a uniform grid at grid point i to the simple example $-u'' + u = f$. Then the error defect at grid point i satisfies the same form of central difference equation with right-hand side give given by the truncation error

$$\tau_i = \frac{h^2}{12}\left(u^{(4)}(x_i)\right) + O(h^4). \tag{23}$$

Retaining only the leading part of the truncation error and evaluating using the differential equation we obtain an algebraic system for the error correction $\mathbf{e}$ to improve $\hat{\mathbf{u}}$.

This concept of constructing a post-processing calculation for the error is related to ideas for constructing coarse grid approximations to a fine grid error in multigrid/multi-level schemes.

3   A POSTERIORI ESTIMATION AND PROJECTION

Most of the research on *a posteriori* error estimation has been directed towards the problem of local grid refinement or redistribution. However, there is a more basic need for local *a posteriori* error estimators simply from the standpoint of reliability and adequacy of a mesh and approximation. Even in this more restrictive setting, efficient and precise local error estimation is a goal that has yet to be attained. Clearly, local post-processing projections play a key role here.

While estimators purport to relate to *a posteriori* analysis and theoretical estimates, indicators have a broader connotation and most finite element codes employ indicators (often deduced from the form of global estimates or bounds). Error indicators can be broadly categorized as feature indicators, indicators based on superconvergence properties or indicators based on other post-processing projections (such as solving a local boundary-value problem).

Richardson extrapolation is a classical procedure for exploiting asymptotes on two grids that can be utilized to construct a local error indicator. Let f_1, f_2 be computational formulas for two grids of characteristic mesh sizes h_1 and h_2 with $h_2 = \gamma h_1$ and satisfying the asymptotic relations

$$f = f_1 + c_1 h_1^p, \qquad f = f_2 + c_2 h_2^p, \tag{24}$$

where f is the analytic quantity of interest and c_1, c_2 are constants. Then

$$f \sim f^* = w f_1 + (1 - w) f_2, \tag{25}$$

with $w = -\gamma^p/(1 - \gamma^p)$ and $f^* - f_1$ or $f^* - f_2$ provide error indicators. Effective use of Richardson extrapolation globally or locally implies the computations are in the asymptotic regime which may not be a valid assumption in an adaptive scheme.

Residuals in finite element analysis are analogous to the truncation error in finite difference approximation and are a natural choice of error indicator. For the linear differential equation $Lu = f$ the error $u - u_h$ satisfies $Le = r$ where $r = f - Lu_h$ is the residual. The local magnitudes of the residual provide an indication of the error distribution and can be used to guide an adaptive refinement scheme. In the least-squares mixed finite element method the minimization functional is constructed from the residuals and this implies that this residual estimator is exact in the associated norm for this variational formulation (Carey and Pehlivanov [11]).

Moreover, it is local and exact in this norm and is computable. However, this property is restricted to the problem norm. This situation is atypical and serves to introduce an important point that is usually overlooked. Normally, the goal of adapting the mesh to equidistribute the error in some norm often does not address the analyst's objectives which may involve more than one variable and choice of norm. For example, in engineering fluid flow and heat transfer applications the analyst may need accurate wall heat fluxes but less accurate interior results (provided these do not pollute the wall values). Similarly, in this case the flow field may be needed to higher accuracy in certain regions. Such coupled problems can also be treated on the same adaptive grid or on separate adaptive grids (although this latter choice appears to be rarely exercised).

4   PROJECTION BETWEEN GRIDS

There are several instances when more than one grid is used in a simulation and solution data or related results must be transferred between grids. Perhaps the most familiar example is the multigrid approach for iteratively solving linear systems. In a typical two-grid V cycle, computed residuals on the fine grid are projected to a coarser grid, and the error correction on this coarse grid is computed. Then this coarse grid error correction is projected back to the fine grid and added to form the next fine grid iterate.

When the fine grid is obtained through an adaptive grid refinement process then the grids are usually fully nested and data transfer is simplified. For example, a standard quadtree or octree refinement strategy will generate a sequence of imbedded meshes with both elements and nodes fully nested. If, instead, refinement is implemented by point insertion in a Delaunay scheme then coarse grid vertex nodes remain common to fine grids but the Delaunay edge swaps imply that the elements (or edges) are no longer nested. Likewise, if a fine grid is coarsened by deleting nodes and then retriangulating, the resulting coarse grid will have imbedding only of vertex nodes and not elements. Obviously this will not be the case for nodes associated with edges or the interior of elements (as is usually the case with higher-degree elements).

In applications to coupled systems such as viscous flow and heat transfer, the respective equations are usually discretized on the same grid even though this may be quite inefficient. For example, fine grids may be required to capture flow structure in the interior near a shock and a thermal layer near a wall. Using two grids here will be more efficient but will also increase the complexity slightly because intergrid projections will no longer be trivial. In a similar vein there are applications where two or more simulations may be essentially decoupled and the solution to one system becomes simply data on a different grid for another system. Coupled flow and transport again provides a good example: the flow field computed on one grid provides the convective velocity approximation for species transport in, say, an environmental model.

4.1 Projection with constraints

In the above discussion we introduced some basic concepts for projections between spaces and associated grids. If the same space is spanned by different choices of bases then the projection is the identity and the characterization simply involves a change of bases. When the projection is the subspace then the subspace basis can be expressed in terms of the basis for the larger space to again determine the projection. If one space is not a subspace of the other then projections can still be devised by, for instance, interpolation or least squares but the previous bases relations no longer apply. In the present investigation we are concerned with constructing projections that satisfy constraints. Of particular interest are projections that satisfy divergence-free properties or similar constraints. Accordingly, let us consider the constrained least-squares problem: given $\mathbf{u}_h \in V_h$ find $\hat{u}_h \in \widehat{V}_h$ such that

$$I = \int_\Omega (\mathbf{u}_h - \hat{\mathbf{u}}_h)^2 \, dx \tag{26}$$

is minimized where $\widehat{V}_h$ is the space of divergence free functions. Let $\psi_j, \hat{\psi}_j$ be the associated bases so that componentwise

$$\mathbf{u}_h = \sum_{j=1}^N \mathbf{u}_{hj} \psi_j(\mathbf{x}), \qquad \hat{\mathbf{u}}_h = \sum_{k=1}^M \hat{\mathbf{u}}_{hk} \hat{\psi}_k(\mathbf{x}). \tag{27}$$

Taking variations in (26), the least-squares projection is defined by

$$\int_\Omega \hat{\mathbf{u}}_h \cdot \delta \hat{\mathbf{u}}_h \, dx = \int_\Omega \mathbf{u}_h \cdot \hat{\mathbf{u}}_h \, dx. \tag{28}$$

That is,

$$\sum_{j=1}^M \left(\int_\Omega \hat{\psi}_i \hat{\psi}_j \, dx \right) \hat{u}_{1j} = \int_\Omega u_{1h} \hat{\psi}_i \, dx \tag{29}$$

$$\sum_{j=1}^M \left(\int_\Omega \hat{\psi}_i \hat{\psi}_j \, dx \right) \hat{u}_{2j} = \int_\Omega u_{2h} \hat{\psi}_i \, dx \tag{30}$$

or, in matrix form, solve

$$\mathbf{M}\hat{\mathbf{u}}_1 = \mathbf{p}_1, \qquad \mathbf{M}\hat{\mathbf{u}}_2 = \mathbf{p}_2, \tag{31}$$

where $\hat{\mathbf{u}}_1$, $\hat{\mathbf{u}}_2$ are the nodal vectors for the divergence-free approximation. Underintegration can be used to approximate (31) by diagonal matrix systems. For the same or similar meshes with the same degree bases the approximation error in the least-squares fit is of the same or better order than in the original computation for $\mathbf{u}_h$. Hence the projection calculation does not degrade accuracy. If the divergence free space is of lower degree or the mesh is coarser then the approximation to the analytic solution u will be degraded.

Of course, it is not necessary to use this least-squares approach to obtain a divergence free approximation $\hat{\mathbf{u}}_h$ since a simple interpolation of $\mathbf{u}_h$ in $\widehat{V}_h$ will suffice.

Let us assume that the meshes for $\mathbf{u}_h$ and $\hat{\mathbf{u}}_h$ are the same and only the local bases differ. Then we simply need to interpolate the necessary degrees of freedom for the expansion in the divergence-free basis.

4.2   Windowing projections and pollution

"Windowing" to a subregion and solving a local problem on a fine grid can be quite effective in providing more accurate results in subregions. The main detractions are that the boundary error on the window subregion pollutes the subgrid solution, and furthermore, the exterior coarse grid solution is not improved. Several projection strategies for iteratively updating the fine subgrid boundary data may be applied to improve this approach. The ideas are also related to the residual and solution projection concepts in domain decomposition and multigrid techniques.

As an illustrative example, consider the simple two-point boundary-value problem

$$-(au')' + bu = f, \qquad 0 < x < 1, \tag{32}$$

with $u(0) = u(1) = 0$, and which has an interior layer in the subregion (α, β). The problem is first solved on a coarse grid to produce an approximate solution u_{h_c} on $(0, 1)$ where h_c is the coarse grid mesh size. This coarse grid approximation is then used to identify the layer and the window (α, β). The "true" window problem is (32) on (α, β) with exact solution values $u(\alpha)$ and $u(\beta)$ specified. The approximate problem utilizes the coarse grid solution to provide end conditions. That is, the improved local approximation u_{h_f} is obtained by solving the window problem on a fine subgrid of mesh size h_f with boundary data

$$u_{h_f}(\alpha) = u_{h_c}(\alpha), \quad u_{h_f}(\beta) = u_{h_c}(\beta). \tag{33}$$

The composite solution u_h^* is then

$$u_h^* = \begin{cases} u_{h_f} & \text{on } (\alpha, \beta), \\ u_{h_c} & \text{elsewhere.} \end{cases} \tag{34}$$

The adjustment of the exterior coarse grid solution to incorporate the effect of the new subgrid information is determined as follows. Introducing (34) into (32), the residual r_h^* corresponding to the composite approximation u_h^* on $(0, 1)$ may be defined and then projected to the original coarse grid as r_{h_c}. The error correction e_{h_c} for this coarse grid residual is then computed, and the new coarse grid solution is

$$\bar{u}_{h_c} = u_{h_c} + e_{h_c}. \tag{35}$$

This solution provides new boundary data for the local subgrid problem. The coupled iteration of coarse and fine grid solutions with residual projection can be repeated to meet a desired solution tolerance. This scheme is equivalent to a block iteration algorithm with hierarchic basis to characterize the error correction on the local subgrid. Of course, in the one dimensional case, local adaptive refinement is superior, and the purpose here is merely to demonstrate the subgrid approach in the simplest setting.

For the higher-dimensional linear elliptic problem

$$Lu = f \quad \text{in } \Omega, \tag{36}$$

with $u = g$ on $\partial\Omega$, the corresponding variational problem is to find $u \in H(\Omega)$ satisfying $u = g$ on $\partial\Omega$ and such that

$$a(u, v) = f(v) \qquad \forall v \in H_0(\Omega) \tag{37}$$

where $a(\cdot, \cdot)$ and $f(\cdot)$ are the corresponding bilinear and linear functionals with $v = 0$ on $\partial\Omega$. Introducing a coarse grid discretization Ω_h, the coarse grid solution u_{h_c} fits the specified data g on $\partial\Omega_h$ and satisfies

$$a(u_{h_c}, v_{h_c}) = f(v_{h_c}) \qquad \forall v_{h_c} \in H_{h_c} \subset H_0(\Omega). \tag{38}$$

Solving the resulting linear system yields u_{h_c}. Let $\widetilde{\Omega}$ be the imbedded subregion $\widetilde{\Omega} \subset \Omega$ with boundary data u_{h_c} on $\partial\widetilde{\Omega}$. Introducing a fine subgrid on $\widetilde{\Omega}$, the imbedded subgrid problem is: find $\tilde{u}_{h_f}$ on $\widetilde{\Omega}$ matching u_{h_c} on $\partial\widetilde{\Omega}$ and satisfying

$$a(\tilde{u}_{h_f}, \tilde{v}_{h_f}) = f(\tilde{v}_{h_f}) \qquad \forall \tilde{v}_{h_f} \in H_{h_f} \subset H_0(\widetilde{\Omega}). \tag{39}$$

The composite solution is again

$$u_h^* = \begin{cases} u_{h_f} & \text{on } \widetilde{\Omega}, \\ u_{h_c} & \text{on } \Omega \setminus \widetilde{\Omega}. \end{cases} \tag{40}$$

Introducing (40) into (36), the composite grid residual is

$$r_h^* = Lu_h^* - f \tag{41}$$

and the composite grid error $e_h^* = u - u_h^*$ satisfies

$$Le_h^* = -r_h^*. \tag{42}$$

From (42), the coarse grid error correction e_{h_c} (due to the subgrid improvement in the solution) is defined by projection to the coarse grid basis as

$$a(e_{h_c}, v_{h_c}) = -r_h^*(v_{h_c}) \qquad \forall v_{h_c} \in H_{h_c}. \tag{43}$$

Solving (43), the adjusted coarse grid solution is

$$\bar{u}_{h_c} = u_{h_c} + e_{h_c} \tag{44}$$

and the procedure may be continued to further fine grid improvements and coarse grid adjustments.

The procedure can also be formulated in terms of the subgrid error and coarse grid corrections. That is, following calculation of u_{h_c}, immediately let $u_h^* = u_{h_c}$ for the composite grid. Then (42) again applies on Ω but now we solve on $\widetilde{\Omega}$

$$a(e_{h_f}, v_{h_f}) = -r_h^*(v_{h_f}) \qquad \forall v_{h_f} \in H_{h_f}, \tag{45}$$

with $e_{h_f} = 0$ on $\partial\widetilde{\Omega}$. The composite solution $u_{h_c} + e_{h_f}$ is, of course, again u_h^* in (40). After solving for the coarse grid error correction in (43), the coarse grid solution can be adjusted as in (44) and then the coarse grid residual is defined by $\bar{r}_{h_c} = L\bar{u}_{h_c} - f$. Projecting to the fine subgrid on $\widetilde{\Omega}$,

$$a(\bar{e}_{h_f}, v_{h_f}) = -\bar{r}_{h_c}(v_{h_f}), \tag{46}$$

with $\bar{e}_{h_f} = 0$ on $\partial\widetilde{\Omega}$. This approach is formally equivalent to the previous scheme. Similar ideas are utilized in multilevel projections as seen next.

4.3   Multigrid projection of residuals

A basic multigrid procedure involves a few iterations on Ω_h, followed by residual projections and similar iterations for error corrections on coarser grids $\Omega_{2h}, \Omega_{4h}, \Omega_{8h}, \ldots$, with an elimination solution at the coarsest level. The various error corrections are then interpolated and smoothed in succession up to the fine grid Ω_h and the fine grid solution is updated. This entire multigrid sequence, or "cycle", is repeated to convergence within a specified tolerance. Since the mid-frequency modes on a given grid become the high-frequency modes on the next coarse grid, the iteration error is damped out rapidly on successive grids which in some sense act as "filters" for different frequency groups.

Since the Galerkin weighted residual method is based on forcing the residual to zero in the continuous problem, it is instructive to examine the residual projection and error equations from this standpoint.

Consider the linear self-adjoint boundary-value problem

$$Lu = f \quad \text{in } \Omega, \tag{47}$$

with

$$u = 0 \quad \text{on } \partial\Omega, \tag{48}$$

where L is a linear elliptic operator on bounded domain Ω with regular boundary $\partial\Omega$.

The residual r for admissible u is defined by $r = Lu - f$. Introducing the test function v in the duality pairing $\langle r, v \rangle = r(v)$ and integrating by parts, the weighted residual condition $r(v) = 0$ implies the weak statement: find $u \in H$ satisfying $u = 0$ on $\partial\Omega$ and such that

$$a(u, v) = f(v) \tag{49}$$

for all $v \in H$ with $v = 0$ on $\partial\Omega$.

Introducing the finite element discretization Ω_h and a basis to define $H^h \subset H$, the approximate problem is: find $u_h \in H^h$ such that

$$a(u_h, v_h) = f(v_h) \quad \text{for all } v_h \in H^h \tag{50}$$

and the resulting algebraic system is

$$\mathbf{A}_h \mathbf{u}_h = \mathbf{b}_h. \tag{51}$$

Assume now that (51) is solved only approximately by a few sweeps of a smoothing iteration to produce $\mathbf{u}_h^*$ as the first step of a multigrid scheme. Now, define the finite element interpolant u_h^* of the grid point values $\mathbf{u}_h^*$, and hence the associated residual $r_h = f - Lu_h^*$. Equivalently, in the weak form define r_h by its "action" as

$$f(v_h) - a(u_h^*, v_h) = r_h(v_h) \tag{52}$$

where u_h^* is the iterate approximation of u_h satisfying (50). Subtracting (52) from (50), the error $e_h^* = u_h - u_h^*$ satisfies

$$a(e_h^*, v_h) = r_h(v_h) \quad \text{for all } v_h \in H^h. \tag{53}$$

That is, the error correction on the fine grid Ω_h satisfies (53) with $r_h(v_h)$ defined by the expression on the left in (52). Since (53) holds for all $v_h \in H^h$ then, in particular, it must apply on Ω_{2h} for the subspace $H^{2h} \subset H^h$, so

$$a(e_h^*, v_{2h}) = r_h(v_{2h}) \quad \text{for all } v_{2h} \in H^{2h}, \tag{54}$$

where now from (52)

$$r_h(v_{2h}) = f(v_{2h}) - a(u_h^*, v_{2h}) \tag{55}$$

is the appropriate projection of the residual to the coarse grid. The best (Galerkin) approximation $e_{2h} \in H^{2h}$ to (54) satisfies

$$a(e_{2h}, v_{2h}) = r_h(v_{2h}) \quad \text{for all } v_{2h} \in H^{2h}, \tag{56}$$

with $r_h(v_{2h})$ defined in (55), since after subtraction $a(e_h^* - e_{2h}, v_{2h}) = 0$ for all $v_{2h} \in H^{2h}$. Now (56) is analogous to (50) and defines a coarse grid approximation e_{2h} to the error e_h^*. Carrying out the integrations in (56) with the coarse grid basis, the matrix problem is

$$\mathbf{A}_{2h}\mathbf{e}_{2h} = \mathbf{r}_{2h}, \tag{57}$$

where $\mathbf{A}_{2h}, \mathbf{r}_{2h}$ are now defined explicitly (and naturally) by the forms in (56) and (55) together with the choice of basis.

In a basic two-level scheme, (57) is then solved by direct elimination to yield e_{2h} satisfying (56). Since e_h^* satisfies (54) and e_{2h} is defined by (56) on H^{2h},

$$a(e_h^*, v_{2h}) = a(e_{2h}, v_{2h}) \quad \text{for all } v_{2h} \in H^{2h}. \tag{58}$$

That is, e_{2h} is the elliptic projection of e_h^* on H^{2h}. The detailed forms of the matrices $\mathbf{A}_h$, $\mathbf{A}_{2h}$ and the projections can be obtained by introducing the respective bases $\{\phi_j^h\}$, $\{\phi_k^{2h}\}$ of rank N and M for the approximation spaces. For the system matrices in (51) and (57)

$$(A_h)_{ij} = a(\phi_i^h, \phi_j^h), \qquad (A_{2h})_{kl} = a(\phi_k^{2h}, \phi_l^{2h}). \tag{59}$$

Setting $v_{2h} = \phi_k^{2h}$ in the residual projection formula (55),

$$r_h(\phi_k^{2h}) = f(\phi_k^{2h}) - a(u_h^*, \phi_k^{2h}). \tag{60}$$

Since $H^{2h} \subset H^h$, we may express ϕ_k^{2h} as a linear combination of fine grid basis functions $\{\phi_j^h\}$,

$$\phi_k^{2h}(\mathbf{x}) = \sum_{j=1}^{N} P_{kj}\phi_j^h(\mathbf{x}). \tag{61}$$

Substituting in (60) and simplifying,

$$r_h(\phi_k^{2h}) = \sum_{j=1}^{N} P_{kj}[f(\phi_j^h) - a(u_h^*, \phi_j^h)] \tag{62}$$

so that, from (52),

$$r_h(\phi_k^{2h}) = \sum_{j=1}^{N} P_{jk}r_h(\phi_j^h). \tag{63}$$

For the Lagrange basis $\phi_j(\mathbf{x}_\ell) = \delta_{j\ell}$ and hence (61) implies $P_{kj} = \phi_k^{2h}(\mathbf{x}_j)$ is the corresponding projection matrix for the fine grid residual vector. Similarly,

$$e_{2h}(\mathbf{x}) = \sum_{k=1}^{M} e_k\phi_k^{2h}(\mathbf{x}) = \sum_{k=1}^{M}\sum_{j=1}^{N} e_k P_{kj}\phi_j^h(\mathbf{x}) \tag{64}$$

and the fine grid (interpolated) values $\tilde{e}_h(x_j)$ can therefore be expressed by the projection as

$$\tilde{e}_j = \sum_{k=1}^{M} e_k P_{kj}, \tag{65}$$

where e_k are the nodal values for e_{2h}. That is, $\tilde{\mathbf{e}}_h = \mathbf{P}^T\mathbf{e}_{2h}$.

In a similar vein, multigrid acceleration can be generalized to a p-type multilevel form. Projection to lower levels then implies simply reducing the degree of the local element basis and projection of error corrections to higher levels involves corresponding increases in the element degree. More specifically, let H^p, H^q with bases $\{\phi_j^p\}$, $\{\phi_j^q\}$ denote the approximation spaces for a discretization of elements of degree p and q respectively with $p < q$. The discrete system on the larger space H^q is

$$\mathbf{A}_q\mathbf{u}_q = \mathbf{b}_q. \tag{66}$$

The residual for iterate $\mathbf{u}_q^*$ of (66) is $\mathbf{r}_q = \mathbf{b}_q - \mathbf{A}_q\mathbf{u}_q^*$ and projecting to the coarse level $p < q$, $\mathbf{r}_p = \mathbf{P}\mathbf{r}_q$. The error equation on H^p becomes

$$\mathbf{A}_p\mathbf{e}_p = \mathbf{r}_p. \tag{67}$$

Solving for $\mathbf{e}_p$ and taking the projection up to H^q, the fine-level correction is

$$\mathbf{e}_q = \widehat{\mathbf{P}}^T\mathbf{e}_p \tag{68}$$

where $\widehat{\mathbf{P}}$ is the projection matrix (65)); e.g., interpolation or elliptic projection.

5   CONCLUDING REMARKS

In this study we have focused on the construction of different types of projections in finite element analysis and applications. Of particular interest are projections in post-processing calculations both as superconvergent extraction formulas to provide more accurate results at little additional cost and as a means for computing error estimators to guide local adaptive refinement. Some additional related ideas concerning projections of residuals in windowing and for multi-level solution strategies are discussed. Projections between grids and particularly non-nested grids are briefly described as well as the question of handling constraints. While a number of advances have been made in these areas, they continue to be topics of strong research interest. There is a pressing need to incorporate these projection and post-processing concepts in engineering applications codes. Part of the problem in, for instance, the adaptive refinement approach is that schemes designed to equidistribute the error do not necessarily address the need for local accuracy in a particular solution variable or subregion.

Acknowledgements   The research on adaptive strategies has been supported by the National Science Foundation and the recent work on projections by the Waterways Experiment Station. I would like to express my appreciation to Varis and Tija for assisting in preparation of this paper at short notice. Some of the ideas and examples here are covered in more detail in the forthcoming monograph Carey [6].

REFERENCES

1. Ainsworth, M., J. Z. Zhu, A. W. Craig, and O. C. Zienkiewicz, *Analysis of the Zienkiewicz–Zhu Error Estimator in the Finite Element Method*, Internat. J. Numer. Methods Engrg. **28** (1989), 2161–2174.
2. Babuška, I. and A. Miller, *The Post-Processing Approach in the Finite Element Method*, Internat. J. Numer. Methods Engrg. **20** (1984), 1085–1129, 2311–2324.
3. Babuška, I. and R. Rodríguez, *The Problem of Selection of an A Posteriori Error Indicator Based on Smoothing Techniques*, Internat. J. Numer. Methods Engrg. **36** (1993), 539–567.
4. Barlow, J., *Optimal Stress Locations in Finite Element Models*, Internat. J. Numer. Methods Engrg. (1976), 243–251.
5. Briggs, W., "Multigrid Tutorial," SIAM, 1987.
6. Carey, G. F., "Computational Grids: Generation, Adaptation and Solution Strategies," Taylor and Francis Publications, 1997.
7. Carey, G. F., *A Mesh Refinement Scheme for Finite Element Computations*, Comp. Methods Appl. Mech. Engrg. **7, 1** (1976), 93–105.
8. Carey, G. F., *Derivative Calculation from Finite Element Solutions*, Comp. Methods Appl. Mech. Engrg. **35** (1982), 1–14.
9. Carey, G. F., S. Chow, and M. Seager, *Approximate Boundary Flux Calculation*, Comp. Methods Appl. Mech. Engrg. **50** (1985), 107–120.
10. Carey, G. F. and B. A. Finlayson, *Orthogonal Collocation on Finite Elements*, J. Ch. Eng. Sci **30, 1** (1974), 69–79.

11. Carey, G. F. and A. I. Pehlivanov, *Local Error Estimation and Adaptive Remeshing Scheme for Least-Squares Mixed Finite Elements*, in press, Comp. Methods Appl. Mech. Engrg. (1997).

12. Carey, G. F. and M. F. Wheeler, C^0-*Collocation-Galerkin Methods*, in "Codes for Boundary Value Problems in Ordinary Differential Equations," Lecture Notes in Computer Science, Springer-Verlag, 1979, pp. 250–256.

13. Chen, M. and Y. Huang, "High Accuracy Theory of Finite Element Methods," Hunan Sci. Tech. Press, Chengdu, 1995.

14. Douglas, J. Jr. and T. Dupont, "Collocation Methods for Parabolic Equations in a Single Space Variable," Springer Lecture Notes, 1974.

15. Dupont, T., *A Unified Theory of Superconvergence for Galerkin Methods for Two-Point Boundary Problems*, SIAM J. Numer. Anal. **13:3** (1976), 362–368.

16. Hinton, E. and J. S. Campbell, *Local and Global Smoothing of Discontinuous Element Functions Using a Least Squares Method*, Internat. J. Numer. Methods Engrg. **8** (1974), 461–480.

17. Kelly, D. W., *The Self-Equilibration of Residuals and Complementary A-Posteriori Error Estimates*, Internat. J. Numer. Methods Engrg. **20** (1984), 1491–1506.

18. Křížek, M. and P. Neittaanmäki, *On Superconvergence Techniques*, Acta Appl. Math. **9** (1987), 175–198.

19. Lazarov, R. D., S. S. Chow, A. I. Pehlivanov, and G. F. Carey, *Superconvergence Analysis of Approximate Boundary-Flux Calculations*, Numerische Mathematik **63** (1992), 483–501.

20. MacKinnon, R. and G. F. Carey, *Superconvergent Derivatives: A Taylor Series Analysis*, Internat. J. Numer. Methods Engrg. **28** (1989), 489–509.

21. MacKinnon, R., G. F. Carey and P. Murray, *A Procedure for Calculating Vorticity Boundary Conditions in the Stream Function-Vorticity Method*, Comm. Numer. Methods Engrg. **6** (1990), 47–48.

22. Oden, J. T., L. Demkowicz, W. Rachowicz, and T. A. Westermann, *A Posteriori Error Analysis in Finite Elements*, Comp. Methods Appl. Mech. Engrg. **82** (1990), 183–203.

23. Pehlivanov, A. I., "Superconvergence and Interior Estimates in the Finite Element Method," Ph.D. thesis, University of Sofia, 1989.

24. Schatz, A. H. and L. B. Wahlbin, *Interior Maximum Norm Estimates for Finite Element Methods*, Math. Comp. **31** (1977), 414–442.

25. Spotz, W. F. and G. F. Carey, *High-Order Compact Schemes for the Stream-Function Vorticity Equations*, Internat. J. Numer. Methods Engrg. **38** (1995), 3497–3512.

26. Wahlbin, L. B., "Superconvergence in Galerkin Finite Element Methods," Springer, 1605, 1995.

27. Wheeler, M. F., *A Galerkin Procedure for Estimating the Flux for Two-Point Problems*, SIAM J. Numer. Anal. **11** (1974), 764–768.

28. Wheeler, M. F. and J. Whiteman, *Superconvergent Recovery of Gradients on Subdomains from Piecewise Linear Finite Element Approximations*, J. Num. Meth. **3:1** (1987), 357–374.

29. Zhu, Q. D. and Q. Lin, "The Hyperconvergence Theory of Finite Elements," Hunan Sci. Tech. Press, Chengdu, 1989.

30. Zienkiewicz, O. C. and J. Z. Zhu, *Superconvergent Derivative Recovery Tech-*

niques and A Posteriori Error Estimation in the Finite Element Method, Internat. J. Numer. Methods Engrg. **33** (1992), 1331–1382.

31. Zlámal, M., *Superconvergence and Reduced Integration in the Finite Element Method*, Math. Comp. **32** (1978), 663–685.

Element analysis method and superconvergence

CHUANMIAO CHEN* Institute of Computer Science, Hunan Normal University, Changsha, Hunan, 410081, People's Republic of China

Abstract   The element analysis method consists of three parts: combining element techniques on almost uniform mesh, piecewise assembly to simulate any domain and locally refined mesh at singularity. Besides, sharp superconvergence estimates on quasiuniform mesh for nonlinear elliptic, linear and nonlinear evolution problems are also derived.

1   INTRODUCTION

Consider the linear elliptic problem to find $u \in H_0^1(\Omega)$ satisfying

$$A(u,v) = (f,v), \quad v \in H_0^1(\Omega), \tag{1}$$

where Ω is a d-dimensional domain with the boundary $\partial\Omega$, and

$$A(u,v) = \int_\Omega (a_{ij}D_iuD_jv + a_0uv)\,dx$$

is bounded and coercive in $H_0^1(\Omega)$. We make the assumption A_p^l: for some integer $l \geq 2$ and $p \geq 2$, (1) has a unique solution $u \in W^{l,p}(\Omega) \cap H_0^1(\Omega)$ and $\|u\|_{l,p,\Omega} \leq c\|f\|_{l-2,p,\Omega}$. Assume that the domain Ω is subdivided into a finite number of elements τ and that the mesh J^h on Ω is quasiuniform, with mesh domain $\Omega_h = \{\tau | \tau \in J^h\}$. Denote the m degree finite element space by

$$S_m^h = \{v \mid v \in C(\Omega_h),\ v|_\tau \in P_m(\tau),\ \tau \in J^h,\ v = 0 \text{ on } \partial\Omega_h\}.$$

Define the finite element approximation $u_h \in S_r^h$ of (1) by

$$A(u_h,v) = (f,v), \quad v \in S_m^h. \tag{2}$$

*Supported by The National Natural Science Funds of China.

Formulae (1) and (2) lead to the basic orthogonal relation for the error $e = u_h - u$,

$$A(e, v) = 0, \quad v \in S_m^h. \tag{3}$$

It is well known that the error estimate

$$\|e\|_{s,\Omega} \leq ch^{m+1-s}\|u\|_{m+1,\Omega}, \quad s = 0, 1, \tag{4}$$

and the negative norm estimate

$$\|e\|_{-s,\Omega} \leq ch^{m+1+s}\|u\|_{m+1,\Omega}, \quad 1 \leq s \leq l - 2, \, l \leq m + 1, \tag{5}$$

hold. From (5), two ideas come, i.e., make the local average and seek superconvergence points.

First, (5) means the approximate orthogonality of e and, because of the arbitrarity of v, which is also a local property (in many cases, unfortunately, the approximate orthogonality is impossible to be valid only on an element) and leads to the local average method, i.e., use the splines on the uniform meshes to construct the kernel function K_h^α with a small support, and makes the convolution $K_h^\alpha * \partial_h^\alpha u_h$ to approximate $D^\alpha u$ with the optimal accuracy order,

$$\|K_h^\alpha * \partial_h^\alpha u_h - D^\alpha u\|_{0,\infty,\Omega_0} = O(h^{2m}), \quad \Omega_0 \subset\subset \Omega.$$

See Bramble and Schatz [4], Thomée [36], Thomée ([37] for linear parabolic case) and Chen ([11] for nonlinear elliptic case). This result is very nice, but valid usually only inside the domain and then its practical application is restricted.

Second, because (e, v) is of higher accuracy than $\|e\|$, the error $e = u_h - u$ has rapidly to change its sign in Ω in order that the cancellation of the positive and negative values makes the integral less. We want to know whether the distribution of zeros of e has certain regulation, and whether we may find their approximate points which are dependent on the shape of elements and independent of the coefficients of A and the behavior of u. It leads to the study of superconvergence points, e.g., the element analysis method and the recent theory by local symmetry proposed by Schatz, Sloan and Wahlbin ([39], 1995). We should point out that the element analysis method generalizes the tensor product [27] and reduced integration [43], and is successful, especially, in studying a lower degree elements.

The basic idea of the element analysis method is to construct an interpolant $u_I \in S_m^h$ of u superclose to the finite element solution $u_h \in S_m^h$ such that

$$D^s(u_h - u_I) = O(h^{m+2-s}), \quad s = 0, 1, \, m + s \geq 2, \tag{6}$$

and then

$$D^s e(x) = D^s(u_I - u)(x) + O(h^{m+2-s}), \quad x \in \tau.$$

So, the change of $D^s(u_I - u)(x)$ approximately describes the behavior of $D^s e(x)$ in an element. In particular, at some special point x_0 one can find that $D^s(u_I - u)(x_0) = O(h^{m+2-s})$. While from (3) we see that the estimate (6) is equivalent to construct the interpolation $u_I \in S_m^h$ having the following approximate orthogonality

$$A(u_I - u_h, v) = A(u_I - u, v) = \begin{cases} O(h^{m+1})\|v\|_{1,p,\Omega}, & v \in S_m^h, \text{ if } m \geq 1; \quad (7) \\ O(h^{m+2})\|v\|_{2,p,\Omega}^*, & v \in S_m^h, \text{ if } m \geq 2. \quad (8) \end{cases}$$

where

$$\|v\|_{2,p,\Omega}^* = \left(\sum_\tau \|v\|_{2,p,\tau}^p \right)^{1/p}.$$

The latter case is a more definite problem to be solved easier. Note that superconvergence estimates in L^∞ and $W^{1,\infty}$ can be implemented by the regularized Green function g_h (see Frehse and Rannacher [28]) and gradient type Green function G_h, $\|G_h\|_{1,1} \le c|\ln h|$ (see Rannacher and Scott [35], and Chen and Liu [22]).

2 THE COMBINING ELEMENT TECHNIQUES

2.1 The one-dimensional case

Define the polynomials in a standard element $\tau = (-h, h)$,

$$M_0 = 1, \quad M_1 = t, \quad M_2 = (t^2 - h^2), ..., M_{m+1} = D_t^{m-1}(t^2 - h^2)^m,$$

which possess the following properties:
1. $l_m(t) = D_t M_{m+1}(t)$ is m-degree Legendre orthogonal polynomial: $(l_m, v) = 0$ if v is a polynomial of degree $\le m - 1$;
2. $M_m(\pm h) = 0$ for $m \ge 2$;
3. $(M_{m+1}, v) = 0$ if v is a polynomial of degree $\le m - 2$.

Expanding $u'(x) = \sum b_{j+1} l_j(x)$ in τ and integrating with respect to x, we get

$$u(x) = u_m + R, \quad u_m = \sum_{j=0}^m b_j M_j(x), \quad R = \sum_{j=m+1}^\infty b_j M_j(x),$$

where $b_0 = (u(-h) + u(h))/2$ and the part sum $u_m(x)$ satisfies the continuity conditions $u_m(\pm h) = u(\pm h)$. It is easy to see that for a special bilinear form $A(u, v) = \int_\Omega u'v' dx$, then $u_h = u_m$ in each τ_j and

$$e(x) = R(x) = b_{m+1} M_{m+1}(x) + O(h^{m+2}),$$
$$e\prime(x) = R'(x) = b_{m+1} l_m(x) + O(h^{m+1}), \tag{9}$$

i.e., they are of higher order accuracy at $m + 1$ order Lobatto points x_j^0, $j = 1, 2, ..., m + 1$, and m order Gauss points x_j', $j = 1, 2, ..., m$, respectively.

In general case of variable coefficients and quasiuniform meshes, there are three kinds of superconvergence:

1*. at each mesh points x_j (Douglas and Dupont [25], [26], 1972–1974),

$$|e(x_j)| \le ch^{2m} \|u\|_{m+1,\Omega}, \quad m \ge 2;$$

2*. at $m + 1$-order Lobatto points x_{ji} within each element τ_j (Chen [7], 1979),

$$e(x_{ji}) = O(h^{m+2}), \quad m \ge 2, \ i = 1, 2, ..., m + 1;$$

3*. at m-order Gauss points x_{ji}' within each element τ_j (Chen [7]),

$$e'(x_{ji}') = O(h^{m+1}), \quad m \ge 1, \ i = 1, 2, ..., m.$$

In order to get the global superconvergence we can make the $m+1$ degree interpolant $I_{m+1}u_h$ by use of values of u_h at $m+2$ successive Lobatto points (or nodal points if $m=1$) (Chen [10]) and then

$$\|I_{m+1}u_h - u\|_{s,\infty,\Omega} = O(h^{m+2-s}), \quad s = 0, 1, \ m+s \geq 2.$$

2.2   The rectangular element

Consider m-degree serendipity family in the standard rectangle $\tau = (-h, h) \times (-k, k)$:

$$Q_\lambda(m) = \sum_{(i,j)\in I_{m,\lambda}} b_{ij}x^iy^j, \quad 1 \leq \lambda \leq m,$$

here the index set $I_{m,\lambda}$ satisfies $\{0 \leq i, \ j \leq m, \ i + j \geq m + \lambda\} \subset I_{m,\lambda} \subset \{0 \leq i, \ j \leq m\}$, $\lambda \geq 1$ or 2. When $\lambda = m$, $Q_m(m)$ is m-degree tensor product. Using one-dimensional expansion, first in x and then in y, we have

$$u(x, y) = \sum_{i,j=0}^{\infty} b_{ij} M_i(x) M_j(y) = u_m + R, \quad u_m = \sum_{(i,j)\in I_{m,\lambda}} b_{ij} M_i(x) M_j(y),$$

and $u_m = u$ at $x = \pm h$ and $y = \pm k$. Therefore,

$$R = b_{r,0}M_r(x) + b_{0,r}M_r(y) + O(h^{m+2}), \quad r = m + 1,$$
$$R_x = b_{r,0}l_m(x) + O(h^r), \quad R_y = b_{0,r}l_m(y) + O(h^r). \tag{10}$$

i.e., the two variables in the main part of the error are separated, and so the zero points of them are the product of one-dimensional superconvergence points.

Consider the square $\Omega = \{0 < x, \ y < 1\}$, nodal points set $Z_h = \{(x_i, y_j) \mid x_i = ih, \ y_j = jk, \ i = 0, 1, ..., L, \ j = 0, 1, ..., N\}$, and rectangular elements $\tau_{ij} = \{x_{i-1} < x < x_i, y_{j-1} < y < y_j\}$. Let $u_I \in S_r^h$ be the piecewise interpolant of u just constructed. Then after integration by parts on each element we have for $R = u_I - u$,

$$A(R, v) = \sum_\tau \int_{\partial\tau} R a_{ij} D_j v \cos(n, x_i) \, ds + \int_\Omega RAv \, dx \, dy.$$

Here the difference from one-dimensional case is that $R \neq 0$ on the boundary $\partial\tau$. So the integrals on boundary $\partial\tau$ do not disappear, and the direct cancellation of each other is impossible. We observe that the integral on an element with constant coefficients:

$$\int_\tau a_{ij} D_i R D_j v \, dx \, dy$$

is still approximately orthogonal if $i = j$. When $a_{12} \neq 0$ the approximate orthogonality is lost. However, using integration by parts and exchanging the order of differentiation, the main part of the integral is transformed to the form

$$\int_\tau a_{12}b_{r,0}l_m(x)v_y \, dx \, dy = -\int_\tau a_{12}b_{r,0}M_r(x)v_{xy} \, dx \, dy = -\int_{\partial\tau} a_{12}b_{r,0}M_r(x)v_x \, dx.$$

$$\tag{11}$$

Therefore, the boundary integrals between two adjacent elements can be still cancelled, and become the higher order quantities.

In general case of variable coefficients there are three kinds of superconvergence:

1*. when $m \geq 1$ and $\lambda \geq 1$, then at m-order Gauss points in each element τ_{ij}, the gradient

$$De(x'_{i\alpha}, y'_{j\beta}) = O(h^{m+1} \ln h), \quad \alpha, \beta = 1, 2, ..., m;$$

2*. when $m \geq 2$ and $\lambda \geq 1$, then at $m + 1$-order Lobatto points on each element τ_{ij},

$$e(x_{i\alpha}, y_{j,\beta}) = O(h^{m+2} \ln h), \quad \alpha, \beta = 1, 2, ..., m + 1;$$

3*. when $m \geq 3$ and $\lambda \geq 2$, $a_{12} = 0$ and the coefficients a_{11} and a_{22} are constants, then at angular nodal points in each element τ_{ij},

$$e(x_i, y_j) = O(h^{m+3} \ln h).$$

Especially, $e(x_i, y_j) = O(h^{2m} \ln h)$, at least, for $m = 1, 2$ and 3.

The above results are proved first in L^2 (without the factor $\ln h$). For 1* and $m = 1, 2$, Zlámal ([42], 1977, the approximate orthogonality) studied the case of $a_{12} = 0$, while the difficult case of $a_{12} \neq 0$ was considered by Chen ([6], 1978, used the technique exhibited here) and by Lesaint and Zlámal ([31], [43], 1979, adopt the reduced integration). For 3*, the case of $A = -\triangle$ is studied by Douglas, Dupont and Wheeler ([27], 1974, used the tensor product which requires $a_{12} = 0$). All results mentioned above are systematically proved by Chen ([9], 1981, the element analysis method), and are extended in the interior maximum norm (by Chen [11]) and the maximum norm (by Zhu [41]).

2.3 The triangular elements

The combining element technique is also applied by Chen ([5], [8], 1978) to study superconvergence of triangular linear element. A simplified proof is formulated as follows.

Assume that Ω is a convex polygon, J^h the uniform mesh of Ω and S^h the linear element subspace. Consider a typical element $\tau = \triangle z_1 z_2 z_3$ with the vertex $z_j = (x_j, y_j)$. Directions of three sides and lengths are S_j and l_j, respectively. Let h be the largest side and $|\tau|$ the area of τ. Let $u_I \in S^h$ be the linear interpolant of u, $R = u_I - u$. For the simplicity, consider $A = -\triangle$. From Green's formula, we have

$$A(R, v) = \sum_\tau \int_{\partial \tau} R \partial_n v \, ds, \quad v \in S^h,$$

where the outward normal derivatives $\partial_n v$ is discontinuous at the boundary $\partial \tau$, the line integrals on the right-hand side cannot be directly cancelled each other. But the tangential derivatives along the boundary $\partial \tau$ is continuous. The key idea is to express the normal derivatives as a linear combination of the tangential derivatives along two sides: $\partial_n v = c_1 \partial_1 v - c_2 \partial_3 v$ on S_1, where the constants c_1 and c_2 are dependent only on the interior angles of τ, and independent of the parameter h. For example, we consider the line integral

$$J_1(\tau) = \int_{S_1} R\partial_n v \, ds_1 = \int_{S_1} R(c_1\partial_1 v - c_2\partial_3 v) \, ds_1.$$

The following two transform formulas are used.

If $R(0) = R(h) = 0$, then after integration by parts twice,

$$\int_0^h R(r) \, dr = \int_0^h \phi(r)u_{rr} \, dr, \quad \phi(r) = r(h-r)/2,$$

where $\phi(r)$ can be extended as a function $\phi(r, s)$: $\phi(r, s) = \phi(r)$ along the straight line $s = $ const. In particular, $\partial_2\phi = 0$ and $\phi(s_3) = s_3(l_3 - s_3)l_1^2/l_3^2$ on S_3.

With the help of area coordinate transform and Green's formula, we have

$$\int_{S_1} F \, ds_1 = l_1 l_3^{-1} \int_{S_3} F \, ds_3 - \frac{l_1 l_2}{2|\tau|} \int_{\tau} \partial_2 F \, dx \, dy,$$

where the last integral is actually a high order disturbance.

Now we can rewrite (simply, denote $r = s_1$, $s = s_2$, $t = s_3$)

$$\begin{aligned}
J_1(\tau) &= \int_{S_1} R\partial_n v \, ds_1 = \int_{S_1} c_1\phi u_{rr}v_r \, dr - \int_{S_1} c_2\phi u_{rr}v_t \, dr \\
&= \int_{S_1} c_1\phi u_{rr}v_r \, dr - \int_{S_3} c_3\phi u_{rr}v_t \, dt + c_4 \int_{\tau} \phi u_{rrs}v_t \, dx \, dy.
\end{aligned} \tag{12}$$

So, $J_1(\tau)$ is decomposed into two line-integrals containing the tangent derivatives v_r and v_t plus a higher order double integral. Because the tangent derivatives are continuous and equal to zero along $\partial\Omega$, then

$$J_1(\Omega) = 0 + 0 + c_4 \int_{\Omega} \phi u_{rrs}v_t \, dx \, dy = O(h^2)\|u\|_{3,p}\|v\|_{1,p'}.$$

Similarly for two other directions. Finally we get the weak estimate (7) with $m = 1$, and uniform superconvergence

$$D(u_h - u_I) = O(h^2 \ln h).$$

This estimate still holds when A is of the variable coefficients.

There are two algorithms to get the applicable results. First, define the average gradient $\overline{D}u_h$ at the midpoint z on common side of two adjacent elements, or at an common vertex z on six adjacent elements, then $\overline{D}u_h(z) - Du(z) = O(h^2 \ln h)$. Another way is to construct the piecewise quadratic interpolant I_2u_h of u_h, then uniformly $\|I_2u_h - u\|_{1,\infty,\Omega} = O(h^2 \ln h)$.

By use of the combining element technique, triangular quadratic element is also studied by Zhu ([40], 1981–1983). At three vertices and midpoints of three sides in each element τ the quadratic finite element solution $u_h \in S_3^h$ attains fourth order accuracy. Let I_3u_h be the three (or four) degree interpolant of u_h, then uniformly

$$\|I_3u_h - u\|_{s,\infty,\Omega} = O(h^4 \ln h), \quad s = 0, 1.$$

For high degree triangular elements, recently we have made some new progress. While some conclusions at symmetrical points can be also derived from theory of local symmetry [39].

3 PIECEWISE ASSEMBLY AND GLOBAL SIMULATION

The above conclusions can be extended to more extensive domains. One of advantages of the element analysis method is a certain flexibility in dividing domain. Chen found that the combining triangular elements can be done under the following *C-condition*: two adjacent element $\triangle z_1 z_2 z_3$ and $\triangle z_3 z_2 z_4$ form an approximate parallelogram $\diamond z_1 z_2 z_4 z_3$, i.e.,

$$|z_1\vec{z_2} - z_3\vec{z_4}| \leq ch^2. \tag{13}$$

Up to now, three kinds of mesh (i.e., C-, PC- and 6PC-meshes) are found. An extensive class of domain can be partitioned by the three kinds of mesh and then superconvergence in whole domain is obtained.

C-Mesh (or The Strongly Regular Mesh, or Almost Uniform Mesh) (Chen [5], [8], 1978) which is quasiuniform and any two adjacent triangles satisfy the C-condition (13).

PC-Mesh (or The Piecewise Strongly Regular Mesh) (Lin and Xu [32], 1985). The domain is first divided by several macro-straight lines into m big triangular subdomains Ω_j, $\overline{\Omega} = \cup_{j=1}^{m} \overline{\Omega}_j$. Let J_j^h be C-mesh of Ω_j, then $J^h = \cup_{j=1}^{m} J_j^h$ is PC-mesh of Ω. The common vertices M of the several subdomains Ω_j are called the meeting points. One inside Ω is called the interior meeting point.

6PC-Mesh (or The Six Pieces Strongly Regular Mesh) (Chen and Liu [22], 1987, Chen and Huang [19], 1989). It is PC-mesh and there are six elements around each interior meeting point which each other satisfy C-condition (13).

It is shown that the C-mesh (for a triangle, sector and quadrilateral), the PC-mesh without interior meeting point (for a convex polygon), PC-mesh (for a general domain) can be easily made. For a general smooth domain, 6PC-mesh can be also constructed by use of 6-pieces smooth and continuous in $\overline{\Omega}$ transformation, Chen [14].

THEOREM   Let J^h be the C-, PC- (without interior meeting points) and 6PC-meshes on Ω, then for the triangular linear element $u_h \in S^h$,

$$\|I_2 u_h - u\|_{1,\infty,\Omega} = O(h^2 \ln h)\|u\|_{3,\infty,\Omega}.$$

If the PC-mesh contains the interior meeting points M, then in the subdomain $\Omega_M = \{x \in \Omega, |x - M| > \gamma > 0\}$,

$$\|I_2 u_h - u\|_{1,\infty,\Omega_M} = (h^2 \ln h).$$

Three kinds of mesh mentioned above are of the significance, which completely break through two previous restrictions: the uniform mesh or the interior of domain. While engineers are often interested in the behaviour near the boundary of the domain. Therefore, the above algorithms seem more applicable.

In order to exhibit the basic idea of proof of piecewise assembly, we consider a typical triangular subdomain $\Omega_1 = \triangle ABC$ having the uniform mesh J_1^h. Applying the expansion (12) to Ω_1, we get

$$J_r(\Omega_1) = \sum_{BC} \int_{S_1} c_1 \phi u_{rr} v_r \, dr - \sum_{AB} \int_{S_3} c_3 \phi u_{rr} v_t \, dt + \sum_{\Omega_1} c_4 \int_\tau \phi u_{rrs} v_t \, dx \, dy. \quad (14)$$

The last term is of order $O(h^2)$ as before. It remains to estimate the line-integrals along $\partial\Omega_1$. Assume that $r_0 = B < r_1 < ... < r_N = C$ is the set of nodal points on the straightline side BC of Ω_1, $h = r_j - r_{j-1}$, then

$$\sum_{BC} \int_{S_1} \phi u_{rr} v_r \, dr = \sum_{j=1}^{N-1} \int_{r_{j-1}}^{r_j} \phi u_{rr} (v_j - v_{j-1}) h^{-1} \, dr$$

$$= -\int_{r_0}^{r_1} \phi u_{rr} v_0 h^{-1} \, dr + \int_{r_{N-1}}^{r_N} \phi u_{rr} v_N h^{-1} \, dr \qquad (15)$$

$$+ \sum_{j=1}^{N-1} \int_{r_{j-1}}^{r_j} \phi(u_{rr}(r+h) - u_{rr}(r)) v_j h^{-1} \, dr$$

$$= r_B + r_C + r^h,$$

where

$$r_B = O(h^2 \max |D^2 u|)|v(B)|,$$
$$r_C = O(h^2 \max |D^2 u|)|v(C)|,$$
$$r^h = O(h^2) \int_{BC} |D^3 uv| \, dr.$$

If $B \in \partial\Omega$, i.e., $v(B) = 0$, then $r_B = 0$ (similarly $r_C = 0$ if $C \in \partial\Omega$). So, by taking $v = G_h(B; z)$ the gradient type Green function, the superconvergence result on PC-mesh without interior meeting points is immediately derived.

If $B \notin \partial\Omega$ is an interior meeting point, in general, $v = G_h(B; z) = O(h^{-1} \ln h)$ and then $J_r(\Omega_1) = O(h \ln h)$, i.e., superconvergence disappears. However, when point z takes away from B, $G_h(B; z) = O(\ln h)$ and then superconvergence is still achieved.

4   LOCALLY REFINED MESHES AT SINGULARITY

In order to get superconvergence in the case of nonsmooth solution, the main idea is to adopt the locally refined meshes at singularity, for example, the graded mesh. The mesh refinement was discussed by Babuška et al [2], [3].

For the simplicity, let Ω be a unit sector with a central angle $\omega = \alpha\pi$, $0.5 < \alpha \leq 2$, and consider a model problem $-\triangle u = f$ in Ω, $u = 0$ on $\partial\Omega$. Assume that the solution u has the regularity estimates $|D^s u| \leq cr^{\beta-s}$, $0 \leq s \leq 3$, $\beta = 1/\alpha \in [0.5, 2)$.

In order to construct a graded mesh of Ω, we first divide Ω into several large subsectors Ω_k with central angle ω_k, $k = 1, 2, ..., m$, and then make the arcs S_j

by radii $r_j = (jh)^\lambda$, $h = 1/N$, $\lambda \geq 1$, $j = 0, 1, ..., N$ in each Ω_k. Finally, join successively the nodal points (r_j, ω_{ij}) in each Ω_k, where $\omega_{ij} = i\omega_k/j$, $i = 0, 1, ..., j$, and get a λ-*graded mesh* (or λ-*PC mesh*), Chen [15]. It has the following properties:

1. it is the regular mesh, i.e., each element τ has the interior angle $\sigma_\tau \geq c_0 > 0$, where the constant c_0 is independent of τ and h;
2. any two adjacent elements τ and τ' in Ω_k satisfy C'-*condition*: $|z_1\vec{z}_2 - z_3\vec{z}_4| \leq ch_j^2/r_j$, where $h_j = r_{j+1} - r_j \approx \lambda r_j/j$;
3. for the element τ_j on j^{th} level, the interior angle σ_j, the length l_j of side and the area $|\tau_j|$ have the difference from the adjacent element τ',

$$[\sigma_j] = O(h_j/r_j), \quad [l_j] = O(h_j^2/r_j), \quad [\tau_j] = O(h_j^3/r_j).$$

Note that if $\lambda = 1$, the above mesh is not the C-mesh, because two adjacent elements around O do not form an approximate parallelogram. However, if $\lambda > 1$, the elements near O are small enough, whether the C-condition is satisfied is already unimportant.

On the element τ_{j+1} of $j + 1^{th}$ level, the error of linear interpolant of u is of

$$R = u_I - u = O(h_j^2)r_j^{\beta-2} = O(h^2)r_j^{\beta-2/\lambda}, \quad DR = O(h_j)r_j^{\beta-2} = O(h)r_j^{\beta-1-1/\lambda},$$

then if $(\beta+1)\lambda \neq 2$ and $\beta\lambda \neq 1$, we have $\|R\|_1 \leq ch^{\min(\lambda\beta,1)}$, $\|R\| \leq ch^{\min((\beta+1)\lambda,2)}$. By the standard argument we also get $\|e\|_1 \leq ch^{\min(\lambda\beta,1)}$, $\|e\| \leq ch^{2\min(\lambda\beta,1)}$. Based on the expansion (12), for λ-PC mesh (if $\lambda\beta > 1$), Chen ([15], 1990) obtained

$$\|u_I - u_h\|_{1,\Omega} \leq ch^2, \quad \|u - I_2 u_h\|_{1,\Omega} \leq ch^2.$$

Using the coordinate transform, similar results in L^∞-norm were obtained by Huang ([29], 1987).

NUMERICAL EXAMPLE   Consider the harmonic solution $u = r^{0.5}\sin 0.5\theta$ in $\Omega = \{0 < r < 1, \ 0 < \theta < 2\pi\}$. Due to the symmetry, it is enough to discuss the upper semi-circle Ω^+ which is divided into three equal sectors Ω_1, Ω_2 and Ω_3. We computed the linear finite element solutions u_N and errors $e_N = u_N - u$, for $\lambda = 1, 2, 4$ and $N = 8, 16, 32$. Let z_{ji} be all interior nodal points on same arc $r = r_j = (jh)^\lambda$, $i = 1, 2, ..., 3j$. Assume that each interior nodal point z in Ω_k is rounded by six adjacent elements τ_i, $i = 1, 2, ..., 6$. We consider the two kinds of average error

$$a^N(z) = \frac{1}{6}\sum_{i=1}^{6}|D_x u_N(\tau_i) - D_x u(z)|, \quad b^N(z) = \frac{1}{6}\sum_{i=1}^{6}(D_x u_N(\tau_i) - D_x u(z)),$$

and the average square root value at $3j$ interior nodal points along the arc $r = r_j$,

$$f_j^N = \left(\frac{1}{3j}\sum_{i=1}^{3j}|f^N(z_{ji})|^2\right)^{1/2}, \quad 1 \leq j \leq N - 1.$$

Table 1   The change of the quotient q_j

q_j	$\lambda = 1$	$\lambda = 2$	$\lambda = 4$
$N = 8$	$2.4 \sim 2$	$4.4 \sim 6$	$12 \sim 7.7$
$N = 16$	$2.5 \sim 2$	$4.5 \sim 10$	$15.6 \sim 14$
$N = 32$	$2.6 \sim 2$	$4.7 \sim 18$	$15.6 \sim 32$

Table 2   Superconvergence of derivatives

r_j	b^8	b^{16}	b^{32}	b^8/b^{16}	b^{16}/b^{32}
0.125	—	9.0E-2	4.0E-2	—	2.25
0.25	6.7E-2	2.9E-2	1.4E-2	2.29	2.10
0.375	3.5E-2	1.6E-2	7.9E-3	2.17	2.07
0.5	2.3E-2	1.1E-2	5.3E-3	2.13	2.05
0.625	1.7E-2	8.2E-3	4.3E-3	2.10	2.05
0.75	1.4E-2	6.6E-3	3.2E-3	2.09	2.04
0.875	1.2E-2	5.5E-3	2.7E-3	2.08	2.04
0.01563	—	2.1E-1	6.5E-2	—	3.17
0.0625	1.1E-1	3.4E-2	1.0E-2	3.22	3.32
0.1406	4.0E-2	1.2E-2	3.6E-3	3.34	3.37
0.25	2.0E-2	5.9E-3	1.7E-3	3.38	3.43
0.3906	1.2E-2	3.5E-3	1.0E-3	3.41	3.46
0.5625	8.0E-3	2.3E-3	6.7E-4	3.43	3.48
0.7656	5.8E-3	1.7E-3	4.9E-4	3.43	3.49
2.441E-4	—	8.3E-1	3.9E-1	—	2.12
3.906E-3	2.8E-1	1.1E-1	3.4E-2	2.44	3.32
0.01978	9.6E-2	3.0E-2	8.1E-3	3.18	3.72
0.0625	4.4E-2	1.2E-2	3.2E-3	3.58	3.87
0.1526	2.4E-2	6.3E-3	1.6E-3	3.76	3.92
0.3164	1.5E-2	3.8E-3	9.6E-4	3.84	3.95
0.5862	9.8E-3	2.5E-3	6.4E-4	3.88	3.96

Note that the change of the quotient $q_j = a_j^N/b_j^N$ exhibits superconvergence effect of b_j^N at interior nodal points when $j = 1 \sim N - 1$ (see Table 1).

Table 2 exhibits superconvergence of derivatives in the case $\lambda = 1, 2, 4$ and along $r_j = (j/8)^\lambda$, $j = 1, 2, ..., 7$. When $\lambda \geq 2$, except a small neighbourhood of O, the superconvergence effects are made.

5   GENERALIZATION TO OTHER PROBLEMS

The above results are also valid for other problems. The following sharp superconvergence estimates hold even on quasiuniform meshes and which means that if some conclusion on superconvergence holds for linear elliptic problem then it is valid for other problems mentioned below. Therefore, the key difficulty in whole study of superconvergence is still to concentrate on the linear elliptic case.

5.1 Nonlinear elliptic case

Let $u_h \in S_m^h$ be the finite element approximation to solution $u \in H_0^1(\Omega)$ of a nonlinear elliptic problem $a(u, v) = 0$, $v \in H_0^1(\Omega)$, where

$$a(u, v) = \int_\Omega \sum_{i=0}^{2} a_i(x, u, Du) D_i v \, dx.$$

Let $u_h^* \in S_m^h$ be the finite element projection of u associated with an auxiliary bilinear form

$$A(u; w, v) = \frac{d}{dt} a(u + tw, v)|_{t=0} = \int_\Omega \sum_{i,j=0}^{2} a_{ij}(x, u, Du) D_i w D_j v \, dx.$$

The following optimal order superconvergence estimate was proved by Chen ([11], [13], 1982–1985)

$$\|u_h - u_h^*\|_{1,\infty,\Omega} = O(h^{2m} \ln h).$$

5.2 The linear and nonlinear evolution problems

Let $u_h(t) \in S_m^h$ be the semidiscrete finite element solution of the linear parabolic problem

$$\begin{aligned}
(u_t, v) + a(u, v) &= (f, v) && \text{in } \Omega \times (0, T], \\
u &= 0 && \text{on } \partial\Omega, \\
u(0) &= u_0 && \text{in } \Omega,
\end{aligned}$$

and $R_h u(t) \in S_m^h$ be the Ritz-projection of u associated with the bilinear form A. Then we have superconvergence estimates (Thomée, Xu and Zhang [38], 1989, and Chen)

$$\begin{aligned}
\|(u_h - R_h u)(t)\|_{1,\infty,\Omega} &= O(h^{m+1} \ln h) && \text{if } m \geq 1, \\
\|(u_h - R_h u)(t)\|_{0,\infty,\Omega} &= O(h^{m+2} \ln h) && \text{if } m \geq 2.
\end{aligned}$$

Let $u_h(t) \in S_m^h$ be the semidiscrete finite element solution of a strongly nonlinear parabolic problem, and $u_h^* \in S_m^h$ be the approximation of u for a linear tangent parabolic operator at u, Chen ([16], 1992) proved the optimal order superconvergence estimate:

$$\|u_h - u_h^*\|_{1,\infty,\Omega} = O(h^{2m} |\ln h|^3).$$

Here the main difficulty is to estimate the gradient type Green function for linear parabolic problem, Chen [18], 1995.

For linear hyperbolic problems, the similar estimates in L^2 are studied by H. Wang (1985), and for weakly nonlinear case by Q. Lin, H. Wang and T. Lin ([33], 1993).

For parabolic integro-differential equation, the similar estimates are also obtained by Chen and Wahlbin (linear case), and by Chen (strongly nonlinear case [17], 1993), cf. also the monograph "Finite Element Methods for Integro-Differential Equations" by Chen and Shih (in English, to appear).

5.3   A successful application to the concrete faced rockfill dams

A successful application to the concrete faced rockfill dams is implemented by Chen and Huang et al ([21], 1993–1995). This is a three dimensional, strongly nonlinear elasto-plastic problem with various media. It leads to solve a nonlinear system of 5000 unknowns. The accuracy of the stresses is improved by superconvergence technique and the actual errors of the displacements and stresses are estimated by an extrapolation algorithm.

REFERENCES

1. Andreev A. B. and Lazarov R. D., *Superconvergence of the gradient for quadratic triangular finite elements*, Numer. Methods Partial Differential Equations **4** (1988), 15–32.

2. Babuška I., *Finite element methods for domains with corners*, Computing **6** (1970), 264–273.

3. Babuška I., Kellogg R. and Pitkäranta J., *Direct and inverse error estimates for finite elements with mesh refinement*, Numer. Math. **33** (1979), 447–471.

4. Bramble J. H. and Schatz A. H., *High order local accuracy by averaging in the finite element method*, Math. Comp. **31** (1977), 94–111.

5. Chen C. M., *Optimal points of the stresses approximated by triangular linear element in FEM*, Natur. Sci. J. Xiangtan Univ. **1** (1978), 77–90.

6. Chen C. M. and Zhu Q. D., *A new estimate for finite element method and optimal point theorem for stresses*, Natur. Sci. J. Xiangtan Univ. **1** (1978), 10–20.

7. Chen C. M., *Optimal points of approximation solution for Galerkin method for two-point boundary value problem*, MR 82e:65086, Numer. Math. J. Chinese Univ. **1:1** (1979), 73–79.

8. Chen C. M., *Optimal points of the stresses for triangular linear element*, MR 83d:65279, Numer. Math. J. Chinese Univ. **2:2** (1980), 12–20.

9. Chen C. M., *Superconvergence of finite element solution and its derivatives*, MR 82m:65100, Numer. Math. J. Univ. **3:2** (1981), 118–125.

10. Chen C. M., "Finite Element Method and its Analysis in Raising Accuracy," Hunan Science and Technique Press, Changsha, 1982.

11. Chen C. M., *Superconvergence of finite element approximations to nonlinear elliptic problems*, MR 85h:65235, in "Proc. of the China–France Symposium on FEM," ed. by Feng Kang and J. Lions, Science Press, Beijing, 1983, pp. 622–640.

12. Chen C. M., *Some estimates for interpolation approximations and their appli-*

cations, MR 86c:41001, Numer. Math. J. Chinese Univ. **6:1** (1984), 35–43.

13. Chen C. M., *Superconvergence and extrapolation of the finite element approximations to quasilinear elliptic problems*, MR 88g:65104, Northeastern Math. J. **2:2** (1986), 228–236.

14. Chen C. M., *High accuracy analysis of finite element methods: superconvergence and postprocessing*, Natur. Sci. J. Xiangtan Univ. **10:4** (1988), 114–123.

15. Chen C. M., *Superconvergence of finite element in polygon with reentrant*, Natur. Sci. J. Xiangtan Univ. **12:4** (1990), 134–141.

16. Chen C. M., *Some estimates for the nonlinear parabolic finite element*, in "Proc. of the First China–Japan Joint Seminar on Numer. Math.," Beijing, Aug. 24–29, 1992, pp. 87–90.

17. Chen C. M., *New estimates of finite element for nonlinear parabolic integro-differential equation*, Natur. Sci. J. Xiangtan Univ. **15:1** (1993), 1–3.

18. Chen C. M., *Gradient type Green function for parabolic problem*, Numer. Math. J. Chinese Univ. **17:1** (1995), 85–90.

19. Chen C. M. and Huang Y. Q., *Extrapolation of triangular linear elements in general domain*, Numer. Math. J. Chinese Univ. **11** (1989), 1–16.

20. Chen C. M. and Huang Y. Q., "High Accuracy Theory of Finite Elements," Hunan Science and Technique Press, Changsha, 1995.

21. Chen C. M., Huang Y. Q., Shu S. and Jin J. C., *The application of high accuracy method in the analysis of concrete faced rockfill dams*, Numer. Math. J. Chinese Univ. **17:4** (1995), 345–351.

22. Chen C. M. and Liu J. G., *Superconvergence of gradient of triangular linear element in general domain*, Natur. Sci. J. Xiangtan Univ. **9** (1987), 114–127.

23. Chen C. M. and Wahlbin L., *Superconvergence for gradient in finite element method for an integro-differential problem*.

24. Dautov R. V., Lapin A. V. and Lyashko A. D., *Some mesh schemes for quasilinear elliptic equations*, USSR Comput. Math. Math. Phys. **20** (1980), 62–78.

25. Douglas J. and Dupont T., *Some superconvergence results for Galerkin methods for the approximate solution of two-point boundary value problems*, in "Topics in Numerical Analysis," Academic Press, 1973, pp. 89–92.

26. Douglas J. and Dupont T., *Galerkin approximations for the two-point boundary problem using continuous, piecewise polynomial spaces*, Numer. Math. **22** (1974), 99–109.

27. Douglas J., Dupont T. and Wheeler M. F., *An L^∞ estimate and a superconvergence result for a Galerkin method for elliptic equations based on tensor products of piecewise polynomials*, RAIRO Anal. Numér. **8** (1974), 61–66.

28. Frehse J. and Rannacher R., *Eine L^1-Fehlerabschatzung diskreter Grundlösungen in der Methode der finiten Elemente*, Tagungsband "Finite Elemente", Bonn. Math. Schrift. **89** (1975), 92–114.

29. Huang Y. Q., "Finite element method: extrapolation and superconvergence," Ph.D. dissertation, Institute of System Science, 1987.

30. Křížek M. and Neittaanmäki P., *Superconvergence phenomenon in the finite element method arising from averaging gradients*, Numer. Math. **45** (1984), 105–116.

31. Lesaint P. and Zlámal M., *Superconvergence of the gradient of finite element solutions*, RAIRO Anal. Numér. **13** (1979), 139–166.

32. Lin Q. and Xu J. C., *Linear finite elements with high accuracy*, J. Comput.

Math. **3** (1985), 115–133.

33. Lin Q., Wang H. and Lin T., *Interpolated FEM for second order hyperbolic equations and their global superconvergence*, Syst. Sci. Math. Sci. **4** (1993), 331–340.

34. Oganesyan L. A. and Rukhovetz L. A., *Study of the rate of convergence of variational difference schemes for second order elliptic equations in a two-dimensional field with a smooth boundary*, USSR Comput. Math. Math. Phys. **9** (1969), 158–183.

35. Rannacher R. and Scott R., *Some optimal estimates for piecewise linear finite element approximations*, Math. Comp. **38** (1982), 437–445.

36. Thomée V., *High order local approximation to derivatives in the finite element method*, Math. Comp. **31** (1977), 652–660.

37. Thomée V., *Negative norm estimates and superconvergence for parabolic problems*, Math. Comp. **34** (1980), 93–113.

38. Thomée V., Xu J. C. and Zhang N. Y., *Superconvergence of the gradient in piecewise linear element approximation to a parabolic problem*, SIAM J. Numer. Anal. **26** (1989), 553–573.

39. Wahlbin L., "Superconvergence in Galerkin finite element methods," Springer, Berlin, 1995, pp. 1–164.

40. Zhu Q. D., *Optimal points for quadratic triangular finite elements*, Natur. Sci. J. Xiangtan Univ. **3** (1981), 36–45.

41. Zhu Q. D., *The uniform superconvergence estimates for derivatives of the finite elements*, Numer. Math. J. Chinese Univ. **5** (1983), 311–318.

42. Zlámal M., *Some superconvergence results in the finite element method*, in "Lecture Notes in Math. 606," Springer, 1977, pp. 353–362.

43. Zlámal M., *Superconvergence and reduced integration in the finite element method*, Math. Comp. **32** (1978), 663–685.

Quadratic interpolation polynomials in vertices of strongly regular triangulations

JOSEF DALÍK*   Department of Mathematics, Technical University of Brno, Žižkova 17, 602 00 Brno, Czech Republic, e-mail: mddal@fce.vutbr.cz

Abstract   The error of interpolation of a given smooth function $u(x) = u(x_1, x_2)$ and of its partial derivatives up to the order two by the Lagrange interpolation polynomial in nodes $a^1, \ldots, a^6$ is studied under the assumption that $a^1, \ldots, a^6$ are vertices of four neighbouring triangles without obtuse angles.

1   INTRODUCTION AND NOTATION

The problem of Lagrange interpolation in n variables for any natural n has been investigated for a long time. See Berezin, Shidkov [2], for example. In Sauer, Xu [7], a satisfactory generalization both of the Newton and the Lagrange forms of the interpolation polynomial from $n = 1$ to arbitrary dimension appears.

In this article, we denote by $\mathcal{P}^2$ the set of polynomials in two variables of degree less or equal to two. Nodes $a^1, \ldots, a^6$ from R^2 are in [7] called *poised* if any given function is interpolated in $a^1, \ldots, a^6$ by exactly one polynomial from $\mathcal{P}^2$. It is well-known that $a^1, \ldots, a^6$ are poised if and only if they are not situated on a unique quadratic curve, but this criterion is hardly applicable in concrete cases. An algorithm from [7] recognizes the poisedness for any given $a^1, \ldots, a^6$.

We say that the nodes $a^1, \ldots, a^6$ are *regular* if the following conditions (a), (b) are satisfied.

(a) The triangles $T_1 = \overline{a^1 a^3 a^5}$, $T_2 = \overline{a^1 a^2 a^3}$, $T_3 = \overline{a^3 a^4 a^5}$, $T_4 = \overline{a^5 a^6 a^1}$ satisfy $T_1 \cap T_2 = \overline{a^1 a^3}$, $T_1 \cap T_3 = \overline{a^3 a^5}$, $T_1 \cap T_4 = \overline{a^5 a^1}$.

(b) There is $\alpha_0 > 0$ such that $\alpha_0 \leq \alpha \leq \pi/2$ for all inner angles α of $T_1, \ldots, T_4$.

We denote by h the maximal length of the sides of $T_1, \ldots, T_4$ and by T the (closed) convex hull of $T_1 \cup \cdots \cup T_4$ in R^2. We reserve the symbol C for a generic positive constant. In what follows, C does never depend on h but, in general, C does depend on α_0.

*Work supported by the Grant Agency of Czech Republic under Grant No. 201/95/0095.

We show that regular nodes are always poised and the related interpolation polynomials are stable in the following way: For given $a^1, \ldots, a^6$ regular and $u \in C^{(3)}(T)$, we describe the basic Lagrange polynomials $L_1(x), \ldots, L_6(x)$ explicitly and sketch the proof of the fact that the interpolation polynomial

$$L(x) = u(a^1)L_1(x) + \cdots + u(a^6)L_6(x)$$

satisfies

$$\left| \frac{\partial^{|m|} u}{\partial x^m}(x) - \frac{\partial^{|m|} L}{\partial x^m}(x) \right| < Ch^{3-|m|}$$

for each multiindex m with $|m| \leq 2$ and for each $x \in T$.

This result plays an important role in our analysis of a numerical method for an approximate solution of the non-stationary two-dimensional convection-diffusion problem with dominating convection, which is an extension of the method from Dalík, Růžičková [3]. But this result is strongly related to other actual topics too. Let us mention the study of superconvergence on non-homogeneous triangulations, see Křížek, Neittaanmäki [6], and the investigation of a posteriori error-estimates. For example, our grad L is a recovery operator in the sense of Ainsworth, Craig [1]. See also Durán, Muschietti, Rodríguez [4] and Durán, Rodríguez [5].

Let $a, b, c, d \in R^2$. We denote by $|ab|$ the length of the segment $\overline{ab}$ and define

$$D(abc) = \frac{1}{2} \begin{vmatrix} a_1 - c_1 & a_2 - c_2 \\ b_1 - c_1 & b_2 - c_2 \end{vmatrix}, \qquad P(abc) = |D(abc)|.$$

Of course, $P(abc)$ is the area of the triangle $\overline{abc}$ and $D(abc) > 0$ or $D(abc) < 0$ whenever the orientation of a, b, c is positive or negative, respectively. We abbreviate $P(T_i)$ by P_i, for $i = 1, \ldots, 4$.

2   THE LAGRANGE INTERPOLATION POLYNOMIAL

CONVENTION   Let $i, j \in \{1, \ldots, 6\}$. If $i + j > 6$ then we denote by a^{i+j} the node a^{i+j-6}.

DEFINITION   We relate the polynomial

$$\begin{aligned}
l_i(x) = {} & D(xa^{4+i}a^{5+i})D(xa^{1+i}a^{2+i})D(a^{3+i}a^{4+i}a^{2+i})D(a^{3+i}a^{5+i}a^{1+i}) \\
& - D(xa^{2+i}a^{4+i})D(xa^{1+i}a^{5+i})D(a^{3+i}a^{4+i}a^{5+i})D(a^{3+i}a^{1+i}a^{2+i})
\end{aligned}$$

to $i = 1, \ldots, 6$.

LEMMA 1   If $i, j \in \{1, \ldots, 6\}$ then we have

$$l_i(a^j) = \begin{cases} 0, & j \neq i, \\ \begin{aligned}(-1)^{i+1}[&D(a^1a^5a^6)D(a^1a^2a^3)D(a^4a^5a^3)D(a^4a^6a^2) \\ &- D(a^1a^3a^5)D(a^1a^2a^6)D(a^4a^5a^6)D(a^4a^2a^3)],\end{aligned} & j = i. \end{cases}$$

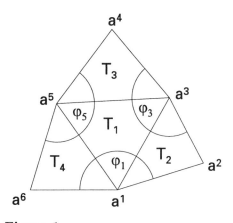

Figure 1

Proof: If $a^j \in \{a^{1+i}, a^{2+i}, a^{4+i}, a^{5+i}\}$ then a value of D, related to a triangle with two vertices identical, appears in each summand of $l_i(a^j)$. If $a^j = a^{3+i}$ then $l_i(a^j)$ is a difference of two members with the same value. Thus $l_i(a^j) = 0$ whenever $j \neq i$. In the case $j = i$, the statement follows by the observation that $l_{i+1}(a^{i+1}) = -l_i(a^i)$ for $i = 1, \ldots, 5$ and $l_1(a^1) = -l_6(a^6)$. $\square$

The well-known standard construction of the base polynomials $L_i(x)$ requires that $l_i(a^i) \neq 0$. The following basic statement tells us more.

PROPOSITION   Let the nodes $a^1, \ldots, a^6$ be regular. Then there exists an index $i \in \{1, 2, 3, 4\}$ such that

$$l_1(a^1) \geq C \cdot P_i P_2 P_3 P_4.$$

Sketch of proof: Let us denote by φ_i the inner angle of the hexagon $\overline{a^1 a^2 a^3 a^4 a^5 a^6}$ at a^i for $i = 1, 3, 5$. See Fig. 1. Instead of the complete proof which is rather technical, we only discuss the concrete forms of the statement in the following cases (a)–(d), characterized by the number of angles among φ_1, φ_3, φ_5 which are less than π:

(a) $\varphi_1 < \pi, \varphi_3 < \pi, \varphi_5 < \pi \implies l_1(a^1) \geq \frac{7}{8} P_1 P_2 P_3 P_4,$

(b) $\varphi_1 < \pi, \varphi_3 < \pi, \varphi_5 \geq \pi \implies l_1(a^1) \geq P_2 P_3 P_4 \cdot \min\{P_3, P_4\},$

(c) $\varphi_1 < \pi, \varphi_3 \geq \pi, \varphi_5 \geq \pi \implies l_1(a^1) \geq \frac{2}{3} \sin^2 \alpha_0 P_2 (P_3)^2 P_4,$

(d) $\varphi_1 \geq \pi, \varphi_3 \geq \pi, \varphi_5 \geq \pi \implies l_1(a^1) \geq P_1 P_2 P_3 P_4.$

The case (c) is critical. Fig. 2 illustrates that, in this case, the regular nodes $a^1, \ldots, a^6$ can be very close to a quadratic curve ($p \cup q$ in Fig. 2). Moreover, we can see from Fig. 2 that the prohibition of obtuse inner angles in $T_1, \ldots, T_4$ cannot be simply omitted in this statement. $\square$

We obtain $l_i(a^i) \neq 0$ for $i = 1, \ldots, 6$ by Proposition and by Lemma 1.

DEFINITION   Let the nodes $a^1, \ldots, a^6$ be regular and let $u = u(x)$ be defined on

Josef Dalík

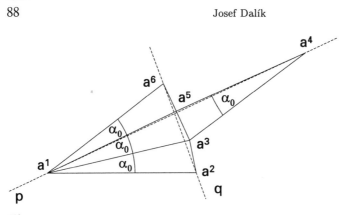

Figure 2

T. Then we put

$$L_i(x) = \frac{l_i(x)}{l_i(a^i)}, \quad \text{for } i = 1, \ldots, 6, \qquad L(x) = \sum_{i=1}^{6} L_i(x)u(a^i).$$

3   UPPER BOUNDS OF THE ERROR OF INTERPOLATION

It is easy to see that

$$P_i > Ch^2, \tag{1}$$

for $i = 1, \ldots, 4$, whenever the nodes $a^1, \ldots, a^6$ are regular. This fact and our Proposition lead to:

LEMMA 2   Let the nodes $a^1, \ldots, a^6$ be regular. Then

$$L_i(x) \leq C, \quad \text{for all } x \in T, \text{ and } i = 1, \ldots, 6.$$

THEOREM 2   Let $a^1, \ldots, a^6$ be regular nodes and $u \in C^{(3)}(T)$. Then

$$|u(x) - L(x)| < Ch^3 \quad \text{for all } x \in T.$$

Proof: Let $x \in T$ be arbitrary. For every $y \in T$, we put

$$P(y) = u(x) + \frac{\partial u}{\partial x_1}(x)(y_1 - x_1) + \frac{\partial u}{\partial x_2}(x)(y_2 - x_2)$$

$$+ \frac{1}{2}\frac{\partial^2 u}{\partial x_1^2}(x)(y_1 - x_1)^2 + \frac{\partial^2 u}{\partial x_1 \partial x_2}(x)(y_1 - x_1)(y_2 - x_2)$$

$$+ \frac{1}{2}\frac{\partial^2 u}{\partial x_2^2}(x)(y_2 - x_2)^2.$$

By the Taylor theorem we obtain

$$u(y) = P(y) + R(y), \quad R(y) = O(h^3).$$

Especially

$$u(a^i) - P(a^i) = O(h^3) \tag{2}$$

for $i = 1, \ldots, 6$. As $P(x) = u(x)$, we have

$$u(x) - L(x) = P(x) - L(x) = \sum_{i=1}^{6} L_i(x)(P(a^i) - u(a^i)) = O(h^3)$$

according to Lemma 2 and (2).   □

In the rest of this section, we denote by a, b, c the vertex a^1, a^3, a^5 of the triangle T_1, respectively.

DEFINITION   Let $a^1, \ldots, a^6$ be regular nodes. If $x \in T$ then we define the real numbers λ_a, λ_b, λ_c by the condition

$$\begin{pmatrix} 1 & 1 & 1 \\ a_1 & b_1 & c_1 \\ a_2 & b_2 & c_2 \end{pmatrix} \begin{pmatrix} \lambda_a \\ \lambda_b \\ \lambda_c \end{pmatrix} = \begin{pmatrix} 1 \\ x_1 \\ x_2 \end{pmatrix}$$

and call them *geometric coordinates* of x in T_1.

LEMMA 3   Let $u \in C^{(3)}(T)$ and $x \in T$. Then

$$
\begin{aligned}
u(x) = {} & \lambda_a u(a) + \lambda_b u(b) + \lambda_c u(c) - \frac{1}{2} \frac{\partial^2 u}{\partial x_1^2}(x) \big[(a_1 - x_1)^2 \lambda_a + (b_1 - x_1)^2 \lambda_b \\
& + (c_1 - x_1)^2 \lambda_c \big] - \frac{\partial^2 u}{\partial x_1 \partial x_2}(x) \big[(a_1 - x_1)(a_2 - x_2) \lambda_a \\
& + (b_1 - x_1)(b_2 - x_2) \lambda_b + (c_1 - x_1)(c_2 - x_2) \lambda_c \big] - \frac{1}{2} \frac{\partial^2 u}{\partial x_2^2}(x) \cdot \\
& \cdot \big[(a_2 - x_2)^2 \lambda_a + (b_2 - x_2)^2 \lambda_b + (c_2 - x_2)^2 \lambda_c \big] + E(u, x)
\end{aligned}
\tag{3}
$$

and $E(u, x) = O(h^3)$. If $u \in \mathcal{P}^2$ then $E(u, x) = 0$.

Proof: By the Taylor theorem we have

$$u(a) = u(x) + \frac{\partial u}{\partial \vec{xa}}(x)|ax| + \frac{1}{2} \frac{\partial^2 u}{\partial \vec{xa}^2}(x)|ax|^2 + e(u, a),$$

$$u(b) = u(x) + \frac{\partial u}{\partial \vec{xb}}(x)|bx| + \frac{1}{2} \frac{\partial^2 u}{\partial \vec{xb}^2}(x)|bx|^2 + e(u, b),$$

$$u(c) = u(x) + \frac{\partial u}{\partial \vec{xc}}(x)|cx| + \frac{1}{2} \frac{\partial^2 u}{\partial \vec{xc}^2}(x)|cx|^2 + e(u, c)$$

and $e(u, y) = O(h^3)$. If $u \in \mathcal{P}^2$ then $e(u, y) = 0$ for $y = a, b, c$. If we multiply these identities by λ_a, λ_b, λ_c consecutively, add them up, and make some simplifications, then we obtain (3).   □

THEOREM 3   Let $a^1, \ldots, a^6$ be regular nodes and $u \in C^{(3)}(T)$. Then

$$\left| \frac{\partial^2 u}{\partial x_i \partial x_j}(x) - \frac{\partial^2 L}{\partial x_i \partial x_j} \right| < Ch$$

for all $x \in T$ and $i, j \in \{1, 2\}$.

Proof: Let $x \in T$ be arbitrary. In (3), we substitute u by $u - L$ and x by $a^m = (b + c)/2$, $b^m = (c + a)/2$, $c^m = (a + b)/2$ consecutively. We obtain the equation

$$
\begin{aligned}
u(y) - L(y) = -\frac{1}{2}\left[\frac{\partial^2 u}{\partial x_1^2}(y) - \frac{\partial^2 L}{\partial x_1^2} \right] &\left[(a_1 - y_1)^2 \lambda_a + (b_1 - y_1)^2 \lambda_b \right. \\
+ (c_1 - y_1)^2 \lambda_c \bigg] &- \left[\frac{\partial^2 u}{\partial x_1 \partial x_2}(y) - \frac{\partial^2 L}{\partial x_1 \partial x_2} \right] \cdot \\
\cdot \big[(a_1 - y_1)(a_2 - y_2)\lambda_a &+ (b_1 - y_1)(b_2 - y_2)\lambda_b \\
+ (c_1 - y_1)(c_2 - y_2)\lambda_c \big] &- \frac{1}{2}\left[\frac{\partial^2 u}{\partial x_2^2}(y) - \frac{\partial^2 L}{\partial x_2^2} \right] \cdot \\
\cdot \big[(a_2 - y_2)^2 \lambda_a + (b_2 - y_2)^2 \lambda_b &+ (c_2 - y_2)^2 \lambda_c \big] + O(h^3)
\end{aligned}
\tag{4}
$$

for $y = a^m$, b^m, c^m. We have $u(y) - L(y) = O(h^3)$ by Theorem 2, $\dfrac{\partial^2 L}{\partial x_i \partial x_j}$ is a constant and $\dfrac{\partial^2 u}{\partial x_i \partial x_j}(y) - \dfrac{\partial^2 u}{\partial x_i \partial x_j}(x) = O(h)$ for all $i, j \in \{1, 2\}$, $y = a^m, b^m, c^m$. Therefore, if we multiply all the equations (4) by -8 and use the facts

$$
\begin{aligned}
\lambda_a = 0 \quad \lambda_b = \tfrac{1}{2} \quad \lambda_c = \tfrac{1}{2} \quad &\text{for } y = a^m \\
\lambda_a = \tfrac{1}{2} \quad \lambda_b = 0 \quad \lambda_c = \tfrac{1}{2} \quad &\text{for } y = b^m \\
\lambda_a = \tfrac{1}{2} \quad \lambda_b = \tfrac{1}{2} \quad \lambda_c = 0 \quad &\text{for } y = c^m
\end{aligned}
$$

then we obtain the following system of equations

$$
\begin{pmatrix}
(b_1 - c_1)^2 & 2(b_1 - c_1)(b_2 - c_2) & (b_2 - c_2)^2 \\
(c_1 - a_1)^2 & 2(c_1 - a_1)(c_2 - a_2) & (c_2 - a_2)^2 \\
(a_1 - b_1)^2 & 2(a_1 - b_1)(a_2 - b_2) & (a_2 - b_2)^2
\end{pmatrix}
\begin{pmatrix}
x_{11} \\ x_{12} \\ x_{22}
\end{pmatrix}
=
\begin{pmatrix}
r_1 \\ r_2 \\ r_3
\end{pmatrix}
\tag{5}
$$

for the unknowns $x_{ij} = \dfrac{\partial^2 (u - L)}{\partial x_i \partial x_j}(x)$ with $r_k = O(h^3)$ for $ij = 11$, 12, 22 and $k = 1, 2, 3$. It is a matter of routine to show that the determinant of the matrix of (5) equals to $2(P(abc))^3$. This result, the Cramer rule and (1) lead to the conclusion that $x_{ij} = O(h)$ for $ij = 11$, 12, 22. $\square$

THEOREM 4   Let $a^1, \ldots, a^6$ be regular nodes and let $u \in C^{(3)}(T)$. Then

$$\left| \frac{\partial u}{\partial x_i}(x) - \frac{\partial L}{\partial x_i}(x) \right| < Ch^2$$

for all $x \in T$ and $i = 1, 2$.

Proof: Let $x \in T$ be arbitrary. Then there exist two vertices among a, b, c, we denote them by a, b, such that $P(abx) \geq \frac{1}{3}P(abc)$. The Taylor theorem and Theorem 3 lead to

$$0 = (u - L)(a) = (u - L)(x) + \frac{\partial(u - L)}{\partial x_1}(x)(a_1 - x_1)$$
$$+ \frac{\partial(u - L)}{\partial x_2}(x)(a_2 - x_2) + O(h^3),$$
$$0 = (u - L)(b) = (u - L)(x) + \frac{\partial(u - L)}{\partial x_1}(x)(b_1 - x_1)$$
$$+ \frac{\partial(u - L)}{\partial x_2}(x)(b_2 - x_2) + O(h^3).$$

Because $(u - L)(x) = O(h^3)$ by Theorem 2, we can rewrite these two equations in the form

$$\frac{\partial(u - L)}{\partial x_1}(x)(a_1 - x_1) + \frac{\partial(u - L)}{\partial x_2}(x)(a_2 - x_2) = r_1,$$
$$\frac{\partial(u - L)}{\partial x_1}(x)(b_1 - x_1) + \frac{\partial(u - L)}{\partial x_2}(x)(b_2 - x_2) = r_2,$$

where $r_i = O(h^3)$ for $i = 1, 2$. Since, moreover,

$$\left| \det \begin{pmatrix} a_1 - x_1 & a_2 - x_2 \\ b_1 - x_1 & b_2 - x_2 \end{pmatrix} \right| = 2P(abx) \geq \frac{2}{3}P(abc),$$

we conclude that $\dfrac{\partial(u - L)}{\partial x_i}(x) = O(h^2)$ for $i = 1, 2$ by the Cramer rule and (1). $\square$

4   AN APPLICATION

Let a triangulation $\mathcal{T}_h$ of a region $\Omega \subseteq R^2$ and $\alpha_0 > 0$ be such that each inner angle α of each triangle from $\mathcal{T}_h$ satisfies $\alpha_0 \leq \alpha \leq \pi/2$. If a, b are vertices of $\mathcal{T}_h$ then we call b a *neighbour* of a whenever $\overline{abc} \in \mathcal{T}_h$ for some vertex c. We denote by $N(a)$ the set of neighbours of a.

In this section we describe a way how to express the values of first partial derivatives of a function u in an inner vertex a as a linear combination of vertices from $N(a) \cup \{a\}$ with an error $O(h^2)$.

Let $a^1, \ldots, a^6$ be regular nodes and $u \in C^{(3)}(T)$. If we put $x = a^1$ in Theorem 4 then we obtain

$$\frac{\partial u}{\partial x_1}(a^1) = u(a^1)\frac{\partial L_1}{\partial x_1}(a^1) + \cdots + u(a^6)\frac{\partial L_6}{\partial x_1}(a^1) + e, \qquad (6)$$

$$e = O(h^2), \quad e = 0 \quad \text{whenever} \quad u \in \mathcal{P}^2.$$

Of course, if $T_1, \ldots, T_4 \in \mathcal{T}_h$ then a^2, a^3, a^5, $a^6 \in N(a^1)$ and $a^4 \notin N(a^1)$. Hence we require that the coefficient $\dfrac{\partial L_4}{\partial x_1}(a^1)$ at $u(a^4)$ in (6) equals to zero.

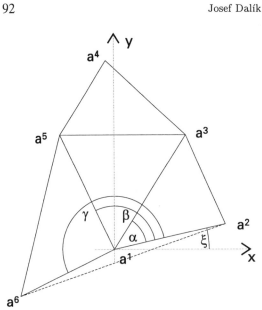

Figure 3

DEFINITION   Let $a^1, \ldots, a^6$ be regular nodes, $\alpha = \sphericalangle a^2 a^1 a^3$, $\beta = \sphericalangle a^2 a^1 a^5$ and $\gamma = \sphericalangle a^2 a^1 a^6$. See Fig. 3. Then we put

$$A_2 = D(a^1 a^3 a^5)D(a^1 a^5 a^6)[D(a^1 a^6 a^2) + D(a^1 a^2 a^3)],$$
$$A_3 = D(a^1 a^5 a^6)D(a^1 a^6 a^2)[D(a^1 a^2 a^3) + D(a^1 a^3 a^5)],$$
$$A_5 = D(a^1 a^6 a^2)D(a^1 a^2 a^3)[D(a^1 a^3 a^5) + D(a^1 a^5 a^6)],$$
$$A_6 = D(a^1 a^2 a^3)D(a^1 a^3 a^5)[D(a^1 a^5 a^6) + D(a^1 a^6 a^2)],$$

and

$$B_1 = A_2|a^1 a^2| \cos 0 - A_3|a^1 a^3| \cos \alpha + A_5|a^1 a^5| \cos \beta - A_6|a^1 a^6| \cos \gamma,$$
$$B_2 = A_2|a^1 a^2| \sin 0 - A_3|a^1 a^3| \sin \alpha + A_5|a^1 a^5| \sin \beta - A_6|a^1 a^6| \sin \gamma.$$

LEMMA 4   Let $a^1, \ldots, a^6$ be regular nodes, and let ξ characterize the rotation of the Cartesian coordinate system in a way obvious from Fig. 3. Then the following assertions (a), (b) are true.

   (a) If $B_1 \neq 0$ or $B_2 \neq 0$ then there exists exactly one $\xi \in (-\pi/2, \pi/2\rangle$ such that
   $$\frac{\partial L_4}{\partial x_1}(a^1) = 0.$$

   (b) If $B_1 = 0$ and $B_2 = 0$ then $\dfrac{\partial L_4}{\partial x_1}(a^1) = 0$ for each $\xi \in (-\pi/2, \pi/2\rangle$.

   Proof: One can easily compute that

$$\frac{\partial L_4}{\partial x_1}(a^1) = \Big[A_2(a^2)_2 - A_3(a^3)_2 + A_5(a^5)_2 - A_6(a^6)_2\Big]/\Big(2 \cdot l_1(a^1)\Big).$$

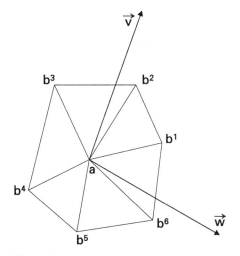

Figure 4

Hence, $\dfrac{\partial L_4}{\partial x_1}(a^1) = 0$ if and only if

$$A_2|a^1 a^2|\sin\xi - A_3|a^1 a^3|\sin(\alpha + \xi)$$
$$+ A_5|a^1 a^5|\sin(\beta + \xi) - A_6|a^1 a^6|\sin(\gamma + \xi) = 0$$

(see Fig. 3) and this condition is equivalent to

$$B_1 \sin\xi + B_2 \cos\xi = 0.$$

The statements (a), (b) follow immediately.   □

COROLLARY   Let $a^1, \ldots, a^6$ be regular nodes and $u \in C^{(3)}(T)$. Then there exists a unit vector $\vec{v}$ such that

$$\frac{\partial L}{\partial \vec{v}}(a^1) = c_1 u(a^1) + c_2 u(a^2) + c_3 u(a^3) + c_5 u(a^5) + c_6 u(a^6),$$

with $c_i = \dfrac{\partial L_i}{\partial \vec{v}}(a^1)$ for $i = 1, 2, 3, 5, 6$, and

$$\left| \frac{\partial (u - L)}{\partial \vec{v}}(a^1) \right| = O(h^2).$$

REMARK   Let a be an inner node of $\mathcal{T}_h$ with neighbours $b^1, \ldots, b^n$. To an arbitrary four-tuple $b^i, \ldots, b^{i+3}$ and $b^j, \ldots, b^{j+3}$ ($j \neq i$) there is a unit vector $\vec{v}$ and $\vec{w}$ such that $\dfrac{\partial u}{\partial \vec{v}}(a)$ and $\dfrac{\partial u}{\partial \vec{w}}(a)$ can be approximated as a linear combination of $u(a)$, $u(b^i), \ldots, u(b^{i+3})$ and $u(a), u(b^j), \ldots, u(b^{j+3})$, respectively, with an error $O(h^2)$ by our Corollary. By a suitable linear combination of these two approximations, we can approximate $\dfrac{\partial u}{\partial \vec{z}}(a)$ for any unit vector $\vec{z}$ with an error $O(h^2)$.

   In Fig. 4, $\dfrac{\partial u}{\partial \vec{v}}(a)$ and $\dfrac{\partial u}{\partial \vec{w}}(a)$ is a linear combination of the values of u in the nodes a, b^1, b^2, b^3, b^4 and a, b^4, b^5, b^6, b^1, respectively.

REFERENCES

1. Ainsworth M., Craig A., *A posteriori error estimators in the finite element method*, Numer. Math. **60** (1992), 429–463.
2. Berezin I.S., Shidkov N.P., "Numerical Methods 1," Nauka, Moskva, 1966.
3. Dalík J., Růžičková H., *An explicit modified method of characteristics for the one-dimensional nonstationary convection-diffusion problem with dominating convection*, Appl. Math. **40** (1995), 367–381.
4. Durán R., Muschietti M.A., Rodríguez R., *On the asymptotic exactness of error estimators for linear triangular finite elements*, Numer. Math. **59** (1991), 107–127.
5. Durán R., Rodríguez R., *On the asymptotic exactness of Bank-Weiser's estimators*, Numer. Math. **62** (1992), 297–303.
6. Křížek M., Neittaanmäki P., *On Superconvergence Techniques*, Acta Appl. Math. **9** (1987), 175–198.
7. Sauer T., Xu Y., *On multivariate Lagrange interpolation*, Math. Comp. **64** (1995), 1147–1170.

Explicit error bounds for a nonconforming finite element method

P. DESTUYNDER   CNAM/IAT, 15 rue Marat, 78210 Saint-Cyr-l'École, France

B. MÉTIVET   EDF/DER, 1 avenue de Général de Gaulle, 92141 Clamart, France

Abstract   Let u be the solution of the following model:

$$\begin{cases} \text{find } u \in H_0^1(\Omega) \text{ such that} \\ -\Delta u = f \quad \text{in } \Omega, \end{cases}$$

where f is a given function in $L^2(\Omega)$ and Ω a bounded set in $\mathbb{R}^2$ with a polygonal boundary. Let u^h be a finite element approximation of u. The goal of this paper is to suggest an explicit bound of the error $(u - u^h)$. Furthermore this bound is obtained without complex computations. The basic idea is to define locally an admissible vector field for the dual model.

1   INTRODUCTION

Let Ω be a bounded open set in $\mathbb{R}^2$ with a boundary Γ which is assumed to be polygonal. For any function f given in the space $L^2(\Omega)$, we set the following model:

$$\begin{cases} \text{find } u \in H_0^1(\Omega) \text{ such that} \\ \forall v \in H_0^1(\Omega), \quad \int_\Omega \nabla u \cdot \nabla v = \int_\Omega fv \end{cases} \tag{1}$$

(∇ denotes the gradient of a function and the dot stands for the scalar product in $\mathbb{R}^2$). It has been proved by several authors that the unique solution u of (1) lies in the space $H^2(\Omega)$ if Ω is convex (see P. Grisvard [?]). Let us now consider a "uniformly regular family of triangulations" of Ω (see P.G. Ciarlet [4] or V. Girault and P.A. Raviart [9]). For each triangulation $\mathcal{T}^h$, we introduce the approximation space

$$V^h = \left\{ v \in C^0(\overline{\Omega}) \cap H_0^1(\Omega) \ ; \ \forall K \in \mathcal{T}^h, \ v|_K \in P_1(K) \right\}.$$

The approximation model consists in finding u^h in V^h such that

$$\forall v \in V^h, \quad \int_\Omega \nabla u^h \cdot \nabla v = \int_\Omega fv. \tag{2}$$

Concerning the error estimation between u and u^h one has (see P.G. Ciarlet [4])

$$\|u - u^h\|_{1,\Omega} \le c \inf_{v \in V^h} \|u - v\|_{1,\Omega} \tag{3}$$

where c is a constant which is independent of both h and u. Furthermore, choosing for v the P1–Lagrange-interpolate of u, one obtains (see again P.G. Ciarlet [4])

$$\|u - u^h\|_{1,\Omega} \le ch|u|_{2,\Omega}. \tag{4}$$

The inequality (4) is not an error bound but an asymptotic error estimate when h tends to zero. Even if the constant c can be roughly bounded (see R. Arcangelli and J.L. Gout [1]), the quantity $|u|_{2,\Omega}$ is not explicit. In order to derive a better (or at least more practical) error bound it is suitable to provide estimates involving u^h instead of u. This is why people use the "a posteriori" terminology.

Let us mention a result due to I. Babuška and Rheinbolt [2] and C. Bernardi, B. Métivet and R. Verfürth [3] that states the following inequality:

$$\|u - u^h\|_{1,\Omega} \le c \left\{ \sum_{K \in \mathcal{T}^h} h^2 \|f + \Delta u^h\|_{0,\Omega}^2 \right\}^{\frac{1}{2}} + \left\{ \sum_{\gamma \in \mathcal{A}^h} h \left\|\left[\frac{\partial u^h}{\partial \nu}\right]\right\|_{0,\gamma}^2 \right\}^{\frac{1}{2}} \tag{5}$$

$$\le c_1 \|u - u^h\|_{1,\Omega},$$

where $\mathcal{A}^h$ is the set of all the internal edges of $\mathcal{T}^h$ and $[\cdot]$ the jump of a function across these edges. Such results have been extensively studied by C. Bernardi, B. Métivet and R. Verfürth [3] for Stokes problem for instance.

Another possibility is obtained using the so-called constitutive relationship error-bound. It is due to P. Ladevéze [10]. This is the method that we employ in this paper. But our results are different from those of L. Gallimard, P. Ladevéze and J.P. Pelle [8] or P. Ladevéze and D. Leguillon [11]. Another approach was suggested by O.C. Zienkiewicz and J.Z. Zhu [14], [15]; but it seems to be also quite different from the one described here.

Let us sketch how this last method works and how we proceed in the following. First of all let us introduce the set

$$H_f(\text{div}, \Omega) = \left\{ q \in (L^2(\Omega))^2 \ ; \ \text{div} \, q + f = 0 \text{ in } \Omega \right\}, \tag{6}$$

where f is the $L^2(\Omega)$ function used in the definition of problem (1). Then one has the following result due to P. Ladevéze [10]

THEOREM 1   Let u be the solution of (1) and $\underset{\sim}{u^h}$ an arbitrary element in the subspace V^h of $H_0^1(\Omega)$. Then one has

$$|u - \underset{\sim}{u^h}|_{1,\Omega} \le \inf_{q \in H_f(\text{div}, \Omega)} \|q - \nabla \underset{\sim}{u^h}\|_{0,\Omega}.$$

Proof: First of all one has

$$|u - \underset{\sim}{u}^h|^2_{1,\Omega} = \int_\Omega \nabla(u - \underset{\sim}{u}^h) \cdot \nabla(u - \underset{\sim}{u}^h)$$

$$= \int_\Omega f(u - \underset{\sim}{u}^h) - \nabla \underset{\sim}{u}^h \cdot \nabla(u - \underset{\sim}{u}^h)$$

$$= \int_\Omega (q - \nabla \underset{\sim}{u}^h) \cdot \nabla(u - \underset{\sim}{u}^h)$$

for any element q lying in the set $H_f(\text{div}, \Omega)$. The proof is then a consequence of the Schwartz inequality.   □

Our goal is to build up a couple $(\underset{\sim}{u}^h, p) \in V^h \times H_f(\text{div}, \Omega)$, where $\underset{\sim}{u}^h$ is an approximation of u, in order to bound the error between u and $\underset{\sim}{u}^h$ by:

$$|u - \underset{\sim}{u}^h|_{1,\Omega} \le \|p - \nabla \underset{\sim}{u}^h\|_{0,\Omega}. \tag{7}$$

Obviously this error bound is meaningful if and only if it satisfies

$$\|p - \nabla \underset{\sim}{u}^h\|_{0,\Omega} \le ch$$

so that it could be compatible with the asymptotic error estimate given by (4). Unfortunately it is impossible to derive such an estimate if $\underset{\sim}{u}^h$ is solution of (2) with an element p which would be built up locally. Hence we suggest the following steps:

Step 1. Definition of a nonconforming finite element method the solution of which is denoted by u^{hh}. It is then proved that the normal derivative of u^{hh} across the boundary between two elements is "almost" continuous.

Step 2. As u^{hh} does not belong to V^h, we cannot use it in the application of Theorem 1. Therefore we define a projection of u^{hh} onto V^h. This projection is computed locally (on clusters of elements having a common vertex). It is denoted by $\underset{\sim}{u}^h$.

Step 3. Because the normal derivative of u^{hh} between two elements is "almost" continuous, it is possible to construct locally (on each element) a vector field, say p, which belongs to the set $H_f(\text{div}, \Omega)$ and almost equal to $\dfrac{\partial u^{hh}}{\partial \nu}$ on the sides of the elements. We use the dual finite elements introduced by P.A. Raviart and J.M. Thomas [**13**].

Step 4. The validity of the method rests upon a final theorem in which we prove that $p - \nabla \underset{\sim}{u}^h$ is bounded in $L^2(\Omega)$-norm by the sum of two terms. One is the error between u^{hh} and $\underset{\sim}{u}^h$, ($\mathcal{O}(h)$ for the first order derivatives). The second one is a local error due to the right hand side of (1) (i.e. f).

Step 5. (Conclusions). Due to the complexity of the method it can be useful to point out that the error between u and $\underset{\sim}{u}^h$ is explicitly bounded by the quantity $\|p - \nabla \underset{\sim}{u}^h\|_{0,\Omega}$ where p is locally computed and $\underset{\sim}{u}^h$ is an element of V^h deduced from the solution u^{hh} of a finite element method, by a local procedure. Furthermore it is proved that this error bound is asymptotically bounded by an $\mathcal{O}(h)$.

Synopsis of the method

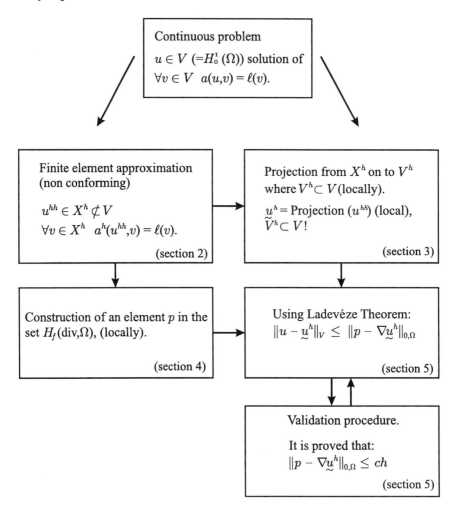

2   THE FINITE ELEMENT METHOD

Let us introduce the approximations space of $H_0^1(\Omega)$, defined by

$$X^h = \left\{ v \mid \forall K \in \mathcal{T}^h,\ v|_K \in P_1(K),\ \forall \gamma \in \mathcal{A}^h\ \int_\gamma [v] = 0,\ \forall \gamma \in \partial\Omega\ \int_\gamma v = 0 \right\} \quad (8)$$

where $\mathcal{A}^h$ and [ ] are defined in the introduction and γ denotes sides of the triangulation. Let us point out that the functions of X^h are not continuous at the boundary between two elements. Then we set for arbitrary elements $v, w \in X^h$

$$a^h(v, w) = \sum_{K \in \mathcal{T}^h} \int_K \nabla v \cdot \nabla w$$

and we use the notation

$$\|v\| = \sqrt{a^h(v, v)}.$$

Then the approximation model of (1) is defined as follows.

$$\begin{cases} \text{Find } u^{hh} \in X^h \text{ such that} \\ \forall v \in X^h, \quad a^h(u^{hh}, v) = \int_\Omega fv. \end{cases} \quad (9)$$

It is quite classical that (9) has a unique solution u^{hh} and furthermore that there exists a constant c, which is independent of both u and h, such that

$$\begin{cases} \|u - u^{hh}\| \le ch|u|_{2,\Omega}, \\ \|u - u^{hh}\|_{0,\Omega} \le ch^2|u|_{2,\Omega}. \end{cases} \quad (10)$$

The proof is similar to the one quoted in P.G. Ciarlet [4] for quadrangles. Let us formulate a very important property of the solution u^{hh} that we use in the following.

THEOREM 2   Let γ be the common side between two neighbour-elements K_1 and K_2. The unit normal along γ is denoted by ν_1 (outside of K_1) or ν_2 (outside of K_2). Let us define by λ_γ the function of X^h which is equal to one at the middle of γ and 0 at the middle of all the other sides of the triangulation (including the boundary of Ω). Then one has

$$\frac{\partial u^{hh}}{\partial \nu_1} - \frac{\int_{K_1} f\lambda_\gamma}{\text{meas}(\gamma)} + \frac{\partial u^{hh}}{\partial \nu_2} - \frac{\int_{K_2} f\lambda_\gamma}{\text{meas}(\gamma)} = 0.$$

If γ is on the boundary of Ω, this relation is meaningless. Hence it is restricted to the sides of $\mathcal{A}^h$.

Proof: Choosing $v = \lambda_\gamma$ in the relation (9) one obtains

$$\int_{K_1} \nabla u^{hh} \cdot \nabla \lambda_\gamma + \int_{K_2} \nabla u^{hh} \cdot \nabla \lambda_\gamma = \int_{K_1 \cup K_2} f\lambda_\gamma$$

and from Stokes formula

$$\int_\gamma \left(\frac{\partial u^{hh}}{\partial \nu_1} + \frac{\partial u^{hh}}{\partial \nu_2} \right) \lambda_\gamma - \int_{K_1 \cup K_2} f\lambda_\gamma = 0$$

(because $\dfrac{\partial u^{hh}}{\partial \nu_i}$ is piecewise constant on ∂K_i and λ_γ, which is linear, is zero at the middle of each side except on γ). Then we deduce

$$\left(\frac{\partial u^{hh}}{\partial \nu_1} + \frac{\partial u^{hh}}{\partial \nu_2} \right) \text{meas}(\gamma) - \int_{K_1} f\lambda_\gamma - \int_{K_2} f\lambda_\gamma = 0,$$

which proves Theorem 2.   □

3   DEFINITION OF A SMOOTHING OF U^{HH}

Let us now introduce a V^h smoothing procedure of u^{hh}. A first attempt consists in setting

$$\underset{v \in V^h}{\text{minimize}} \frac{1}{2} \int_\Omega (u^{hh} - v)^2 \tag{11}$$

or else

$$\begin{cases} \text{find } \underset{\sim}{u_0^h} \in V^h \text{ such that} \\ \forall v \in V^h, \quad \int_\Omega \underset{\sim}{u_0^h} v = \int_\Omega u^{hh} v. \end{cases} \tag{12}$$

But unfortunately the solution of (12) cannot be computed locally. It is necessary to solve a global problem. Even if the associated-matrix is well conditioned (Jacobi algorithm works perfectly) it is too expensive for our purpose. Therefore we suggest another V^h smoothing which is based on a reduced integration (leading to the lumped matrix associated to (12)). For any continuous function v on K, we define the approximation

$$\int_K v \simeq \frac{\text{meas}(K)}{3} \big[v(S_1^K) + v(S_2^K) + v(S_3^K) \big], \tag{13}$$

where S_i^K are the three vertices of the triangle K.

Then we introduce (instead of $\underset{\sim}{u_0^h}$), the element $\underset{\sim}{u^h}$ of V^h solution of the following smoothing procedure:

$$\underset{v \in V^h}{\min} \frac{1}{2} m^h(u^{hh} - v, u^{hh} - v), \tag{14}$$

where $m^h(\cdot, \cdot)$ is defined by

$$m^h(u, v) = \sum_{K \in T^h} \frac{\text{meas}(K)}{3} \big[uv(S_1^K) + uv(S_2^K) + uv(S_3^K) \big]. \tag{15}$$

Therefore $\underset{\sim}{u^h}$ is solution of the linear system

$$\begin{cases} \text{find } \underset{\sim}{u^h} \in V^h \text{ such that} \\ \forall v \in V^h, \quad m^h(\underset{\sim}{u^h}, v) = m^h(u^{hh}, v). \end{cases} \tag{16}$$

It is worth noting that the matrix of this linear system is diagonal. More precisely, choosing in (16) a function $v \in V^h$, equal to one at one vertex S inside Ω and to zero at the other vertices, we obtain (cf. Figure 1)

$$\underset{\sim}{u^h}(S) = \sum_{K \in C(S)} \frac{\text{meas}(K) u_K^{hh}(S)}{\sum\limits_{K \in C(S)} \text{meas}(K)}, \tag{17}$$

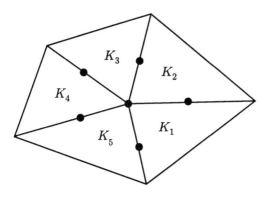

$C(S)=\{K_1,K_2,K_3,K_4,K_5\}$

Figure 1   Definition of $\underset{\sim}{u}^h$.

where $C(S)$ is the cluster of S (i.e. the set of all the triangles which have S as a vertex). Furthermore, $u_K^{hh}(S)$ is the value of u^{hh} on the element K at point S.

The goal of the next section is to estimate the error between u^{hh} and $\underset{\sim}{u}^h$. First of all, let us notice that from the definition of $\underset{\sim}{u_0^h}$ (see (11)), one can write

$$\forall v \in V^h \quad \|u^{hh} - \underset{\sim}{u_0^h}\|_{0,\Omega} \leq \|u^{hh} - v\|_{0,\Omega}.$$

But from the triangle inequality we deduce that

$$\forall v \in V^h \quad \|u^{hh} - \underset{\sim}{u_0^h}\|_{0,\Omega} \leq \|u^{hh} - u\|_{0,\Omega} + \|u - v\|_{0,\Omega},$$

where we recall that u is the solution of (1). Then choosing for instance $v = \Pi u$ (Lagrange interpolate of u) in V^h, we obtain (we also use (10))

$$\|u^{hh} - \underset{\sim}{u_0^h}\|_{0,\Omega} \leq ch^2 |u|_{2,\Omega}. \tag{18}$$

In order to evaluate $\|u^{hh} - \underset{\sim}{u}^h\|_{0,\Omega}$ we need the next property.

THEOREM 3   Let v and w be two elements in the space X^h. We set

$$m(v,w) = \int_\Omega vw.$$

Then there exists a constant c which is independent of h, v and w such that

$$|m^h(v,w) - m(v,w)| \leq ch^2 \|\|uv\|\| \, \|\|w\|\|.$$

Proof: Let us denote by Π_0 the linear operator from $L^2(\Omega)$ into X^h defined by

$$\forall v \in L^2(\Omega), \forall K \in \mathcal{T}^h, \quad \Pi_0 v_K = \frac{1}{\text{meas}(K)} \int_K v.$$

Then from Lemma 2 (see the annex)

$$\|v - \Pi_0 v\|_{0,\Omega} \le ch\|v\|.$$

Observing that

$$\forall v, w \in L^2(\Omega), \quad \int_\Omega (v - \Pi_0 v)\Pi_0 w = 0,$$

we deduce that

$$\forall v, w \in L^2(\Omega), \quad m(v - \Pi_0 v, w) = m(v - \Pi_0 v, w - \Pi_0 w),$$

and then

$$\forall v \in L^2(\Omega), \forall w \in X^h, \quad m^h(\Pi_0 v, w) = m(\Pi_0 v, w).$$

Finally for any $v \in X^h$ one has

$$m(v, v) - m^h(v, v) = m(v - \Pi_0 v, v) + m^h(\Pi_0 v - v, v)$$

and

$$m(v, v) - m^h(v, v) = m(v - \Pi_0 v, v - \Pi_0 v) - m^h(v - \Pi_0 v, v - \Pi_0 v).$$

Let us now set ($v \in X^h$!)

$$w = v - \Pi_0 v, \quad \left(\int_K w = 0 \quad \forall K \in \mathcal{T}^h \right).$$

Then, from the definition of $m^h(\cdot, \cdot)$, one has

$$m^h(w, w) = \sum_{K \in \mathcal{T}^h} \frac{\text{meas}(K)}{3} \left[w^2(S_1^K) + w^2(S_2^K) + w^2(S_3^K) \right]$$

and from $\int_K w = 0$, we deduce that

$$w(S_1^K) + w(S_2^K) + w(S_3^K) = 0 \quad (w \text{ is linear on } K),$$

and therefore

$$m^h(w, w) = \sum_{K \in \mathcal{T}^h} \frac{\text{meas}(K)}{3} \left[\left(w(S_1^K) + w(S_2^K) \right)^2 + \left(w(S_2^K) + w(S_3^K) \right)^2 \right.$$

$$\left. + \left(w(S_3^K) + w(S_4^K) \right)^2 \right].$$

Thus for any w in X^h, satisfying $\int_K w = 0$, one has $m^h(w, w) = 4m(w, w)$.

(One has $\int_K w^2 = \frac{\text{meas}(K)}{12}[(w(S_1^K) + w(S_2^K))^2 + (w(S_2^K) + w(S_3^K))^2 + (w(S_1^K) + w(S_3^K))^2]$.)

Finally we conclude that for any element v lying in the space X^h, one has

$$|m(v, v) - m^h(v, v)| \le 3\|v - \Pi_0 v\|_{0,\Omega}^2 \le 3ch^2\|v\|^2.$$

The proof of Theorem 3 is then completed by observing that

$$\begin{cases} m(v, w) = \dfrac{1}{4}\left[m(v + w, v + w) - m(v - w, v - w) \right] \\[2mm] m^h(v, w) = \dfrac{1}{4}\left[m^h(v + w, v + w) - m^h(v - w, v - w) \right]. \end{cases}$$

$\square$

The errors between the solutions u^{hh} and u^h of (9) and (16), are then upper-bounded in the following result.

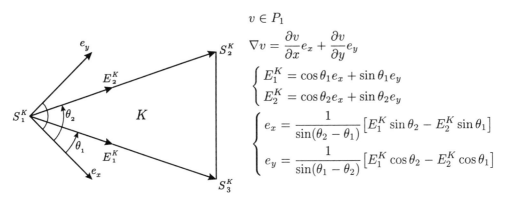

Figure 2   Gradient of a linear function on K.

THEOREM 4   There exists a constant c which is both independent of h and u and such that

$$\|u^{hh} - u^h\| \leq ch|u|_{2,\Omega}$$

and

$$\|u^{hh} - u^h\|_{0,\Omega} \leq ch^2|u|_{2,\Omega}.$$

Proof: Let v be an element in the space X^h. From the definition of the bilinear form m^h one has

$$m^h(v,v) = \sum_{K \in T^h} \frac{\text{meas}(K)}{3} \left[v^2(S_1^K) + v^2(S_2^K) + v^2(S_3^K) \right].$$

But from a simple calculus one has, using the fact that the family of triangulations is uniformly regular so that the angles of each triangle would be under-bound independently of h (see Figure 2)

$$|v|_{1,K}^2 \leq c \left[\left(v(S_1^K) - v(S_2^K) \right)^2 + \left(v(S_1^K) - v(S_2^K) \right)^2 \right],$$

which leads to a so-called inverse inequality: for any element v in the space X^h

$$\|v\|^2 = \sum_{K \in T^h} |v|_{1,K}^2 \leq \frac{c}{h^2} m^h(v,v). \qquad (19)$$

But from the definition of $\underset{\sim}{u}^h$ given at (14), one deduces

$$m^h(u^{hh} - \underset{\sim}{u}^h, u^{hh} - \underset{\sim}{u}^h) \leq m(u^{hh} - \underset{\sim}{u}_0^h, u^{hh} - \underset{\sim}{u}_0^h)$$

or else, using Theorem 3

$$m^h(u^{hh} - \underset{\sim}{u}^h, u^{hh} - \underset{\sim}{u}^h) \leq \|u^{hh} - \underset{\sim}{u}_0^h\|_{0,\Omega}^2 + ch^2\|u^{hh} - \underset{\sim}{u}_0^h\|^2.$$

The classical inverse inequality between $L^2(K)$ and $H^1(K)$ for linear functions lead to the following bound (cf. P.G. Ciarlet [4]):

$$m^h(u^{hh} - \underset{\sim}{u}^h, u^{hh} - \underset{\sim}{u}^h) \leq c\|u^{hh} - u_0^h\|_{0,\Omega}^2$$

and finally, from (18),

$$m^h(u^{hh} - \underset{\sim}{u}^h, u^{hh} - \underset{\sim}{u}^h) \leq ch^4|u|_{2,\Omega}^2 \tag{20}$$

or else using (19)

$$\|u^{hh} - \underset{\sim}{u}^h\| \leq ch|u|_{2,\Omega}.$$

Let us now prove the $L^2(\Omega)$ error estimate between u^{hh} and $\underset{\sim}{u}^h$.

Let us first notice that for any linear function on the element K, one has

$$\int_K v^2 = \frac{\text{meas}(K)}{12}\left[\left(v(S_1^K) + v(S_2^K)\right)^2 + \left(v(S_2^K) + v(S_3^K)\right)^2 + \left(v(S_1^K) + v(S_3^K)\right)^2\right]$$

or else

$$\int_K v^2 \leq \frac{\text{meas}(K)}{3}\left[v^2(S_1^K) + v^2(S_2^K) + v^2(S_3^K)\right],$$

which implies for any function in X^h

$$m(v,v) \leq m^h(v,v),$$

and finally from (20)

$$\|u^{hh} - \underset{\sim}{u}^h\|_{0,\Omega}^2 = m(u^{hh} - \underset{\sim}{u}^h, u^{hh} - \underset{\sim}{u}^h) \leq m^h(u^{hh} - \underset{\sim}{u}^h, u^{hh} - \underset{\sim}{u}^h) \leq ch^4|u|_{2,\Omega}^2,$$

which completes the proof of Theorem 4.   □

REMARK   It is worth noting that the error between u^{hh} and $\underset{\sim}{u}^h$ is compatible with the one between u^{hh} and u.

4   DEFINITION OF AN EQUILIBRIUM VECTOR FIELD

In this section, we build up an element p in the space $H_f(\text{div}, \Omega)$ which is quite close to ∇u^h, so that we obtain an explicit error bound between u and $\underset{\sim}{u}^h$ which is compatible with the theoretical error estimates obtained in a finite element method. Furthermore, we point out, one more time, that this element must be built up locally in order to be operational. The main difficulty is to ensure the continuity of the normal flux across the boundary between two elements. Following the results mentioned in Theorem 2 we use the quantity

$$\frac{\partial u^{hh}}{\partial \nu} - \frac{\int_K f\lambda_\gamma}{\text{meas}(\gamma)},$$

which satisfies the right continuity condition for $p \cdot \nu$ along ∂K (see Theorem 2).

Let us then set on K

$$
\begin{cases}
p \in (L^2(K))^2, \quad \operatorname{div} p + f = 0 \quad \text{in } K \\[2mm]
\text{and} \\[2mm]
p \cdot \nu = \dfrac{\partial u^{hh}}{\partial \nu} - \dfrac{\int_K f \lambda_\gamma}{\operatorname{meas}(\gamma)} \qquad \text{on } \partial K.
\end{cases}
\tag{21}
$$

Such an element p would belong to the set $H_f(\operatorname{div}, \Omega)$.

First of all we introduce the element p^h such that (x and y are the Cartesian coordinates)

$$
p^h = \begin{cases}
a_K + b_K x \\
c_K + b_K y
\end{cases} \qquad \text{in } K
\tag{22}
$$

and such that on each side γ of ∂K

$$
p^h \cdot \nu|_\gamma = \frac{\partial u^{hh}}{\partial \nu} - \frac{\int_K f \lambda_\gamma}{\operatorname{meas}(\gamma)} \qquad \forall \gamma \subset \partial K.
\tag{23}
$$

Then p^h is in $H(\operatorname{div}, \Omega)$. This vector field is perfectly determined. The idea of such a polynomial approximation is due to P.A. Raviart and J.M. Thomas [13]. We refer the reader, if necessary, to the previous authors for additional explanations concerning this kind of finite elements. Let us nevertheless underline that

$$
\int_K \operatorname{div} p^h = 2 \operatorname{meas}(K) b_K = \int_{\partial K} p^h \cdot \nu
$$

$$
= \int_{\partial K} \frac{\partial u^{hh}}{\partial \nu} - \int_K f(\lambda_{\gamma_1} + \lambda_{\gamma_2} + \lambda_{\gamma_3})
$$

$$
= \int_{\partial K} \frac{\partial u^{hh}}{\partial \nu} - \int_K f
$$

(γ_1, γ_2 and γ_3 denotes the three sides of ∂K).

But u^{hh} is linear on K. Hence

$$
\Delta u^{hh} = 0 \quad \text{on } K
$$

and therefore

$$
\int_{\partial K} \frac{\partial u^{hh}}{\partial \nu} = 0.
$$

Finally

$$
b_K = -\frac{1}{2 \operatorname{meas}(K)} \int_K f.
\tag{24}
$$

Let us now set

$$
\delta p = p - p^h.
$$

One has

$$
\begin{cases}
\delta p \cdot \nu = 0 & \text{on } \partial K, \\[2mm]
\operatorname{div} \delta p = -(f + 2b_K) = -\left(f - \dfrac{1}{\operatorname{meas}(K)} \int_K f \right) & \text{in } K,
\end{cases}
\tag{25}
$$

so that in K, one can choose for δp the gradient of the function δu_K defined by

$$
\begin{cases}
-\Delta \delta u_K = \left(f - \dfrac{1}{\text{meas}(K)} \displaystyle\int_K f \right) & \text{in } K, \\[2mm]
\dfrac{\partial}{\partial \nu}(\delta u_K) = 0 & \text{on } \partial K, \\[2mm]
\displaystyle\int_K \delta u_K = 0.
\end{cases}
\tag{26}
$$

Furthermore δu_K can be explicitly calculated using Green integrals.

Existence and uniqueness of δu_K are standard. Then we define p on each K, as follows:

$$
p = p^h + \nabla \delta u_K,
\tag{27}
$$

and clearly this element satisfies all the required properties in order to belong to the set $H_f(\text{div}, \Omega)$.

Let us complete this section by estimating the term $\nabla \delta u_K$ in the definition of p.

THEOREM 5   Let δu_K be the solution of the problem (26) and let us assume that f lies in the space $H^1(K)$. Then there exists a constant c which is both h and f independent and such that

$$
|\delta u_K|_{1,K} \le ch^2 \|f\|_{1,K} \quad (\le ch\|f\|_{0,K} \text{ if we only have } f \in L^2(K)).
$$

Proof: From the definition of δu_K (system (26)), one has

$$
- \int_K \Delta \delta u_K \delta u_K = \int_K |\nabla \delta u_K|^2 = \int_K (f + 2b)\delta u_K = \int_K f \delta u_K
$$

because

$$
\int_K \delta u_K = 0.
$$

Therefore

$$
|\delta u_K|^2_{1,K} \le \|f - \Pi_0 f\|_{0,K}\|\delta u_K\|_{0,K},
$$

and from Lemma 1 and 2 (see the annex)

$$
|\delta u_K|_{1,K} \le ch^2 \|f\|_{1,K}, \quad \text{(if } f \text{ is smooth enough)}
$$
$$
\big(\text{or } |\delta u_K|_{1,K} \le ch\|f\|_{0,K} \text{ if } f \text{ is only } L^2(\Omega)\big).
$$

$\square$

Therefore the term $p - p^h$ is $\mathcal{O}(h^2)$ in $L^2(\Omega)$ norm. The computation of δu_K on each element, can appear to be a little bit tough. Theorem 5 indicates that if f is smooth enough, this term can be omitted in the definition of p, because it is only a second order term. But in such a choice the error, bound is no more exact. Let us also point out that is also easy to upper bound $|\delta u_K|_{1,K}$ by an a priori estimate which can be made explicit and such that

$$
|\delta u_K|_{1,K} \le ch\|f\|_{0,K}.
$$

5 COMPATIBILITY BETWEEN THE EXPLICIT ERROR BOUND AND THE CLASSICAL ESTIMATES

First of all we recall that we have built up two terms, u^h at (16), and p at (27) satisfying (according to Theorem 1)

$$|u - u^h|_{1,\Omega} \leq \|p - \nabla u^h\|_{0,\Omega}.$$

Our goal is now to bound the right hand side of the above inequality. Let us set

$$q = p - \nabla u^h.$$

From the triangle inequality we deduce that

$$\|q\|_{0,\Omega} \leq \left[\sum_{K \in \mathcal{T}^h} \int_K |p^h + \nabla \delta u_K - \nabla u^{hh}|^2 \right]^{\frac{1}{2}} + \|u^{hh} - u^h\|$$

and from (10) Theorem 4

$$\|q\|_{0,\Omega} \leq \left[\sum_{K \in \mathcal{T}^h} \int_K |p^h + \nabla \delta u_K - \nabla u^{hh}|^2 \right]^{\frac{1}{2}} + ch|u|_{2,\Omega}.$$

Let us consider in K the vector

$$k = p^h + \nabla \delta u_K - \nabla u^{hh} = \begin{pmatrix} k_1 \\ k_2 \end{pmatrix}.$$

It satisfies

$$\text{curl}(k) = \partial_1 k_2 - \partial_2 k_1 = 0 \quad \text{in } K,$$

and necessarily

$$\int_{\partial K} k \cdot \tau = 0.$$

Then there exists a function φ_K on K such that

$$k = \nabla \cdot \varphi_K \quad \text{and} \int_K \varphi_K = 0.$$

But φ_K also satisfies

$$\text{div } k = \Delta \varphi_K = \text{div } p^h + \Delta \delta u_K = -f \quad \text{on } K,$$

and

$$\frac{\partial \varphi_K}{\partial \nu} = p^h \cdot \nu + \frac{\partial \delta u_K}{\partial \nu} - \frac{\partial u^{hh}}{\partial \nu} = -\frac{\int_K f \lambda_\gamma}{\text{meas}(\gamma)} \quad \text{on } \gamma \subset \partial K.$$

From the definition of φ_K one has

$$|u - u^h|_{1,\Omega} \leq \|p - \nabla u^h\|_{0,\Omega} \tag{28}$$

$$\leq ch|u|_{2,\Omega} + \left[\sum_{K \in \mathcal{T}^h} |\varphi_K|^2_{1,K} \right]^{\frac{1}{2}},$$

where φ_K is solution of

$$\begin{cases} -\Delta \varphi_K = f & \text{on } K, \quad \int_K \varphi_K = 0, \\ \dfrac{\partial \varphi_K}{\partial \nu} = -\dfrac{\int_K f \lambda_\gamma}{\text{meas}(\gamma)} & \text{on } \gamma \subset \partial K. \end{cases} \tag{29}$$

The basic point is now to prove the following result.

THEOREM 6   Let φ_K be the unique solution of (29). Then there exists a constant c, which is both h and f independent and such that

$$\|\varphi_K\| = \left[\sum_{K \in \mathcal{T}^h} |\varphi_K|_{1,K}^2 \right]^{\frac{1}{2}} \leq ch\|f\|_{0,\Omega}.$$

Proof: First of all one has

$$-\int_K \Delta\varphi_K \varphi_K = \int_K f\varphi_K = -\int_{\partial K} \frac{\partial \varphi_K}{\partial \nu} + \int_K |\nabla\varphi_K|^2$$

or else

$$|\varphi_K|_{1,K}^2 = \int_K f\varphi_K + \sum_{i=1,2,3} \frac{\int_K f\lambda_{\gamma_i}}{\text{meas}(\gamma_i)} \int_{\gamma_i} \varphi_K.$$

Then one has, because φ_K is such that (see Lemma 1 in the annex)

$$\int_K \varphi_K = 0,$$

(i)
$$\left| \int_K f\varphi_K \right| \leq \|f\|_{0,K} \|\varphi_K\|_{0,K} \leq ch\|f\|_{0,K} |\varphi_K|_{1,K}$$

and

(ii) $$\left| \sum_{i=1,2,3} \left(\frac{\int_K f\lambda_{\gamma_i}}{\text{meas}(\gamma_i)} \int_{\gamma_i} \varphi_K \right) \right| \leq \|f\|_{0,K} \sqrt{\frac{\text{meas}(K)}{3}} \sum_{i=1,2,3} \frac{1}{\text{meas}(\gamma_i)} \left| \int_{\gamma_i} \varphi_K \right|,$$

and from the Schwartz inequality and because of the uniform regularity of the triangulation (see the annex)

$$\left| \sum_{i=1,2,3} \frac{1}{\text{meas}(\gamma_i)} \int_{\gamma_i} \varphi_K \right| \leq \sum_{i=1,2,3} \frac{1}{\sqrt{\text{meas}(\gamma_i)}} \|\varphi_K\|_{0,\gamma_i}$$

$$\leq c|\varphi_K|_{1,K}.$$

Finally we proved that

$$|\varphi_K|_{1,K} \leq ch\|f\|_{0,K}$$

and the proof is complete.   □

A direct consequence of Theorem 6 is the next one.

THEOREM 7   Let f be an element of the space $L^2(\Omega)$. Then u^h and $\underset{\sim}{p}$ being the elements defined respectively at (16) and (27); then one has (assuming that u solution of (1) belongs to $H^2(\Omega)$)

$$|u - u^h|_{1,\Omega} \leq \|\underset{\sim}{p} - \nabla u^h\|_{0,\Omega} \leq ch.$$

Proof: The proof is straight-forwards from the previous results.   □

It is possible to make more precise the estimate given in Theorem 7 when f is $L^\infty(\Omega)$.

THEOREM 8   Let f be an element of the space $L^\infty(\Omega)$, $\underset{\sim}{u}^h$ and p being defined in (16) and (27). Then

$$|u - \underset{\sim}{u}^h|_{1,\Omega} \le \|p - \nabla \underset{\sim}{u}^h\|_{0,\Omega}$$

$$\le \|u^{hh} - \underset{\sim}{u}^h\| + ch^2 \left(\|f\|_{1,\Omega} + \left(\sum_{K \in \mathcal{T}^h} \|f\|_{0,\infty,K}^2 \right)^{\frac{1}{2}} \right).$$

Proof: Let us just modify the proof of Theorem 6 (estimate (ii)), by using the uniform regularity of the triangulation and Lemma 3 in the annex

$$\left| \sum_{i=1,2,3} \left(\frac{\int_K f \lambda_{\gamma_i}}{\text{meas}(\gamma_i)} \int_{\gamma_i} \varphi \right) \right| \le \frac{\text{meas}(K)}{3} \|f\|_{0,\infty,K} \left(\sum_{i=1,2,3} \frac{1}{\text{meas}(\gamma_i)} \left| \int_{\gamma_i} \varphi \right| \right)$$

$$\le ch^2 \|f\|_{0,\infty,K} |\varphi|_{1,K}.$$

Furthermore one has

$$\int_K f\varphi = \int_K (f - \Pi_0 f)\varphi \quad \left(\text{because} \int_K \varphi = 0 \right).$$

Therefore (see the annex)

$$\left| \int_K f\varphi \right| \le \|f - \Pi_0 f\|_{0,K} \|\varphi\|_{0,K} \le ch^2 \|f\|_{1,K} |\varphi|_{1,K}$$

and the proof is established.   □

The interest of Theorem 8 is that when f has a support restricted to a small number of elements of $\mathcal{T}^h$, then the term

$$\left(\sum_{K \in \mathcal{T}^h} \|f\|_{0,\infty,K}^2 \right)^{\frac{1}{2}}$$

can be assumed to be quite small and the error due to f is locally in h^2 (as far as h is not too small). Then the main contribution to the error between u and $\underset{\sim}{u}^h$ is the one between u^{hh} and $\underset{\sim}{u}^h$. This remark is valuable for a mesh refinement algorithm. More precisely, if $\underset{\sim}{f} \equiv 0$ on the element K, then $\varphi_K = 0$ (φ_K is defined at (29)). One deduces that

$$\|p - \nabla \underset{\sim}{u}^h\|_{0,K} = |u^{hh} - \underset{\sim}{u}|_{1,K}.$$

As far as one is interested by an error indicator, the term $\|p - \nabla \underset{\sim}{u}^h\|_{0,K}$ is a nice candidate. The advantage is that it is fully explicit (with local computations).

REMARK   We proved previously that the term $\nabla \delta u_K$ is $O(h^2)$ in $L^2(K)$ norm when f is $H^1(K)$. Therefore it is well justified to use the following approximation of the error, mainly if one is concerned with mesh refinement algorithms

$$\varepsilon = \left[\sum_{K \in \mathcal{T}^h} \|p^h - \nabla \underset{\sim}{u}^h\|_{0,K}^2 \right]^{\frac{1}{2}},$$

because one has, assuming that f belongs to the space $H^1(\Omega)$, (see Theorem 5)

$$|u - \underset{\sim}{u}^h|_{1,\Omega} \leq \varepsilon + ch^2\|f\|_{1,\Omega}.$$

6 CONCLUSION

We showed that for a non-conforming finite element method it is possible to bound the error by an explicit term. The indicator of the error (local contribution to the global error) can be easily computed element by element and a mesh refinement algorithm can be derived. The strategy seems to be quite general for second order elliptic problems. For instance, it can be extended to elasticity problem without major difficulties. Furthermore, the extension to conforming finite element methods can also be carried out, but using a slightly different strategy and other assumptions (see Destuynder and Métivet [6]).

REFERENCES

1. R. Arcangelli and J.L. Gout, *Sur l'évaluation de l'erreur d'interpolation de Lagrange dans un ouvert de $\mathbb{R}^n$*, Revue Française d'Automat. Informat., R.O. série rouge: Anal. Numér. **10** (1976), 5–27.

2. I. Babuška and W.C. Rheinbolt, *Error estimates for adaptive finite element computation*, SIAM J. Numer. Anal. **15**, n° 4 (1978), 736–754.

3. C. Bernardi, B. Métivet and R. Verfürth, *Analyse numérique d'indicateurs d'erreur*, note EDF–M172/93062, 1993.

4. P.G. Ciarlet, "The finite element method for elliptic problems," North-Holland, 1978.

5. Ph. Destuynder and B. Métivet, *Explicit error bounds for a conforming finite element method*, (1996) (to appear).

6. Ph. Destuynder and B. Métivet, *Estimation explicite de l'erreur pour une méthode d'éléments finis non conforme*, note aux CRAS, soumise (1996).

7. K. Erikson and C. Johnson, *An adaptive finite element method for linear elliptic problems*, Math. Comp. **50** (1988), 361–383.

8. L. Gallimard, P. Ladevéze and J.P. Pelle, *La méthode des erreurs en relation de comportement appliquée au contrôle des calculs non linéaires*, École CEA EDF-INRIA. Cours du 18–21 septembre 1995 sur le calcul d'erreur a posteriori et adaptation de maillage (1995).

9. V. Girault and P.A. Raviart, "Finite element methods for Navier–Stokes equations. Theories and algorithms," Springer, Berlin, 1986.

10. P. Ladevéze, "Comparaison de modèles de milieux continus," Université Paris VI, Paris.

11. P. Ladevéze and D. Leguillon, *Error estimate procedure in the finite element method and applications*, SIAM J. Numer. Anal. **20**, n° 3 (1983), 485–509.

12. R. Verfürth, *A posteriori error estimation and adaptive mesh-refinement techniques*, J. Comput. Appl. Math. **50** (1994), 67–83.

13. P.A. Raviart and J.M. Thomas, *A mixed finite element method for second order elliptic problems*, in "Proc. of the Symposium on the Mathematical Aspects of

the Finite Element Methods," Rome, 1975, Lecture Notes in Mathematics 606, Springer, 1977, pp. 292–315.

14. J.C. Zhu and O.C. Zienkiewicz, *Adaptive techniques in the finite element method*, Comm. Appl. Numer. Methods **4** (1988), 197–204.

15. O.C. Zienkiewicz and J.Z. Zhu, *A simple error estimator and adaptive procedure for practical engineering analysis*, Internat. J. Numer. Methods Engrg. **24** (1987), 337–357.

ANNEX

In the paper we used several classical results quite well known by people from numerical analysis. We recall them here after just for sake of brevity.

LEMMA 1   Let K be a triangle of a mesh family assumed to be regular (see P.G. Ciarlet [4] and V. Girault and P.A. Raviart [9]). Then let h be the maximum length of the sides of K. For any function $\varphi \in H^1(K)$, satisfying

$$\int_K \varphi = 0,$$

there exists a constant (say c) with is both h and φ independent and such that

(i) $$\|\varphi\|_{0,K} \leq ch|\varphi|_{1,K}.$$

LEMMA 2   (Same hypothesis as in Lemma 1). Let $\Pi_0\varphi$ be defined by

$$\Pi_0\varphi = \frac{1}{\text{meas}(K)} \int_K \varphi.$$

Then there exists a constant c which is both h and φ independent and such that

$$\|\varphi - \Pi_0\varphi\|_{0,K} \leq ch|\varphi|_{1,K}.$$

LEMMA 3   (Same hypothesis as in Lemma 1 but we also assume that the family of triangulation is uniformly regular as described in V. Girault and P.A. Raviart [9]). There exists a constant c which is both h and φ independent and such that

$$\|\varphi\|_{0,\partial K} \leq \sqrt{h}\|\varphi\|_{1,K}.$$

Analysis of the average efficiency of an error estimator

PEDRO DÍEZ, JUAN JOSÉ EGOZCUE, and ANTONIO HUERTA   Departamento de Matemática Aplicada III, E.T.S. de Ingenieros de Caminos, Universitat Politècnica de Catalunya, Campus Nord C–2, E–08034 Barcelona, Spain

1   INTRODUCTION

The error estimator introduced in [1] gave accurate estimates in all the analyzed examples. This estimator has been proved to underestimate the actual error but a complete theoretical analysis has not been carried out.

The effectivity index, that is, the ratio between the estimated and the actual error, and its asymptotic behavior are the main tools in the analysis of error estimators (see references [2]–[6]). Local and global effectivity indices can be evaluated in concrete examples using a more accurate reference solution or the exact solution if available. The theoretical analysis of the error estimators usually finds lower and upper bounds of the global effectivity index depending on unknown constants. The estimator is said to be asymptotically exact if both the lower and upper bounds tend to one when the characteristic size of the mesh, h, goes to zero.

In [1], the effectivity index was used in order to evaluate the performance of the estimator in the application examples. Both global and local effectivity indices were close to one and, therefore, the estimate was considered accurate. Nevertheless, since the estimator undervaluates the actual error, it is worth to quantify this undervaluation and, consequently, to improve the estimate. In order to give an a priori computed and practical correction factor, we study the accuracy of the estimator from a probabilistic point of view. The efficiency of the estimator is described by the mean value of the effectivity index of the estimator applied to a randomized error. Thus, the average behavior of the estimator is described and the bias of the estimate can be corrected.

From a theoretical point of view, the study of the asymptotic exactness is important [3] but it is useless for correction purposes. On the other hand, in reference [6], Babuška and co-workers provide a general tool for the practical analysis of error estimators. This process uses the estimator as a black-box and finds the extreme values of the computed effectivity indices in a family of problems with known ana-

113

lytical solution. Nevertheless, the pessimistic values of the effectivity index cannot be used in the correction of the bias associated with the systematic undervaluation.

In Section 2, the projection strategy used in [1] is described. This strategy approximates the error function e using a composition of two projections of e on subspaces V and U. Consequently, we call it a multiprojection strategy. Then, the effectivity index of the projection strategy is defined, as well as its mean value, ϕ, which represents the average behavior of the estimator. In Section 3, a simple example with geometric interpretation is used to illustrate the whole process. In this example we easily find a relation between ϕ and the angle, α, between V and U. Here, α has the classical geometric definition of dihedral angle.

The remainder of the paper is devoted to the generalization of the relationship between ϕ and α for arbitrary subspaces V and U. Section 4 deals with the problem of evaluating the mean effectivity index ϕ for a single projection. In Section 5 a general concept of angle between subspaces is presented. This allows to obtain a magnitude α analogous to the dihedral angle. This generalized angle is related with the efficiency of the multiprojection strategy.

Finally, Section 6 presents a numerical example of the theoretical results introduced in the previous sections and their application to the study and correction of the error estimator introduced in [1].

2   THE MULTIPROJECTION STRATEGY

The goal of any error estimator is to approximate (a measure of) the error e given by $e := u - u_h$, where u is the exact solution of the problem and u_h is the Finite Element approximation to u in the interpolation space $\mathcal{V}_h$. The space $\mathcal{V}_h$ is generated by a mesh of characteristic size h. In this section we describe the error estimator introduced in [1] and we introduce the tools used to quantify its performance.

Let us consider a mesh of characteristic size $\tilde{h}$, smaller than h ($\tilde{h} << h$), generating an interpolation space $\mathcal{V}_{\tilde{h}}$. Since the exact solution u cannot be found and the solution obtained with this mesh, $u_{\tilde{h}}$, is more accurate than u_h, $u_{\tilde{h}}$ is taken as a reference solution. Accordingly, a reference error is defined by $e_r := u_{\tilde{h}} - u_h$. Thus, the goal of the error estimation is, in fact, to approximate e_r. However, the computation of e_r requires the solution of a global refined problem. The direct computation of $u_{\tilde{h}}$ (and then, the evaluation of e_r) is, in general, unaffordable. The error estimator presented in [1] approximates e_r with a multiprojection strategy. It is important to observe, however, that e_r is the projection of e on $\mathcal{V}_{\tilde{h}}$, therefore, the energy norm of e_r is always lower than the norm of e ($\|e_r\| \leq \|e\|$).

As previously said, instead of computing e_r by solving a global problem, the error estimation procedure introduced in [1] uses the projections of e on a set of subspaces in $\mathcal{V}_{\tilde{h}}$. A family of orthogonal subspaces V_k is defined, each one associated with one element Ω_k of the mesh generating $\mathcal{V}_h$. Each subspace V_k is generated by the interpolation functions of $\mathcal{V}_{\tilde{h}}$ having their support in Ω_k. The so-called interior approximation ε is obtained by projecting e on the space V which is the sum of all subspaces V_k. Since the subspaces V_k are orthogonal, ε is the sum of the local projections of e on each subspace V_k. The cost of computing ε is obviously lower than the cost of solving a global problem. On the other hand, since V is included in $\mathcal{V}_{\tilde{h}}$, ε is also the projection of e_r on V. Therefore, the norm of ε is lower than the norm of e_r ($\|\varepsilon\| \le \|e_r\| \le \|e\|$).

However, ε is forced to vanish along the interelement boundaries and, consequently, it may be a poor approximation to e. Then, an enrichment of the estimate is performed by adding the contribution of a new set of projections of e. A new family of subspaces U_l is defined, each one associated with one patch Λ_l. The patches are geometrical subdomains partitioning again the global domain Ω. In reference [1] various choices for the shape of the patches are shown. A global subspace U in $\mathcal{V}_{\tilde{h}}$ is generated as the sum of all subspaces U_l. In order to obtain an enriched estimate we do not freely project e on U but we impose the projection to be orthogonal to ε. The orthogonality constraint is imposed via the Lagrange multipliers technique. The patch estimate, that is, the projection of e on U orthogonal to ε, is denoted by η. The projection η vanishes on the boundary of the patches but it is nonnull in most of the points where ε is forced to be zero.

Finally, $\varepsilon + \eta$ is an error estimate which approximates e_r and which always undervaluates its norm ($\|\varepsilon + \eta\| \le \|e_r\| \le \|e\|$).

Notice that there are few nodes of the reference mesh lying on the intersection of the boundaries of the elements and the patches, then the sum of V and U, $W := V + U$, is not exactly $\mathcal{V}_{\tilde{h}}$. Those are called hidden points because the approximation to the error, $\varepsilon + \eta$, is forced to vanish on them. In reference [1] it is shown how the hidden points can be a priori chosen by selecting suitable definitions of patches. Then, in fact, the estimate $\varepsilon + \eta$ can only approximate $\hat{e}_r$, the projection of e_r on W ($\|\varepsilon + \eta\| \le \|\hat{e}_r\| \le \|e_r\| \le \|e\|$).

Thus, the error estimator described above is composed by two computational phases, the interior projection and the patch projection. Nevertheless, in the numerical analysis performed here the complete process (i.e. the transformation from e to $\varepsilon + \eta$) is splitted in three steps, each one corresponding to a different projection. The first one is the projection of e on $\mathcal{V}_{\tilde{h}}$ (from e to e_r). The second one accounts for the hidden points effect, it is the projection of e_r on $W = V + U$ (from e_r to $\hat{e}_r$). And finally, the third step reflects the computation itself, that is the two computational phases: the interior estimate ε (projection of $\hat{e}_r$ on V) and the patch estimate η (projection of $\hat{e}_r$ on U, orthogonal to ε). Thus, this last step corresponds to the transformation from $\hat{e}_r$ to the complete estimate $\varepsilon + \eta$.

Of course, these steps do not correspond to the computational practice but they are useful for the theoretical analysis of the total undervaluation. Indeed, each one of these steps introduces an undervaluation and they can be studied separately.

The first step is easily studied because it needs of the the standard Finite Element a priori estimates. Indeed, assume that an a priori error estimate stands

$$\|e\| = C h^p, \tag{1}$$

where C is an unknown constant and p is related to the maximum degree of the complete polynomial included in the interpolation. Then, the Richardson extrapolation can be used to determine the asymptotic relation between $\|e_r\|$ and $\|e\|$ and to obtain an reliable approximation of the factor relating them:

$$\|e\|^2 \approx \left[1 - \left(\frac{\tilde{h}}{h} \right)^{2p} \right]^{-1} \|e_r\|^2. \tag{2}$$

Thus, if we know an approximation to the norm of e_r we can easily correct it and obtain a better approximation to the norm of e by means of Eq. (2).

The undervaluation associated with the second step is discussed, as a particular case, in Section 4. In fact, this step is a single projection on W and, then, its study is analogous to the study of the first phase of the complete multiprojection strategy. Therefore, the second step is treated as a particular case of the third step.

The study of the undervaluation associated with the third step, the multiprojection strategy, is the main goal of this paper. The space W is decomposed as a vectorial sum of two subspaces V and U. An element $\hat{e}_r$ in W is approximated by projecting it on V and U separately. We analyze the efficiency of this approximation by using its effectivity index, that is, the ratio of the norm of the computed approximation $\varepsilon + \eta$ and the norm of $\hat{e}_r$.

As previously said, the effect of the hidden points, the second step, is a single projection on W. This can be analyzed as a particular case of the third step because if U is included in V, then the second phase, the projection on U, has null effect.

Thus, in the remainder of the text, we concentrate on the study of the undervaluation introduced by the third step. The error which must be approximated is, then, the output of the second step, $\hat{e}_r$. In the following we do not mention the original Finite Element context and we use a general notation: the space W is an Euclidean space with a scalar product denoted by $\langle \cdot, \cdot \rangle$. Moreover, in order to remark that $\hat{e}_r$, ε and η are vectors in W, from now on, they will be boldfaced and denoted by $\hat{\mathbf{e}}_r$, $\boldsymbol{\varepsilon}$ and $\boldsymbol{\eta}$, respectively. Thus, the multiprojection strategy is defined as follows. First, we project $\hat{\mathbf{e}}_r$ on V and we obtain the interior approximation $\boldsymbol{\varepsilon}$. Then, we project $\hat{\mathbf{e}}_r$ on U imposing orthogonality to $\boldsymbol{\varepsilon}$. That is, we project $\hat{\mathbf{e}}_r$ on $\widetilde{U} := U \cap (\mathrm{span}\{\boldsymbol{\varepsilon}\})^{\perp}$ and we obtain $\boldsymbol{\eta}$ ($\perp$ stands for orthogonal subspace).

The vector $\hat{\mathbf{e}}_r$ is approximated using the sum of $\boldsymbol{\varepsilon}$ and $\boldsymbol{\eta}$. Since $\boldsymbol{\varepsilon}$ and $\boldsymbol{\eta}$ are orthogonal the squared norm of $\boldsymbol{\varepsilon} + \boldsymbol{\eta}$ is $\|\boldsymbol{\varepsilon}\|^2 + \|\boldsymbol{\eta}\|^2$. Using this to approximate $\hat{\mathbf{e}}_r$, we are undervaluating its actual norm because $\|\hat{\mathbf{e}}_r\|^2 \geq \|\boldsymbol{\varepsilon}\|^2 + \|\boldsymbol{\eta}\|^2$. The aim of this study is to quantify this undervaluation depending on the "relative position" of the spaces V and U.

Let us define the effectivity index of the multiprojection strategy applied to $\hat{\mathbf{e}}_r$ as the scalar function

$$\nu(\hat{\mathbf{e}}_r) := \frac{[\|\boldsymbol{\varepsilon}\|^2 + \|\boldsymbol{\eta}\|^2]^{1/2}}{\|\hat{\mathbf{e}}_r\|}. \tag{3}$$

Given V and U, $\nu(\hat{\mathbf{e}}_r)$ can be evaluated for every $\hat{\mathbf{e}}_r$. Since the error e and, consequently, $\hat{\mathbf{e}}_r$ are unknown, $\hat{\mathbf{e}}_r$ can be fitted with random character. Then, in order to quantify the average efficiency of the projection strategy we can study the mean value of $\nu(\hat{\mathbf{e}}_r)$.

Since projections are linear operators, the value of $\nu(\hat{\mathbf{e}}_r)$ does not depend on the norm of $\hat{\mathbf{e}}_r$ but only on its direction. Indeed, for any nonnull scalar β,

$$\nu(\hat{\mathbf{e}}_r) = \nu(\beta \hat{\mathbf{e}}_r). \tag{4}$$

Consequently, the study of the average efficiency of the projection strategy can be reduced to an hypersphere in W. We define hypersphere of radius R as being the set of vectors in W of modulus R. Denoting by n the dimension of W, the hypersphere of radius R is denoted by S_R^n.

Then, the mean value of $\nu(\hat{\mathbf{e}}_r)$ is

$$E(\nu(\hat{\mathbf{e}}_r)) = \int_{S_R^n} \nu(\hat{\mathbf{e}}_r) f(\hat{\mathbf{e}}_r) dS, \tag{5}$$

where $f(\cdot)$ is the probability density of $\hat{\mathbf{e}}_r$ over S_R^n. Since we do not have any a priori information about the distribution of $\hat{\mathbf{e}}_r$, we assume $\hat{\mathbf{e}}_r$ to be uniformly distributed and therefore

$$f(\hat{\mathbf{e}}_r) = \frac{1}{\mu(S_R^n)}, \tag{6}$$

where $\mu(S_R^n)$ stands for the measure of S_R^n, which is a general concept including the length of the circle S_R^2 and the area of the sphere S_R^3. Then, our goal is to evaluate

$$\phi := \frac{1}{\mu(S_R^n)} \int_{S_R^n} \nu(\hat{\mathbf{e}}_r) dS, \tag{7}$$

which depends on the dimensions of subspaces V and U and their relative position but does not depend on R. The value of ϕ is lower than or equal to 1. If ϕ is equal to 1 the projection strategy is said to be optimal because, for every $\hat{\mathbf{e}}_r$,

$$\|\hat{\mathbf{e}}_r\|^2 = \|\boldsymbol{\varepsilon}\|^2 + \|\boldsymbol{\eta}\|^2. \tag{8}$$

If ϕ is lower than 1, this value can be used as a correction factor of the obtained estimate.

If instead of considering $\nu(\hat{\mathbf{e}}_r)$ we study the average value of $[\nu(\hat{\mathbf{e}}_r)]^2$, the goal is now to evaluate

$$\psi := \frac{1}{\mu(S_R^n)} \int_{S_R^n} [\nu(\hat{\mathbf{e}}_r)]^2 dS, \tag{9}$$

which is different from ϕ^2. The squared value $[\nu(\hat{\mathbf{e}}_r)]^2$ is easier to handle and the derived expressions for ψ are simpler, as it is shown in Sections 4 and 6. The optimality of the multiprojection strategy is also equivalent to $\psi = 1$.

Both ϕ and ψ can be used in the evaluation of the average behavior of the projection strategy and the correction of the estimate.

3   GEOMETRIC INTERPRETATION

The projection strategy described above as well as the final goal of this study, can be easily understood exploring a simple particular case with geometric interpretation. Indeed, if W has dimension 3 and V and U have both dimension 2, then V and U can be seen as secant planes in a three-dimensional space W. Here, $\tilde{U}$ is a straight line in U which depends on the position of $\boldsymbol{\varepsilon}$, and therefore on $\hat{\mathbf{e}}_r$. Fig. 1 shows the geometric representation of the projection strategy in this simple case. Here the relative position of subspaces V and U is measured by the dihedral angle α. The dihedral angle is characterized by its intersection with a plane orthogonal to the common edge of the two faces V and U (plane angle). If α is null, V and U are the same plane. If α is a right angle, V and U are perpendicular. Note that here,

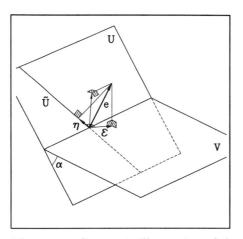

Figure 1   Geometric illustration of the projection strategy.

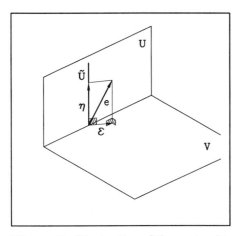

Figure 2   Illustration of the optimality of the projection strategy for perpendicular planes.

perpendicularity is different from orthogonality. Two subspaces are orthogonal if every vector in one of them is orthogonal to all the vectors in the other. The orthogonal space of a plane in a three-dimensional space is represented by a straight line. Two planes cannot be orthogonal but they can be perpendicular.

Let us study for this simple case the mean undervaluation ϕ and its dependence on the dihedral angle α. The objective is to obtain the relation $\phi(\alpha)$ for α between 0 and $\pi/2$ (right angle).

If $\alpha = \pi/2$, i.e., V and U are perpendicular, the projection strategy is optimal. This fact is illustrated in Fig. 2. In this case the space $V^\perp$ is included in U. Then $V^\perp$ is included in $\widetilde{U}$ because $\widetilde{U}$ is the intersection of U and $<\varepsilon>^\perp$ and both include $V^\perp$. Since $\hat{e}_r - \varepsilon$ is in $V^\perp$, $\boldsymbol{\eta} = \hat{e}_r - \varepsilon$, the projection strategy is optimal and we deduce that $\phi(\pi/2) = 1$.

In this particular three-dimensional case the expression of $\nu(\hat{e}_r)$ and, hence, that

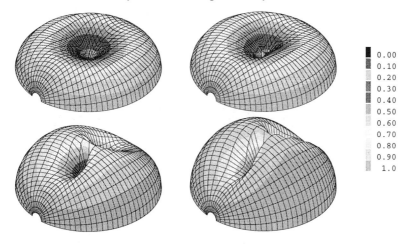

■	0.00
■	0.10
▨	0.20
▨	0.30
■	0.40
▨	0.50
▨	0.60
⋯	0.70
⋯	0.80
⋯	0.90
▨	1.0

Figure 3 Representation of $\nu(\hat{\mathbf{e}}_r)$ over the sphere for four values of α: 0, 30°, 60° and 75° (from left to right and from top to bottom).

of $\phi(\alpha)$ and $\psi(\alpha)$ can be obtained for any α between 0 and $\pi/2$.

In this case each vector $\hat{\mathbf{e}}_r$ can be represented by a point of the three-dimensional space. Let us then consider the points of a sphere. The representation of the values of $\nu(\hat{\mathbf{e}}_r)$ on the sphere can be found in Fig. 3. The sphere is deformed multiplying each radius by the value of $\nu(\hat{\mathbf{e}}_r)$ and shrinking it. We also trace the curves corresponding to equal values of $\nu(\hat{\mathbf{e}}_r)$. In any case (except $\alpha = \pi/2$) we find a "hole" associated with vectors which have a value of $\nu(\hat{\mathbf{e}}_r)$ close to zero (bad approximation). For large values of α the hole is narrow and, then undervaluation affects to a small number of vectors.

The curves plotting ϕ, ψ and $\sqrt{\psi}$ versus α can be seen in Fig. 4. Note that ϕ starts with a minimal value in $\alpha = 0$ of approximately 0.785 and it raises to 1 in $\alpha = \pi/2$, as expected, while ψ starts in $\alpha = 0$ at approximately 0.667.

On the other hand, it is worth remarking that $\sqrt{\psi}$ is greater than ϕ. This stands as a general result because $\psi - \phi^2$ is the variance of $\nu(\hat{\mathbf{e}}_r)$ and, therefore, it is positive.

Thus, the relative position of V and U is measured by α and we can find the value of the mean undervaluation ϕ (or ψ) as a function of α. Of course, this only applies for a very restricted and simple case with $n = 3$. Nevertheless, also in a general case, the average efficiency of the projection strategy must depend only of the relative position of V and U. The objective of this work is to generalize the definition of dihedral angle and to explore its relationship with the mean value of $\nu(\hat{\mathbf{e}}_r)$. We want to define a parameter, analogous to the dihedral angle, for general cases, to measure it and to relate it with the mean value of $\nu(\hat{\mathbf{e}}_r)$.

This process is divided in two parts. Fist, in Section 4, we study the average efficiency of one projection. This allows to a priori determine the starting left value of the curve in Fig. 4. Second, in Section 5, we relate the generalized angle with the projection strategy and deduce an approximated expression for the curve appearing in Fig. 4.

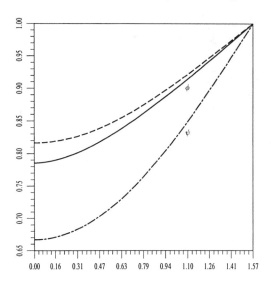

Figure 4   Curves plotting ϕ (solid line), ψ (dashed-dotted line) and $\sqrt{\psi}$ (dashed line) versus α.

4   EFFICIENCY OF A SINGLE PROJECTION

The average efficiency of the projection on a subspace V in W depends only on both the dimensions of V and W. Let us denote by n the dimension of W and by $n - m$ the dimension of V. We want to evaluate the average effect of "loosing" m dimensions in a projection. In this section we give the expressions of the values of ϕ and ψ corresponding to the single projection on V. These values must be functions of n and m and they are denoted by ϕ_m^n and ψ_m^n, respectively.

The results obtained in this section can be used in two of the steps we define in Section 2 in order to analyze the estimator. The average undervaluation introduced in the second step (single projection on W) can be directly evaluated by these results. Moreover, the single projection is a limit case of the multiprojection strategy if V and U are embedded (V is included in U or U is included in V).

The expression giving the mean undervaluation ϕ_m^n depends on the parity of both m and n:

$$\phi_m^n = \frac{(n - m) \cdot (n - m + 2) \cdots (n - 2)}{(n - m + 1) \cdot (n - m + 3) \cdots (n - 1)} \quad \text{for even } m, \tag{10}$$

while the expression for odd m must be splitted:

$$\phi_m^n = \frac{2}{\pi} \frac{[2 \cdot 4 \cdots (n - m - 1)]^2 (n - m + 1) \cdot (n - m + 3) \cdots (n - 2)}{[3 \cdot 5 \cdots (n - m - 2)]^2 (n - m) \cdot (n - m + 2) \cdots (n - 1)} \quad \text{for even } n, \tag{11}$$

and

$$\phi_m^n = \frac{\pi}{2} \frac{[3 \cdot 5 \cdots (n - m - 1)]^2 (n - m + 1) \cdot (n - m + 3) \cdots (n - 2)}{[2 \cdot 4 \cdots (n - m - 2)]^2 (n - m) \cdot (n - m + 2) \cdots (n - 1)} \quad \text{for odd } n. \tag{12}$$

Note that in Eq. (12) both numerator and denominator have $m/2$ terms.

The expression giving the mean squared undervaluation ψ_m^n is much simpler:

$$\psi_m^n = \frac{n-m}{n} = 1 - \frac{m}{n}.$$ (13)

5   EVALUATION OF THE AVERAGE EFFICIENCY OF THE MULTIPROJECTION STRATEGY

This section is devoted to find the relationship between the average underestimation, ϕ or ψ, and the relative position of spaces V and U. This relative position is described by the singular values of a rectangular matrix. The set of these singular values is interpreted as a vector of cosinus of the "angles" between the spaces V and U. We also find an approximate expression for ψ as a function of the singular values of A. The goodness of this approximate expression is tested in Section 6.

It is worth remarking that η is used as an enrichment of the interior estimate ϵ. Then, the norm of the complete estimate is greater than the interior one and the related effectivity index is closer to 1. Therefore, the second projection has an associated value of the mean efficiency which is greater than the value given by ϕ_m^n or ψ_m^n. The amount of the improvement associated with the second projection increases if the angle between the spaces V and U is large (close to the right angle).

In the simple case of Section 3, using a suitable basis, the matrix of the scalar product is

$$[<\cdot,\cdot>]_B = \begin{pmatrix} 1 & 0 & \cos\alpha \\ 0 & 1 & 0 \\ \cos\alpha & 0 & 1 \end{pmatrix}.$$ (14)

Thus, the cosinus of the dihedral angle appears in the corners of the matrix. Moreover, the expression of ψ in this particular case (which is represented in the curve of Fig. 4) has a very simple expression:

$$\psi = 1 - \frac{1}{3}\cos\alpha.$$ (15)

In the general case, by selecting a suitable basis, the matrix of the scalar product is expressed by a simple block shape:

$$[<\cdot,\cdot>]_B = \begin{pmatrix} I_{n_v} & 0_{n_v \times n_c} & A \\ 0_{n_c \times n_v} & I_{n_c} & 0_{n_c \times n_u} \\ A^T & 0_{n_u \times n_c} & I_{n_u} \end{pmatrix}.$$ (16)

The rectangular matrix A contains all the information regarding the relative position of V and U. Moreover, this information is syntheticly resumed in the singular values of A. The set of the singular values of A, $\sigma_1, \dots, \sigma_r$, is interpreted as a set of cosinus describing the "angle" between V and U. In order to give a simple expression of ψ as a function of these singular values, we propose the following approximation:

$$\psi \approx 1 - \frac{1}{n}\sum_{i=1}^{r}\sigma_i.$$ (17)

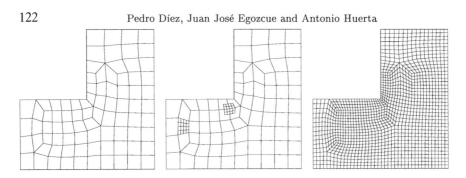

Figure 5 L-shaped domain and computational mesh generating $\mathcal{V}_h$ (left), elementary submesh and node-centered patch submesh (center) and reference mesh generating $\mathcal{V}_{\bar{h}}$ (right).

The validation via numerical simulation of this approximation is performed in Section 6.

Unfortunately, the mean value of $\nu(\hat{\mathbf{e}}_r)$, ϕ, is much more involved and we did not find an expression approximating ϕ in a general case. Nevertheless, ϕ increases monotonously with the singular values of A, as well as ψ does. Recall that $\sqrt{\psi}$ is greater than ϕ. Moreover, for large values of the dimension of W, n, the difference between ϕ_m^n and $\sqrt{\psi_m^n}$ is very small for any m. Thus, we can assume that for large values of n, ϕ and $\sqrt{\psi}$ are practically equal.

6 NUMERICAL EXAMPLE

In this section we show a numerical example coming from the error estimation procedure described in [1]. This example allows both to test the results obtained in the previous sections by numerical simulation and to study the undervaluation of the error estimate.

The results introduced above are applied to a simple case showed in [1]. The Poisson equation is solved in the L-shaped domain of Fig. 5 with a source term and Dirichlet boundary condition on the whole boundary such that the exact solution is $u(x,y) = x^2 + y^2$. The error estimator is applied with elementary submeshes of 16 elements (the refinement parameter $h/\bar{h}$ is equal to 4) and node-centered patches, as shown in Fig. 5. The reference mesh generating $\mathcal{V}_{\bar{h}}$ is build up by assembling all the elementary submeshes and it is also shown in Fig. 5.

The analysis of the obtained estimate has been carried out by examining global and local effectivity indices. Note that, since in this case the actual error is available, the effectivity indices can be easily evaluated. The global effectivity index is equal to 0.919 and the values of the local effectivity index are represented by the histogram of Fig. 6. This histogram plots a bar representing the number of elements with local effectivity index in the associated range. Let us remark that the bias of the estimate can be corrected by dividing the estimate by the global effectivity index. This operation is equivalent to shift the whole histogram to the right and center it around the unity. Since the dispersion of the local effectivity indices is not important, the corrected local estimate is also very accurate. Of course, this can only be done because the actual value of the error is known. However, we can a

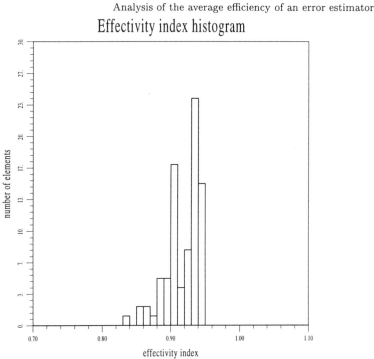

Figure 6 caption:

Figure 6   Histogram representing the distribution of local effectivity indices.

priori assess this correction factor and improve the estimate by applying the results introduced in the previous sections.

In the analyzed example, the elements are bilinear quadrilaterals and, then, the coefficient p of Eq. (2) is equal to 1. Thus, the undervaluation factor associated with the first step (projection on $\mathcal{V}_{\tilde{h}}$) is

$$\frac{\|e_r\|}{\|e\|} \approx \left[1 - \left(\frac{\tilde{h}}{h}\right)^2\right]^{1/2} \approx 0.968. \tag{18}$$

Note that, since this result is used for correction purposes, it is expressed with the same precision of the available effectivity index.

In order to study the undervaluation associated with the second step (effect of the hidden points), it is necessary to carry out some accounting on the topology of the related meshes. The original mesh, see Fig. 5, has 84 elements, 105 nodes and 188 edges, 40 of them in the boundary. Thus, the mesh has 148 interior edges, each one associated with a remaining hidden point. On the other hand, the refined reference mesh of Fig. 5 has 1425 nodes. Thus, the undervaluation associated with this second step is due to force the approximation $\varepsilon + \eta$ to be zero in 148 hidden nodes over a total of 1425. Then, by using Eqs. (14), (15) and (16), $\phi_{148}^{1425} = 0.946628$ and, using Eq. (17), $\psi_{148}^{1425} = 0.89614$. Let us remark that, as it is indicated in the previous section $\sqrt{\psi} = 0.946647$ is very close to ϕ. Thus, the correction factor associated with this step is 0.947.

In order to assess the undervaluation factor associated with the third step (multiprojection strategy) we have to measure the angle between the subspaces V and

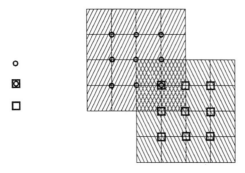

Figure 7   Synthetic mesh generating V_1 and U_1. The nodes marked with a circle are associated with V_1 and the nodes marked with a square are associated with U_1.

U. Let us remark that, in this case, the dimension of space V is 756 (9×84), the dimension of $V \cap U$ is 336 (4×84) and the dimension of U is 697 ($1425 - 40 \times 4 - 148 - 756 + 336$). Thus, in this case, A is a 420×361 rectangular matrix. Obviously, the cost of finding the singular values of this matrix must be precluded. However, as it is indicated in Section 2, both V and U are generated by a family of orthogonal subspaces:

$$V = V_1 \oplus V_2 \oplus \cdots \oplus V_m \quad \text{and} \quad U = U_1 \oplus U_2 \oplus \cdots \oplus U_{m'}, \tag{19}$$

where m is the number of elements and m' is the number of patches (that is, in this case, the number of nodes). Moreover, each subspace V_k is only nonorthogonal to the subspaces U_l such that the element Ω_k and the patch Λ_l have nonnull intersection. Such subspaces will be called connected subspaces. Then, the relative position of V and U can be studied by examining the relative position of all connected subspaces. Nevertheless, each couple of connected subspaces has approximately the same structure and, therefore, it can be assumed that the study of the global relative position of V and U can be reduced to the study of one standard couple of connected subspaces. Moreover, assuming that the distorsion of the elements is not significant we study the angle formed by two connected subspaces V_1 and U_1 generated by the interpolation functions associated with the nodes of the mesh of Fig. 7. The nodes marked with a circle are in the interior of an element and generate V_1 while the nodes marked with a square are in the interior of a patch and generate U_1. In this example there is just one node belonging to both V_1 and U_1. Indeed, the dimension n of $V_1 + U_1$ is 17, $n_v = 8$, $n_c = 1$ and $n_u = 8$. Thus, the matrix A is a 8×8 square rank 3 matrix with the following singular values: $\sigma_1 = 0.23932$, $\sigma_2 = 0.11452$ and $\sigma_3 = 0.01276$. Then, applying Eq. (17) we obtain $\psi \approx 0.978435$. Thus, the approximation of ϕ is $\phi \approx \sqrt{\psi} \approx 0.989159 \approx 0.989$.

In order to verify the goodness of the approximation of Eq. (75), a numerical simulation is performed by randomly generating 100,000 vectors and obtaining the corresponding value of ϕ. That leads to experimental values of

$$\phi \approx 0.990121 \quad \text{and} \quad \psi \approx 0.980736,$$

which are the main values of ν and ν^2, respectively. Thus, the approximated value of the standard deviation of ν is 0.019924. The standard deviation of the employed statistics is small enough to rely on 3 decimal positions in these results. Note that, if ν is assumed to be normally distributed, the value of ν is larger than 0.950273 with a probability of 95%. Then, the multiprojection strategy is not expected to furnish an undervaluation under 95.03%.

Thus, following Eq. (17) the mean undervaluation factor associated with the third step is taken to be 0.989.

The total a priori computed undervaluation factor is then $0.968 \times 0.947 \times 0.989 = 0.907$ which is a good approximation of the global effectivity index 0.919. Indeed, if we correct the estimate using this factor, we obtain a improved global effectivity index of $0.919/0.907 = 1.013$. Thus the resulting error in the error estimate is now of 1.3% instead of the 8.1% of the former estimation. The histogram representing the local effectivity index distribution associated to the corrected estimate does not practically increase its dispersion and it is centered around one.

7 CONCLUSIONS

In this paper, a method for assessing the quality of an error estimator and improving it has been presented. The originality of the analysis is to use a randomized error function in order to a priori evaluate the average behavior of the estimator. The systematic bias introduced in the estimate can, hence, be corrected and an unbiased estimate is obtained.

When it is applied to the error estimator introduced in [1], this analysis requires to introduce mathematical tools which are not usual in the domain of the Finite Element error estimation. Thus, we have developed a set of hyperspherical coordinates allowing to integrate in the hyperspheres. We also introduced the generalized concept of angle between subspaces and we use it to synthetically describe the relative position of two subspaces. The average efficiency of the error estimator has been proved to depend on this synthetic information. Moreover, we give an simple expression that approximately describes this average behavior.

The application to a simple example of error estimation in a bidimensional Poisson equation with analytical solution shows that this theory can be used in both the analysis and the practical use of the error estimator introduced in [1].

REFERENCES

1. Díez, P., J.J. Egozcue, and A. Huerta, *A Posteriori error Estimation for Standard Finite Element Analysis*, Publication nº75 CIMNE, Barcelona, 1995.
2. Babuška, I., T. Strouboulis and C.S. Upadhyay, *A model study of a posteriori error estimators for linear elliptic problems. Error estimation in the interior of patchwise uniform grids of triangles*, Comp. Methods Appl. Mech. Engrg. **114** (1994), 307–378.
3. Durán, R. and R. Rodríguez, *On the asymptotic exactness of Bank–Weiser's estimator*, Numer. Math. **62** (1992), 297–303.
4. Zienkiewicz, O.C. and J.Z. Zhu, *A simple error estimator and adaptive proce-*

    dure for practical engineering analysis, Internat. J. Numer. Methods Engrg. **24** (1987), 337–357.

5. Rodríguez, R., *Some remarks on Zienkiewicz–Zhu estimator*, Numer. Methods Partial Differential Equations **10** (1994), 625–635.

6. Babuška, I. , T. Strouboulis, C.S. Upadhyay, S.K. Gangaraj and K. Copps, *Validation of a posteriori error estimators by numerical approach*, Internat. J. Numer. Methods Engrg. **37** (1994), 1073–1123.

On the mesh for difference schemes of higher accuracy for the heat-conduction equation

ISTVÁN FARAGÓ*   Eötvös Loránd University, Department of Applied Analysis, H-1088 Múzeum krt. 6–8, Budapest, Hungary, e-mail: faragois@ludens.elte.hu

1   INTRODUCTION

In domain $\Omega = [0, 1] \times \mathbb{R}_0^+$ we consider the initial-boundary value problem for the heat-conduction equation

$$\frac{\partial w}{\partial t} = \frac{\partial^2 w}{\partial x^2}; \quad x \in (0, 1),\ t > 0,$$

with initial condition $w(x, 0) = w_0(x)$ and first homogeneous boundary condition $w(0, t) = w(1, t) = 0$.

We assume that w_0 is sufficiently smooth. Then, there exists a unique solution $w : \Omega \mapsto \mathbb{R}$ which is sufficiently smooth and has several special characteristic properties, like contractivity in time in different norms, preservation of both the nonnegativity and shape of the initial function, regular exponential convergence and others. (The regular exponential convergence means that by increasing time, the solution of the problem tends to the solution of the corresponding elliptic problem without oscillation and the convergence rate in l_2-norm is the same as the convergence rate of an exponential function. See, e.g., [2], [4], [8], [10].)

As typical, the continuous problem cannot be solved analytically. Therefore, we apply some numerical process which can be described in the following way. First, we semi-discretize the continuous problem with respect to the space variable and obtain a Cauchy problem for a system of ordinary differential equations. For the time integration of this problem the so-called θ-method is applied. It is easily implementable. For both methods the meshes are assumed uniform and they are characterized by the step-sizes h and τ, respectively. Altogether, it results in that on the uniform mesh $\Omega_{h,\tau} \subset \Omega$ we define an approximation $w_{h,\tau}$ to the solution w. Obviously, the basic question is the convergence, that is, when refining h and τ,

*This work was supported by the Hungarian National Science Foundation (OTKA) under grant No. T19460.

the sequence $w_{h,\tau}$ should be convergent to w in the sense of a certain norm. This problem is widely investigated in the literature and, as typical, the requirement of the convergence results in conditions for the choice of the step-sizes h and τ. Usually, we distinguish two cases, namely,

(a) there is a bound for the choice of $q = \tau/h^2$ (the so-called *conditionally stable methods*);

(b) there is no bound for the choice of q (the so-called *unconditionally stable methods*).

(See, e.g., [3], [9], [10].)

It is not less important to require the preservation of the discrete analogues of the basic qualitative properties of the continuous solution mentioned above at certain fixed numerical solution (or at all of them). This requirement results in conditions (upper and lower bounds) for the choice of q, even for unconditionally stable schemes, too. That is, aiming at a qualitatively adequate numerical solution, we lose the main advantage of the unconditionally stable methods: the independent choice of the parameters h and τ.

Our aim is to minimize the computational work. Therefore, it is suitable to choose numerical methods of higher accuracy.

2   THE ACCURACY AND CONVERGENCE OF THE (σ, θ)-METHOD

On a fixed mesh $\Omega_{h,\tau}$, the realization of the numerical process given above yields a one-step iteration process with respect to the unknown vectors $\mathbf{y}^j$, $j = 1, 2, \ldots$. Here the components of the vectors $\mathbf{y}^j \in \mathbb{R}^n$ denote the approximation to the solution $w(x, j\tau)$ at the points of the space-discretization with $h = 1/n$, $n \in \mathbf{N}$.

Assume that $\sigma \in [0, 0.25)$ and $\theta \in [0, 1]$ are given numbers and we denote the uniformly continuant tridiagonal matrices $tridiag[-1, 2, -1]$ by $\mathbf{Q}$ and $tridiag[\sigma, 1 - 2\sigma, \sigma]$ by $\mathbf{M}$ in $\mathbb{R}^{n \times n}$, respectively.

Now, $\mathbf{y}^j$ (the approximation to the solution $w(x, t)$ at the time level t_j) is defined by the iteration

$$\mathbf{M}(\mathbf{y}^{j+1} - \mathbf{y}^j) = \theta q \mathbf{Q} \mathbf{y}^{j+1} + (1 - \theta) q \mathbf{Q} \mathbf{y}^j, \tag{2.1}$$

$$\mathbf{y}^0 \text{ is given.} \tag{2.2}$$

DEFINITION   A numerical method leading to the approximation (2.1), (2.2) is called (σ, θ)-*method*.

By refining the mesh we assume that q is constant. Then, at the point (x_i, t_j) the local error of the (σ, θ)-method is

$$\psi_i^j = K_1 h^2 + K_2 h^4 + O(h^6), \tag{2.3}$$

where

$$K_1 = [q(\theta - 0.5) + 1/12 - \sigma]\frac{\partial^4 w(x_i, t_j + 0.5\tau)}{\partial x^4},$$

$$K_2 = \left[\frac{1}{12}q^2 + \frac{1}{360} - \frac{\sigma}{12} + \frac{\theta - 0.5}{12}q\right]\frac{\partial^6 w(x_i, t_j + 0.5\tau)}{\partial x^6}.$$

Analogously, for the global error at the time-level t_j we obtain the estimate

$$\|\mathbf{z}^j\| \le \|\mathbf{z}^0\| + \frac{1}{1 - 4\sigma}\tau\sum_{m=1}^{j}\|\mathbf{\Psi}^m\|, \qquad (2.4)$$

where $\mathbf{\Psi}^m$ is the vector with coordinates ψ_i^m, $i = 1, \ldots, n$. Hence, the approximation of the initial function should have an order corresponding to the order of the (σ, θ)-method.

By simple analysis of the eigenvalues, it is easy to see that the (σ, θ)-method is convergent for all n if and only if the condition

$$q \le \begin{cases} \dfrac{1 - 2\sigma}{2(1 - 2\theta)} & \text{when } \theta \in [0, 1/2), \\[2mm] \text{is arbitrary} & \text{when } \theta \in [1/2, 1] \end{cases} \qquad (2.5)$$

is satisfied. It means that in case of (2.5) the order of the convergence is equal to the order of the local error.

3 SPECIAL SCHEMES OF HIGHER ACCURACY

Furthermore, we assume that $\mathbf{y}^0$ approximates the initial function w_0 according to the accuracy of the method. For arbitrary values of the parameters σ and θ the (σ, θ)-method has an accuracy $O(h^2 + \tau)$, that is, if q is constant then $O(h^2)$. On the other hand, as we will see, by special choice of the parameters we are able to construct schemes of higher accuracy, too.

Let us consider these cases.

1. The order of accuracy is $O(h^4)$ either
   - $\theta = 1/2$ and $\sigma = 1/12$, or
   - $q = \dfrac{\sigma - 1/12}{\theta - 1/2}$ (where $\theta \in [0, 1/2)$ and $\sigma \in [0, 1/12)$ or $\theta \in (1/2, 1]$ and $\sigma \in (1/12, 1/4)$ are arbitrary).

2. If
   $$\theta = \frac{3 - \sqrt{5}}{6} + 2\sigma\sqrt{5}, \quad q = \frac{\sqrt{5}}{10}, \quad \sigma \in \left[0, \frac{3\sqrt{5} + 5}{60}\right],$$

   then the order of accuracy is $O(h^6)$. We notice that we can guarantee the methods to be explicit by a special choice of the parameter q.

3. For the above case 1 the $(\sigma, 6\sigma)$-method with $q = 1/6$ is explicit and convergent for all $\sigma \in [0, 1/6]$.

It is remarkable, that for all σ these explicit methods (that is the $(\sigma, 6\sigma)$-methods with $q = 1/6$) are equivalent, that is, the only explicit method with the accuracy $O(h^4)$ is the method $(0,0)$ with $q = 1/6$. This latter method is known from the theory of finite difference methods [9], [12].

We also remark that for $\sigma = 0$ (that is for the finite difference method) the case 2 is known [1], [11]. On the other hand, with the above choice of q the finite element method with $\theta = (3 + \sqrt{5})/6$ and the $\sigma = 1/12$-method with $\theta = 1/2$ have also this order of accuracy.

4   QUALITATIVE PROPERTIES OF THE SCHEMES OF HIGHER ACCURACY

In order to preserve the main qualitative properties of the solution of the continuous problem, the bounds for the choice of q are usually not sufficient. We have to replace the condition (2.5) by an other, a stronger one. A detailed analysis shows that one of the basic requirements is the preservation of the nonnegativity, that is, when for any nonnegative vector $\mathbf{y}^0$ the iteration process (2.1) results in a nonnegative vector-sequence $(\mathbf{y}^j; j = 1, 2, \dots)$. In order to formulate the conditions of the nonnegativity preservation and the regular exponential convergence we recall the following results.

THEOREM   [6].   The (σ, θ)-method $((\sigma, \theta) \neq (0,0))$ is nonnegativity preserving for all number n of space division if and only if the condition

$$\frac{\sigma}{\theta} \leq q \leq \frac{\frac{1}{2} - \sigma}{1 - \theta}, \quad \theta \geq 2\sigma, \tag{4.1}$$

is satisfied. For the (0,0)-method the condition is

$$q \leq \frac{1}{2}. \tag{4.2}$$

REMARK   For the finite difference and finite element schemes a more detailed analysis is given in [5].

THEOREM   [7].   The (σ, θ)-method has the regular exponential convergence if and only if the condition

$$q < \begin{cases} \dfrac{1 - 2\sigma}{2(1 - \theta)} & \text{when } \theta \in [0, 1/2 + \sigma], \\[2mm] \dfrac{2\sigma}{2\theta - 1} & \text{when } \theta \in (1/2 + \sigma, 1], \end{cases} \tag{4.3}$$

holds.

REMARK   For the finite difference and finite element schemes a more detailed analysis is given in [7].

Let us formulate the conditions both of the nonnegativity preservation and regular exponential convergence for the schemes of higher accuracy given in previous section.
1. The $(1/12, 1/2)$-method is nonnegativity preserving and it has a regular exponential convergence if and only if the condition

$$1/6 \leq q < 5/6 \tag{4.4}$$

holds. The (σ, θ)-method with $q = \dfrac{\sigma - 1/12}{\theta - 1/2}$ is nonnegativity preserving and it has a regular exponential convergence if and only if the condition

$$\begin{aligned} \sigma \in [0, 1/12) & \quad \text{and} \quad \theta \in [6\sigma, (6\sigma + 2)/5], \\ \sigma \in (1/12, 1/6] & \quad \text{and} \quad \theta \in [(6\sigma + 2)/5, 6\sigma], \\ \sigma \in (1/6, 1/4) & \quad \text{and} \quad \theta \in [(6\sigma + 2)/5, 1], \end{aligned} \tag{4.5}$$

Table 1 Accuracy of the methods

Method	Accuracy	$n = 10$	$n = 50$	$n = 100$
FDM	$O(h^2)$	7.08×10^{-4}	1.54×10^{-5}	2.78×10^{-6}
FEM	$O(h^2)$	3.55×10^{-4}	7.68×10^{-6}	1.39×10^{-6}
1.	$O(h^4)$	9.73×10^{-7}	9.72×10^{-10}	4.48×10^{-11}
2.	$O(h^6)$	6.58×10^{-9}	1.02×10^{-11}	1.91×10^{-12}
3.	$O(h^4)$	9.73×10^{-7}	9.72×10^{-10}	4.48×10^{-11}

holds.

2. The $(\sigma, \frac{3-\sqrt{5}}{6} + 2\sqrt{5}\sigma)$-method with $q = \frac{\sqrt{5}}{10}$ preserves the nonnegativity and it has a regular exponential convergence.

3. The $(\sigma, 6\sigma)$-method with $q = 1/6$ is nonnegativity preserving and it has a regular exponential convergence.

We remark that under the above conditions of the nonnegativity preservation, a number of further qualitative properties (like regular and exponential convergence, preservation of shape of the initial function, contractivity, etc.) are preserved, too [4], [6].

5 NUMERICAL EXAMPLES

In this chapter we analyze the numerical solution of the problem

$$\frac{\partial w}{\partial t} = \frac{\partial^2 w}{\partial x^2}, \qquad x \in (0,1), \ t > 0, \tag{5.1}$$

$$w(x,0) = \sin(\pi x), \qquad x \in (0,1), \tag{5.2}$$

$$w(0,t) = w(1,t) = 0. \tag{5.3}$$

This problem has the exact solution

$$w(x,t) = \exp(-\pi^2 t) \sin \pi x. \tag{5.4}$$

As before, we assume, as earlier, that q is constant. We give our analysis at the time level $T = 0.1$.

The Table 1 shows the accuracy of the methods examined previously.

The finite difference method and the finite element method are defined by the parameter $\theta = 2/3$. For all methods (with the exception of the case 2) the parameter $q = 1/6$ is chosen, that is, for these schemes the calculation is performed on the same mesh. Since the required value of q in the case 2 is greater then $1/6$, therefore, for this method the mesh is less dense with respect to the time discretization. It means that the amount of the computations of the method 2 is less then it is for the other implicit methods.

In Table 2 and Table 3 we give an analysis of the same method (namely, for the finite element method) with different values of θ. We can see that in a certain neighbourhood of q, that guarantees a higher accuracy, the methods are more accurate. (For instance, for $q = 1$ the $(1/6, 3/5)$-method is more accurate then the

Table 2   Accuracy of $(1/6, 2/3)$-method (FEM with $\theta = 2/3$)

q	$n = 10$	$n = 50$	$n = 100$	$n = 150$
1	4.86×10^{-4}	1.15×10^{-5}	2.08×10^{-6}	7.63×10^{-7}
2/3	1.60×10^{-4}	3.82×10^{-6}	6.95×10^{-7}	2.54×10^{-7}
1/2	8.68×10^{-6}	8.74×10^{-9}	4.04×10^{-10}	6.61×10^{-11}
1/3	1.80×10^{-4}	3.84×10^{-6}	6.96×10^{-7}	2.55×10^{-7}
1/6	3.55×10^{-4}	7.68×10^{-6}	1.39×10^{-6}	5.09×10^{-7}
1/12	4.44×10^{-4}	9.60×10^{-6}	1.74×10^{-6}	6.36×10^{-7}

Table 3   Accuracy of $(1/6, 3/5)$-method (FEM with $\theta = 3/5$)

q	$n = 10$	$n = 50$	$n = 100$	$n = 150$
1	6.47×10^{-5}	2.26×10^{-6}	4.15×10^{-7}	1.52×10^{-7}
5/6	2.79×10^{-5}	2.82×10^{-8}	1.30×10^{-9}	2.13×10^{-10}
2/3	1.23×10^{-4}	2.32×10^{-6}	4.18×10^{-7}	1.53×10^{-7}
1/2	2.22×10^{-4}	4.62×10^{-6}	8.35×10^{-7}	3.05×10^{-7}
1/3	3.23×10^{-4}	6.91×10^{-6}	1.25×10^{-6}	4.58×10^{-7}
1/6	4.27×10^{-4}	9.21×10^{-6}	1.67×10^{-6}	6.11×10^{-7}
1/12	4.79×10^{-4}	1.04×10^{-5}	1.88×10^{-6}	6.87×10^{-7}

Table 4   Accuracy of $(1/6, 1)$-method for different mesh-size

q	$n = 50$	$n = 100$	$n = 150$	$n = 200$
1	5.75×10^{-5}	1.04×10^{-5}	3.82×10^{-6}	1.87×10^{-6}
1/2	2.30×10^{-5}	4.17×10^{-6}	1.53×10^{-6}	7.47×10^{-7}
1/3	1.15×10^{-5}	2.09×10^{-6}	7.64×10^{-7}	3.73×10^{-7}
1/6	9.72×10^{-10}	4.49×10^{-11}	7.34×10^{-12}	2.03×10^{-12}
1/12	5.76×10^{-6}	1.04×10^{-6}	3.82×10^{-7}	1.87×10^{-7}

$(1/6, 1/2)$-method. The opposite happens when $q = 1/3$.) It is remarkable that sufficiently far from the region of attractivity of the higher accuracy the methods give almost the same accuracy. These tables also show that at fixed h the further refinement of the mesh (that is the decrease of τ) does not necessarily imply the decrease of the global error. In fact, it shows a slightly increasing character.

Table 4 gives the characterization of the accuracy by different n and q for the $(1/6, 1)$-method. As we can see, if one takes both the accuracy at the required computational work into consideration, the schemes of higher accuracy are especially efficient when the number of the space division is not too large.

Finally we remark that analogous treatment can be carried out for the (σ, θ)-methods with increasing q-s (e.g., for the Cranc–Nicolson $(0, 1/2)$-method with the choice $h = \tau$). In this case, the qualitative properties, however, cannot be preserved for the numerical solution.

Acknowledgement   The author wishes to thank M.N. Spijker for stimulating discussions about the topic of this paper. The author is also indebted to K. Balla and R. Horváth for reading preliminary versions of this paper and making many valuable suggestions concerning the presentations.

REFERENCES

1. Axelsson, O., Steihaug, T., *Some computational aspects in the numerical solution of parabolic equations*, J. Comp. Appl. Math. **4** (1978), 129–142.
2. Bers, L., John, F., Schechter, M., "Partial differential equations," Interscience Publ., New York, 1964.
3. Dekker, K., Verwer, J.G., "Stability of Runge–Kutta methods for stiff nonlinear differential equations," North-Holland, Amsterdam, 1984.
4. Faragó, I., Pfeil, T., *Preserving concavity in initial-boundary value problems of parabolic type and its numerical solution*, Periodica Math. Hung. **30** (1995), 135–139.
5. Faragó, I., *Nonnegativity of the difference schemes*, Pure Math. Appl. **6** (1996), 38–50.
6. Faragó, I., *One-steps methods of solving a parabolic problem and their qualitative properties*, Publ. Appl. Analysis, ELTE, 1996.
7. Faragó, I., *Regular and exponential convergence of difference schemes for the heat-conduction equation*, Comp. Math. Appl. (to appear).
8. Friedmann, A., "Partial differential equations of parabolic type," Prentice-Hall, Englewood Cliffs, NJ, 1964.
9. Richtmyer, R.D., Morton, K.W., "Difference methods for initial-value problems," Interscience Publ., New York, 1967.
10. Samarskii, A.A., "Theory of the difference schemes," Nauka, Moscow, 1977. (in Russian)
11. Thomée, V., "Galerkin finite element methods for parabolic problems," Lect. Notes in Math. 1054, 1984.
12. Thomée, V., *Finite difference methods for linear parabolic equations*, in "Handbook of numerical analysis," (ed. by Ciarlet, P.G. and Lions, J.L.), North-Holland, Amsterdam, 1990.

Shape design sensitivity formulae approximated by means of a recovered gradient method

IVAN HLAVÁČEK and JAN CHLEBOUN Mathematical Institute, Academy of Sciences of the Czech Republic, Žitná 25, 115 67 Praha 1, Czech Republic, e-mail: chleboun@math.cas.cz

Abstract   A recovered gradient method is applied to some smooth optimal shape design problems. Attention is payed to the gradient of a cost functional, i.e, to certain boundary integrals which result from a shape design sensitivity analysis. To obtain sufficiently accurate approximation of the sensitivity formulae, a recovered gradient method is recommended.

INTRODUCTION

In smooth optimal shape design problems with an elliptic state equation, the gradient of a cost functional with respect to design variables exists and can be derived with the aid of the shape design sensitivity analysis. Various sensitivity analysis techniques are used in computational mathematics. Purely numerical ones suffer from limited reliability due to a loss of accuracy unless they are sufficiently sophisticated. In any case, one can expect either extensive calculations or extensive programming.

Semi-analytical methods are a compromise which can save man's or computer's labour but under some circumstances their accuracy is questionable or even insufficient.

Among analytical approaches, the material derivative method [1], [6], [7], [15] is quite popular. It gives directional derivatives of an integral cost functional. The derivatives are expressed by either domain or boundary integrals. The latter involve integrals of the boundary flux of the state solution. The resulting formulae are easy to program and they need minimal, if any, interference into a state problem solver. Thus an efficient method for computing both the state solution and the boundary flux is highly desirable. However, the direct differentiation of a finite element solution does not yield satisfactory results, as they are inaccurate.

To take advantage of boundary design sensitivity formulae one can either utilize

135

a postprocessing technique which recovers the gradient of the finite element state solution, or use a special method to solve the state problem (see [4], [9], for instance). Then the approximation of the boundary flux can be sufficiently accurate to enable applicable evaluation of sensitivity formulae.

On the basis of a postprocessing method proposed in [11], [12] we present an application to shape optimization in this paper, more detailed treatment can be found in [10]. In Section 1, we start with some optimal shape design problems and corresponding design sensitivity formulae. Section 2 focuses on the discretization by linear finite elements. The final Section 3 is devoted to both the technique of recovered gradients and its application to the design sensitivity evaluation.

1   SHAPE OPTIMIZATION PROBLEMS, DESIGN SENSITIVITY FORMULAE

In the paper, the symbol $W_p^k(\Omega)$ stands for the standard Sobolev space of functions the derivatives of which in the sense of distributions up to the order k belong to the space $L^p(\Omega)$. The corresponding norm is denoted by $\| \cdot \|_{k,q,\Omega}$ and, if $q = 2$, by $\| \cdot \|_{k,\Omega}$.

Let us choose a simple model domain (see Fig. 1)

$$\Omega(\theta) = \{(x_1, x_2) \in \mathbb{R}^2 \,; \; 0 < x_1 < 1, \; 0 < x_2 < \theta(x_1)\}$$

the shape of which is controlled by a design function θ.

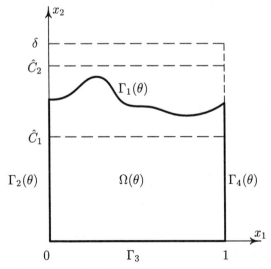

Figure 1   An admissible domain $\Omega(\theta)$.

To set *a family $\mathcal{U}_{ad}$ of admissible design functions*, first we define

$$Q = \Big\{ \alpha \in \mathbb{R}^{n+1}; \ \hat{C}_1 \leq \alpha_i \leq \hat{C}_2, \ i = 0, \ldots, n,$$

$$|\alpha_{i+1} - \alpha_i| \leq \hat{C}_3/n, \ i = 0, \ldots, n-1, \ \sum_{i=0}^{n} \alpha_i = \hat{C}_4(n+1) \Big\},$$

$$\theta(s) = F(\alpha)(s) = \sum_{i=0}^{n} \alpha_i \beta_i^n(s), \quad \beta_i^n(s) = \binom{n}{i} s^i (1-s)^{n-i}, \tag{1.1}$$

where $n > 1$ is a fixed integer and $\hat{C}_1$, $\hat{C}_2$, $\hat{C}_3$, $\hat{C}_4$ are given positive parameters. We suppose $Q \neq \emptyset$.

Thus θ is a Bézier curve (see [2]) and the shape of $\Omega(\alpha)$ is generated by the x_2-coordinates of the control points

$$\{B_i\}_{i=0}^n, \quad B_i = \Big(\frac{i}{n}, \alpha_i \Big) \in \mathbb{R}^2.$$

The formula (1.1) determines the mapping $F \colon \mathbb{R}^{n+1} \to C([0,1])$ and, if we set $\mathcal{U}_{ad} = F(Q)$, it can be proved (see [4], [10]) that

$$\mathcal{U}_{ad} \subset \hat{\mathcal{U}}_{ad} \equiv \Big\{ \theta \in C^{(0),1}([0,1]); \ \hat{C}_1 \leq \theta(x_1) \leq \hat{C}_2,$$

$$|\theta'(x_1)| \leq \hat{C}_3, \ \int_0^1 \theta(x_1)\, \mathrm{d}x_1 = \hat{C}_4 \Big\},$$

where $C^{(0),1}([0,1])$ denotes the space of Lipschitz-continuous functions and the prime stands for $(\mathrm{d}/\mathrm{d}x_1)$.

Any domain $\Omega(\theta)$, $\theta \in \mathcal{U}_{ad}$, is contained in the rectangle $\Omega_\delta = (0,1) \times (0,\delta)$, $\delta > \hat{C}_2$. We distinguish four parts ("sides") of $\partial\Omega(\theta)$, i.e.,

$$\partial\Omega(\theta) = \Gamma_1(\theta) \cup \Gamma_2(\theta) \cup \Gamma_3 \cup \Gamma_4(\theta)$$

(see Fig. 1), and we define

$$\hat{\Gamma}_2 = \{(x_1, x_2); \ x_1 = 0, x_2 \in (0,\delta)\}, \quad \hat{\Gamma}_4 = \{(x_1, x_2); \ x_1 = 1, x_2 \in (0,\delta)\}.$$

Let a symmetric matrix $\mathcal{A} = (a_{ij})_{i,j=1}^2$, $a_{ij} \in W_2^2(\Omega_\delta)$, and a function $\sigma \in L^\infty(\hat{\Gamma}_2 \cup \hat{\Gamma}_4)$, $\sigma \geq 0$, determine the following bilinear form

$$a(\theta; y, w) = \int_{\Omega(\theta)} \mathcal{A}\nabla y \cdot \nabla w \,\mathrm{d}x + \int_{\Gamma_2(\theta) \cup \Gamma_4(\theta)} \sigma y w \,\mathrm{d}\gamma, \quad y, w \in W_2^1(\Omega(\theta)). \tag{1.2}$$

We suppose that a constant $a_0 > 0$ exists such that

$$\mathcal{A}(x)t \cdot t \geq a_0 \|t\|_{\mathbb{R}^2}^2 \quad \forall x \in \Omega_\delta, \ \forall t \in \mathbb{R}^2,$$

i.e., if we introduce the space

$$V(\theta) = \big\{ w \in W_2^1(\Omega(\theta)); \ w = 0 \text{ on } \Gamma_1(\theta) \cup \Gamma_3 \big\},$$

then the form $a(\theta; \cdot, \cdot)$ is $V(\theta)$-elliptic.

We are now in a position to define *the state problem*: Find $y(\theta) \in V(\theta)$ such that

$$a(\theta; y(\theta), w) = \int_{\Omega(\theta)} fw\,\mathrm{d}x + \int_{\Gamma_2(\theta)\cup\Gamma_4(\theta)} gw\,\mathrm{d}\gamma \quad \forall w \in V(\theta), \tag{1.3}$$

where $f \in L^2(\Omega_\delta)$ and $g \in L^2(\hat{\Gamma}_2 \cup \hat{\Gamma}_4)$ are given functions.

To evaluate the unique solution $y(\theta)$ of the state equation (1.3) the following cost functionals are considered

$$J_1(\theta) = \int_{\Omega(\theta)} \big(y(\theta) - y_p\big)^2 \,\mathrm{d}x, \tag{1.4}$$

$$J_2(\theta) = \int_{\Omega(\theta)} fy(\theta)\,\mathrm{d}x, \tag{1.5}$$

$$J_3(\theta) = \int_{\Omega(\theta)} |\nabla y(\theta)|^2 \,\mathrm{d}x, \tag{1.6}$$

where $y_p \in L^2(\Omega_\delta)$ is a given function.

We are ready to define *the shape design optimization problems* $\mathcal{P}_j$, $j \in \{1, 2, 3\}$: Find $\alpha_{\mathrm{opt}}^j \in Q$ such that

$$\alpha_{\mathrm{opt}}^j = \arg\min_{\alpha \in Q} J_j\big(F(\alpha)\big), \tag{1.7}$$

where $y\big(F(\alpha)\big)$ solves the state problem (1.3) for $\theta = F(\alpha)$.

Following the proof of [8, Lemma 2.1], for instance, we can prove that the problem $\mathcal{P}_j$, $j \in \{1, 2, 3\}$, has at least one solution.

As explained in the introduction, sensitivity analysis should be a part of an efficient real-life computation. We will present Gâteaux differentials of the cost functionals provided that the coefficients a_{ij} are constant.

The basic idea of the material derivative method consists in introducing a mapping $x \mapsto x + tv$, $x \in \overline{\Omega(\theta)}$, which determines a change of the shape of the domain $\Omega(\theta)$ with respect to a given velocity field $v \in \mathbb{R}^2$ and an artificial time t, see [7], [15] for details. Taking both the definition of the domain $\Omega(\theta)$ and the problem (1.7) into account, we can set

$$v = (0, v_2), \quad v_2(x_1, x_2) = x_2\hat{\theta}(x_1)/\theta(x_1),$$

where $\hat{\theta} \in C^{(0),1}([0, 1])$. We have $v_2 = 0$ on Γ_3 and $v_2 = \hat{\theta}$ on $\Gamma_1(\theta)$.

On the basis of [7, Chapter 3] and supposing certain regularity for functions involved in calculation, we can derive after some labour (see [10]) that

$$
\begin{aligned}
DJ_1(\theta, \hat{\theta}) &= \lim_{t \to 0} \frac{J_1(\theta + t\hat{\theta}) - J_1(\theta)}{t} \\
&= \int_0^1 \Big(\frac{\partial y(\theta)}{\partial \nu}\big(\theta(x_1)\big)\frac{\partial z_1(\theta)}{\partial \nu_A}\big(\theta(x_1)\big) + y_p^2\big(\theta(x_1)\big)\Big)\hat{\theta}(x_1)\,\mathrm{d}x_1,
\end{aligned} \tag{1.8}
$$

where $\partial/\partial\nu_A \equiv \sum_{i,j=1}^2 a_{ij}\nu_i\partial/\partial x_j$ and ν denotes the unit outward normal along $\Gamma_1(\theta)$ in this particular case. The function $z_1(\theta) \in V(\theta)$ is the unique solution of the following adjoint equation to (1.3), (1.4)

$$a(\theta; z_1(\theta), w) = 2 \int_{\Omega(\theta)} (y(\theta) - y_p)w \, dx \quad \forall w \in V(\theta).$$

Similar calculation leads to

$$DJ_2(\theta, \hat{\theta}) = \int_0^1 \frac{\partial y(\theta)}{\partial \nu}(\theta(x_1)) \frac{\partial z_2(\theta)}{\partial \nu_A}(\theta(x_1))\hat{\theta}(x_1) \, dx_1, \qquad (1.9)$$

where $z_2(\theta) \in V(\theta)$ solves the adjoint equation to (1.3), (1.5)

$$a(\theta; z_2(\theta), w) = \int_{\Omega(\theta)} fw \, dx \quad \forall w \in V(\theta).$$

Finally, the equation

$$a(\theta; z_3(\theta), w) = 2 \int_{\Omega(\theta)} \nabla y(\theta) \cdot \nabla w \, dx \quad \forall w \in V(\theta)$$

is adjoint to the problem (1.3), (1.6), and

$$DJ_3(\theta, \hat{\theta}) = \int_0^1 \left[\frac{\partial y(\theta)}{\partial \nu} \frac{\partial z_3(\theta)}{\partial \nu_A} - \left(\frac{\partial y(\theta)}{\partial \nu}\right)^2 \right](\theta(x_1))\hat{\theta}(x_1) \, dx_1. \qquad (1.10)$$

The formulae (1.8)–(1.10) can be somewhat modified by virtue of the equality

$$\nabla w|_{\Gamma_1(\theta)\cup\Gamma_3} = \frac{\partial w}{\partial \nu}\nu,$$

which is valid for any $w \in V(\theta)$.

To pass from DJ_j to $\partial J_j/\partial\alpha_k$, $k = 0, 1, \ldots, n$, $j = 1, 2, 3$, we can utilize the equalities

$$\frac{\partial J_j}{\partial \alpha_k}\big(F(\alpha)\big) = DJ_j\big(F(\alpha), F'(\alpha, \varepsilon_k)\big), \quad F'(\alpha, \varepsilon_k) = \beta_k^n,$$

where $F'(\alpha, \varepsilon_k)$ is the derivative of the mapping F at the point α with respect to the unit $\mathbb{R}^{n+1}$-coordinate vector ε_k, cf. (1.1).

REMARK 1.1 Besides the problem (1.3), similar equations with different boundary conditions are considered in [10], too. The resulting sensitivity formulae are identical or similar (simpler) to the above cited, see [10]. □

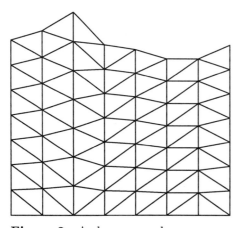

Figure 2 A chevron mesh.

2   DISCRETIZED PROBLEMS

The set $\mathcal{U}_{\mathrm{ad}}$ will be approximated by a set $\mathcal{U}_{\mathrm{ad}}^h$ of continuous and piecewise linear functions. To this end we choose a positive integer N and, introducing $h = 1/N$, we define the set

$$U^h = \{\theta_h \in C^{(0),1}([0,1]);\ \theta_h|_{e_j} \in P_1(e_j),\ j = 1,\ldots,N\},$$

where $P_1(e_j)$ is the set of linear functions on $e_j = [(j-1)h, jh]$.

Let the mapping $F_h : Q \to U^h$ be defined by the relations

$$F_h(\alpha)(jh) = F(\alpha)(jh), \quad j = 0, 1, \ldots, N.$$

Then we set

$$\mathcal{U}_{\mathrm{ad}}^h = \{\theta_h \in U^h;\ \exists \alpha \in Q,\ F_h(\alpha) = \theta_h\}.$$

The mapping F_h is injective provided h is sufficiently small ($h \leq 1/n$).

The domain $\Omega(\theta_h)$, $\theta_h \in \mathcal{U}_{\mathrm{ad}}^h$, is to be subdivided into triangles. Keeping a superconvergence postprocessing technique in mind, we should use triangulations which are images of a reference triangulation of a square grid under a W_∞^2-diffeomorphism, see [11], [12], [14]. The reference mesh is either a chevron mesh or a uniform slope one on the unit square $S = [0,1]^2$.

To construct a proper mapping Φ_θ which maps S onto $\overline{\Omega(\theta)}$, $\theta \in \mathcal{U}_{\mathrm{ad}}$, we put $\Phi_\theta(X_1, X_2) = (X_1, X_2\theta(X_1))$. The mapping Φ_θ can be extended to a W_∞^2-diffeomorphism $\hat{\Phi}_\theta$ defined on a domain $\hat{\Omega}$, $S \subset \hat{\Omega}$, i.e., $\hat{\Phi}_\theta$ is injective, $\hat{\Phi}_\theta(\hat{\Omega})$ is an open set, $\hat{\Phi}_\theta$ and $\hat{\Phi}_\theta^{-1}$ belong to the space W_∞^2 on $\hat{\Omega}$ and $\hat{\Phi}_\theta(\hat{\Omega})$, respectively.

Attention should be payed to the slope of diagonals dividing quadrangles into triangles. In order to avoid triangles with obtuse inner angle we make the slope dependent on the behaviour of the function $\theta = F(\alpha)$ or, which is equivalent, $\theta_h = F_h(\alpha)$. A detailed algorithm is presented in [10], its idea is obvious from Fig. 2.

On the basis of the mapping Φ_{θ_h}, we can easily construct a strongly regular family

$$\mathcal{T} = \{\mathcal{T}_h(\theta_h);\ \theta_h \in \mathcal{U}_{\mathrm{ad}}^h,\ h \to 0_+\}$$

of triangulations $\mathcal{T}_h(\theta_h)$. Let us remind that triangulations are strongly regular if there exist constants $C_1 > 0$, $C_2 > 0$, independent of h and θ_h, and such that $C_1 h^2 \leq \text{meas}(K)$, $\text{diam}(K) \leq C_2 h$ for any triangle $K \in \mathcal{T}_h(\theta_h)$.

The state equation (1.3) is discretized by the finite element method via the following spaces

$$X_h(\theta_h) = \{w_h \in C(\overline{\Omega(\theta_h)});\ w_h|_K \in P_1(K)\ \forall K \in \mathcal{T}_h(\theta_h)\} \subset W_2^1(\Omega(\theta_h)),$$
$$V_h(\theta_h) = \{w_h \in X_h(\Omega(\theta_h));\ w_h = 0 \text{ on } \Gamma_1(\theta_h) \cup \Gamma_3\}.$$

Then the approximate state equation reads: Find $y_h(\theta_h) \in V_h(\theta_h)$ such that

$$a\big(\theta_h; y_h(\theta_h), w_h\big) = \int_{\Omega(\theta_h)} f w_h \, dx + \int_{\Gamma_2(\theta_h) \cup \Gamma_4(\theta_h)} g w_h \, d\gamma \quad \forall w_h \in V_h(\theta_h). \quad (2.1)$$

The solution $y_h(\theta_h)$ is evaluated by a cost functional $J_{hj}\big(F_h(\alpha)\big)$ which arises from J_j by substitution $y_h(\theta_h)$ for $y(\theta)$.

The approximate domain optimization problem $\mathcal{P}_h^j$ for fixed h and $j \in \{1, 2, 3\}$: Find $\alpha_{\text{opt},h}^j \in Q$ such that

$$\alpha_{\text{opt},h}^j = \arg\min_{\alpha \in Q} J_{hj}\big(F_h(\alpha)\big). \quad (2.2)$$

It can be shown that the problem $\mathcal{P}_h^j$ has at least one solution, see [10].

The following theorem gives a link between solutions of the problems $\mathcal{P}_h^j$ and $\mathcal{P}$. The solutions of the equations (1.3), (2.1) are extended by zero in $\Omega(\delta) \setminus \Omega(\theta)$ and $\Omega(\delta) \setminus \Omega(\theta_h)$, respectively.

THEOREM 2.1   Let $\{\alpha_{\text{opt},h}^j\}$, $h \to 0_+$, be a sequence of solutions of the approximate domain optimization problems $\mathcal{P}_h^j$ (2.2), $j \in \{1, 2, 3\}$. Then a subsequence $\{\alpha_{\text{opt},\widehat{h}}^j\}$ exists such that

$$\alpha_{\text{opt},\widehat{h}}^j \to \alpha_{\text{opt}}^j \qquad\qquad \text{in } \mathbb{R}^{n+1},$$
$$\theta_{\text{opt},\widehat{h}}^j \equiv F_{\widehat{h}}(\alpha_{\text{opt},\widehat{h}}^j) \to \theta_{\text{opt}}^j \equiv F(\alpha_{\text{opt}}^j) \quad \text{in } C([0,1]),$$
$$y_{\widehat{h}}(\theta_{\text{opt},\widehat{h}}^j) \to y(\theta_{\text{opt}}^j) \qquad\qquad \text{in } W_2^1(\Omega_\delta),$$

hold as $\widehat{h} \to 0_+$, where $y_{\widehat{h}}(\theta_{\text{opt},\widehat{h}}^j)$ solves (2.1) on $\Omega(\theta_{\text{opt},\widehat{h}}^j)$, $y(\theta_{\text{opt}}^j)$ solves (1.3) on $\Omega(\theta_{\text{opt}}^j)$ and α_{opt}^j is a solution of the problem $\mathcal{P}_j$ (1.7).

Proof: The theorem can be proved by a slight modification of the proof of [3, Theorem 7.1], or the proof of [4, Theorem 4.6] can be taken over almost literally. The above cited references together with [10] also show how to prove corresponding auxiliary lemmas. $\qquad\qquad\square$

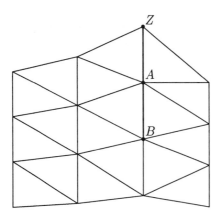

Figure 3

3 RECOVERED GRADIENT METHOD

As indicated in the previous sections, we propose to approximate the shape design sensitivity formulae by means of a recovered gradient technique. Unlike piecewise constant approximation of $\nabla y(\theta)$ which results from a direct differentiation of the finite element solution $y_h(\theta_h)$, the recovered gradient is continuous, piecewise linear and gives a more accurate approximation.

In our particular case, we can confine ourselves to the second component of the recovered gradient $G_h y_h(\theta_h)$ at a grid point Z (mesh node). Following [11], [12], [13], we set (θ_h is omitted)

$$[G_h y_h]_2(Z) = \eta y_h(A) - (\eta + \xi)y_h(Z) + \xi y_h(B), \qquad (3.1)$$

where A and B are the two nearest grid points in the x_2-direction (see Fig. 3) and

$$\eta = \frac{b}{a(b - a)}, \qquad \xi = \frac{a}{b(a - b)},$$
$$a = x_2(A) - x_2(Z), \quad b = x_2(B) - x_2(Z).$$

Having values at nodes, we define $[G_h y_h]_2$ on the whole domain $\Omega(\theta_h)$ by piecewise linear interpolation.

To give a theoretical link to a sensitivity formula approximation we consider a simplified state problem on a polygonal domain $\Omega \equiv \Omega(\theta_h)$. In (1.2), (1.3), we set $\sigma = 0$, $g = 0$ and $V(\theta_h) \equiv H_0^1(\Omega)$, i.e., as a consequence, $V_h(\theta_h)$ is also restricted to functions with zero traces on $\partial\Omega$. Moreover, we assume that

    i) domains $\mathcal{G}_1 \subset\subset \mathcal{G}_2$ exist (which are not contained in Ω, see Fig. 4) such that the solution $y \equiv y(\theta_h)$ of the state problem (1.3) belongs to the subspace $W_q^3(\Omega_1) \cap W_2^3(\Omega_2)$, where $\Omega_j = \mathcal{G}_j \cap \Omega$, $j = 1, 2$, and $q > 2$;

    ii) a parameter $\gamma \in (0, 1]$ exists such that $y \in W_2^{1+\gamma}(\Omega)$ for any $f \in L^2(\Omega)$ and

$$\|y\|_{1+\gamma,\Omega} \le C\|f\|_{0,\Omega}. \qquad (3.2)$$

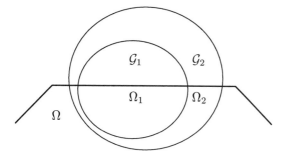

Figure 4

Then, according to [11, Theorem 6.4], a local error estimate up to the boundary has the form

$$\left\|\frac{\partial y}{\partial x_2} - [G_h y_h]_2\right\|_{0,\Omega_0} \leq C(\Omega_0, \Omega_1) h^{2\gamma} (\|y\|_{3,\Omega_2} + \|y\|_{3,q,\Omega_1} + \|f\|_{0,\Omega}) \qquad (3.3)$$

for any sufficiently small h and any subdomain $\Omega_0 = \mathcal{G}_0 \cap \Omega$, $\mathcal{G}_0 \subset\subset \mathcal{G}_1$.

Let us remark that under the regularity assumption considered in (3.2) the standard finite element theory [5] guarantees only the h^γ rate of convergence for $\|\partial y/\partial x_2 - \partial y_h/\partial x_2\|_{0,\Omega_0}$.

The second component of ∇y on $\Gamma_1(\theta_h)$ is approximated by $q_{2h} = [G_h y_h]_2 \in C(\Gamma_1(\theta_h))$, see (3.1).

To approximate the first component, we employ the homogeneous boundary condition. The optimized boundary $\Gamma_1(\theta_h)$ consists of segments $\Gamma_{1j}(\theta_h)$, $j = 1, \dots, N$. Along $\Gamma_1(\theta_h)$, we have

$$\frac{\partial y_h(\theta_h)}{\partial t_j} = 0, \quad j = 1, \dots, N, \qquad (3.4)$$

where the vector $t_j = (t_{j1}, t_{j2}) = \big(h, \theta_h(jh) - \theta_h((j-1)h)\big)$ is tangential to the segment $\Gamma_{1j}(\theta_h)$.

In accordance with (3.4), we prescribe

$$q_{1h} = -q_{2h} t_{j2}/t_{j1} = -q_{2h}[\theta_h(jh) - \theta_h((j-1)h)]/h$$

on $\Gamma_{1j}(\theta_h)$, $j = 1, \dots, N$.

The vector $q_h = (q_{1h}, q_{2h})$ can be used to approximate $\partial y/\partial \nu$, $\partial y/\partial \nu_A$ in the sensitivity formulae (1.8)–(1.10).

The same approach can be employed to obtain approximations of $\partial z_i/\partial \nu$ and $\partial z_i/\partial \nu_A$, $i = 1, 2, 3$.

Theorem 6.4 in [11] was proved under rather restrictive assumptions. Some of them can be met by virtue of properties of our mesh which takes advantages of the shape of the domain $\Omega(\theta)$, $\theta \in \mathcal{U}_{\mathrm{ad}}$.

The recovered gradient formula (3.1) lies in weighting function values at points Z, A and B. If A and B coincide with the nodes of $\mathcal{T}_h(\theta_h)$ for all mesh nodes $Z \in \partial\Omega_1 \cap \partial\Omega$ (see Fig. 3) then we arrive at the rate $h^{2\gamma}$ as used in (3.3).

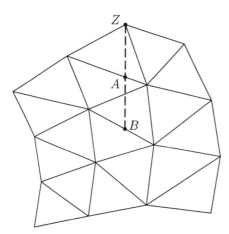

Figure 5

If the mesh $\mathcal{T}_h(\theta_h)$ is distorted in the both coordinates we can get (3.3) with h^ρ on the assumption that the points A and B are placed as depicted in Fig. 5. In this case, however, $\rho = \min\{2\gamma, 3/2\}$, see [11, Theorem 5.9].

In general, estimation of $\|\nabla y - G_h y_h\|_{0,\Omega_0}$ is split into two main steps originating from the triangle inequality

$$\|\nabla y - G_h y_h\|_{0,\Omega_0} \leq \|\nabla y - G_h y\|_{0,\Omega_0} + \|G_h y - G_h y_h\|_{0,\Omega_0}.$$

The first term is estimated on the basis of the formula (3.1) which can be applied to a function on a quite irregular triangulation, see [12], [13] for details. To estimate the second term and not to lose a higher convergence rate, special quasiuniform meshes are supposed, see [11], [12], [14].

REMARK 3.1   Though, strictly speaking, the accuracy of the recovered gradient along the boundary $\Gamma_1(\theta)$ is not assessed by (3.3) the estimate gives at least some approval to the boundary formulae approximation.

Let us remark that the recovered gradient method can also be applied to design sensitivity formulae expressed by domain integrals.

Acknowledgements   The work was supported by Grant A 1019601 of the Grant Agency of the Academy of Sciences of the Czech Republic. This support is gratefully acknowledged. The authors also thank the journal *Applications of Mathematics* for the kind permission to reprint Fig. 1 and Fig. 2.

REFERENCES

1. J.S. Arora, *An exposition of the material derivative approach for structural shape sensitivity analysis*, Comput. Methods Appl. Mech. Engrg. **105** (1993), 41–62.
2. R.H. Bartels, J. C. Beatty and B.A. Barsky, "An Introduction to Splines for use in Computer Graphics and Geometric Modelling," Morgan Kaufmann, Los Altos, 1987.

3. D. Begis, R. Glowinski, *Application de la méthode des éléments finis à l'approximation d'un probléme de domaine optimal*, Appl. Math. Optim. **2** (1975), 130–169.

4. J. Chleboun, R.A.E. Mäkinen, *Primal hybrid formulation of an elliptic equation in smooth optimal shape problems*, Adv. in Math. Sci. and Appl. **5** (1995), 139–162.

5. P.G. Ciarlet, "Basic error estimates for elliptic problems, Handbook of Numerical Analysis II (P.G. Ciarlet, J.L. Lions eds.)," North-Holland, Amsterdam, 1991.

6. J. Haslinger, P. Neittaanmäki, "Finite Element Approximation for Optimal Shape Design, Theory and Applications," John Wiley, Chichester, 1988.

7. E.J. Haug, K.K. Choi and V. Komkov, "Design Sensitivity Analysis of Structural Systems," Academic Press, Orlando, London, 1986.

8. I. Hlaváček, *Optimization of the domain in elliptic problems by the dual finite element method*, Apl. Mat. **30** (1985), 50–72.

9. I. Hlaváček, *Shape optimization by means of the penalty method with extrapolation*, Appl. Math. **39** (1994), 449–477.

10. I. Hlaváček, J. Chleboun, *A recovered gradient method applied to smooth optimal shape problems*, Appl. Math. **41** (1996), 281–297.

11. I. Hlaváček, M. Křížek, *Optimal interior and local error estimates of a recovered gradient of linear elements on nonuniform triangulations*, J. Comput. Math. **14** (1996), 345–362.

12. I. Hlaváček, M. Křížek and V. Pištora, *How to recover the gradient of linear elements on nonuniform triangulations*, Appl. Math. **41** (1996), 241–267.

13. M. Křížek, *Higher Order Global Accuracy of a Weighted Averaged Gradient of the Courant Elements on Irregular Meshes*, in "Finite Elements Methods: Fifty Years of the Courant Element," M. Křížek, P. Neittaanmäki, R. Stenberg (eds), Lecture Notes in Pure and Applied Mathematics 164, Marcel Dekker, New York, 1994, pp. 267–276.

14. N. Levine, *Superconvergent recovery of the gradient from piecewise linear finite element approximations*, IMA J. Numer. Anal. **5** (1985), 407–427.

15. J. Sokolowski, J.P. Zolesio, "Introduction to Shape Optimization: Shape Sensitivity Analysis," Springer, Berlin, 1992.

A posteriori error estimates for three-dimensional axisymmetric elliptic problems

IVAN HLAVÁČEK and MICHAL KŘÍŽEK*   Mathematical Institute, Academy
of Sciences, Žitná 25, CZ–115 67 Praha 1, Czech Republic, e-mail: krizek@math.cas.cz

Abstract   We propose and examine the primal and dual finite element method
for solving an axially symmetric elliptic problem with mixed boundary conditions.
We derive a posteriori error estimates and apply the hypercircle method.

1   MOTIVATION

In many theoretical books on the finite element method we meet the so-called *a
priori error estimates* which are usually of the form (see, e.g., [**9**], [**10**], [**11**])

$$\|u - u_h\|_1 \le Ch^k |u|_{k+1}, \quad k \in \{1, 2, ...\},$$

where u is a sufficiently smooth weak solution of some second order elliptic problem, u_h is its finite element approximation, h is the usual discretization parameter,
$\| \cdot \|_k$ and $| \cdot |_k$ are, respectively, the standard norm and seminorm in the Sobolev
space $H^k(\Omega)$. Such a priori error estimates show the asymptotic behaviour of the
error for $h \to 0$. Anyway, we often do not know a realistic bound for the constant
$C > 0$, which depends on several other constants coming from the Céa lemma, the
Bramble-Hilbert lemma, the trace theorem, Friedrichs' or Poincaré's inequality, the
regularity of used triangulations etc. Some of these constants are mostly unknown
and, moreover, we do not know the seminorm $|u|_{k+1}$, since u is usually unknown.
Moreover, this seminorm need not be defined in the case of mixed boundary conditions, reentrant corners, jumps in coefficients, etc. The right-hand side of the
corresponding a priori error estimate then contains an unknown constant C multi-
plied by "infinity". Therefore, we are not able to establish "how far" is the finite
element solution u_h from the true solution u for a given triangulation.

*The work was supported by grant No. A1019601 of the Grant Agency of the Academy of Sciences
of the Czech Republic.

The aim of this paper is to introduce the so-called *a posteriori error estimates* which give us some information about the error after computation. From a practical point of view, they are more valuable and useful, since real-life problems are often nonlinear, large, ill-conditioned, unstable, with multiple solutions, singularities, etc. We generalize the method of [6] to axially symmetric elliptic problems. This method is based on simultaneous application of the primal and dual finite element method, see also [11]. An application of the dual finite element method to axisymmetric optimal shape design problems is given in [8].

In [1], [2], [3], [5], [9], [15], [20], several different approaches to a posteriori error estimation are presented. Various superconvergence phenomena and averaging techniques (see, e.g., [12], [13], [16], [18]) can also be employed to a posteriori estimation. These techniques enable us to get a higher order approximation of the true solution, which is then used for adaptive mesh refinements. A posteriori error estimates for some nonlinear problems are derived in [4], [19].

2 FORMULATION OF THE PROBLEM

In this section we introduce a three-dimensional elliptic boundary value problem which is axisymmetric with respect to the axis x_3. Then the most suitable approach is to use cylindrical coordinates (r, θ, z) for which

$$
\begin{aligned}
x_1 &= r \cos \theta, \\
x_2 &= r \sin \theta, \\
x_3 &= z.
\end{aligned}
\tag{2.1}
$$

Assume that a domain $\Omega \subset R^3$ is generated by the rotation of a bounded polygonal connected set

$$
D \subset \{(r, z) \in R^2 \mid r \geq 0\}
$$

about the axis $x_3 = z$. Assume that Ω has a Lipschitz boundary $\partial \Omega$ (see [15]) so that the set $\partial D \cap \sigma$, where $\sigma = \{(r, z) \in R^2 \mid r = 0\}$, does not contain isolated points. The set D is called the *meridian section* of Ω. Let

$$
\partial D = \overline{\Gamma}_\sigma \cup \overline{\Gamma}_u \cup \overline{\Gamma}_g,
\tag{2.2}
$$

where Γ_σ, Γ_u, Γ_g have finite number of components, are mutually disjoint, and are open sets in the boundary ∂D, $\Gamma_u \neq \emptyset$ and Γ_σ is the interior of the set $\partial D \cap \sigma$.

In terms of the cylindrical coordinates (2.1), we define for any $U = U(x_1, x_2, x_3)$ the function

$$
u(r, \theta, z) = U(r \cos \theta, r \sin \theta, z).
\tag{2.3}
$$

A function U is said to be *axisymmetric* with respect to the axis $x_3 = z$ if the corresponding function u is independent of θ. In this case we write only $u = u(r, z)$.

Consider the boundary value problem

$$
-\sum_{i=1}^{3} \frac{\partial}{\partial x_i} \left(A_i \frac{\partial U}{\partial x_i} \right) = F \quad \text{in } \Omega,
$$

$$
U = \overline{U} \quad \text{on } \partial \Omega_u,
\tag{2.4}
$$

$$
\sum_{i=1}^{3} N_i A_i \frac{\partial U}{\partial x_i} = G \quad \text{on } \partial \Omega_g,
$$

where $\partial\Omega_u$ and $\partial\Omega_g$ are generated, respectively, by rotation of Γ_u and Γ_g about the axis x_3, $N = (N_1, N_2, N_3)^T$ is the unit outward normal to $\partial\Omega$, and for variational formulation we assume that $A_i \in L^\infty(\Omega)$, $A_1 = A_2$, $F \in L^2(\Omega)$, $G \in L^2(\partial\Omega_g)$ and $\overline{U} \in H^1(\Omega)$. Let a positive constant C exist such that

$$A_i(x) \geq C \quad \text{a.e. in } \Omega, \quad i = 1, 2, 3. \tag{2.5}$$

Finally, let all the data F, G, $\overline{U}$ and A_i be axisymmetric. Then the problem (2.4) can be formally reduced via (2.3) to the two-dimensional problem

$$
\begin{aligned}
-\frac{1}{r}\frac{\partial}{\partial r}\left(r a_r \frac{\partial u}{\partial r}\right) - \frac{\partial}{\partial z}\left(a_z \frac{\partial u}{\partial z}\right) &= f \quad \text{in } D, \\
u &= \overline{u} \quad \text{on } \Gamma_u, \\
n \cdot \operatorname{cograd} u &= g \quad \text{on } \Gamma_g,
\end{aligned}
\tag{2.6}
$$

where n is the unit outward normal to ∂D and

$$\operatorname{cograd} u = \left(a_r \frac{\partial u}{\partial r}, a_z \frac{\partial u}{\partial z}\right)^T,$$
$$a_r = A_1, \quad a_z = A_3.$$

3   PRINCIPLE OF MINIMUM OF COMPLEMENTARY ENERGY

If an axisymmetric function U belongs to $L^2(\Omega)$ then, by the substitution theorem, the corresponding function u (cf. (2.3)) belongs to the weighted Lebesque space $L_r^2(D)$ which consists of all measurable functions v defined on D such that

$$\int_D v^2(r, z) r \, dr \, dz < \infty.$$

Note that $L^2(D) \subset L_r^2(D)$. Introduce the weighted Sobolev space

$$W_r^{1,2}(D) = \left\{ v \in L_r^2(D) \,\middle|\, \frac{\partial v}{\partial r}, \frac{\partial v}{\partial z} \in L_r^2(D) \right\}$$

equipped with the norm

$$\|v\|_{1,r,D} = \left(\int_D \left(v^2 + \left(\frac{\partial v}{\partial r}\right)^2 + \left(\frac{\partial v}{\partial z}\right)^2 \right) r \, dr \, dz \right)^{1/2}$$

and seminorm

$$|v|_{1,r,D} = \left(\int_D \left(\left(\frac{\partial v}{\partial r}\right)^2 + \left(\frac{\partial v}{\partial z}\right)^2 \right) r \, dr \, dz \right)^{1/2}.$$

Let $L_r^2(\Gamma)$ denote the space of measurable functions φ defined on Γ such that

$$\int_\Gamma \varphi^2 r \, ds < \infty,$$

where Γ is a measurable part of $\partial D \setminus \Gamma_\sigma$.

LEMMA 3.1 Let $\Gamma_1 \subset \partial D \backslash \Gamma_\sigma$, meas $\Gamma_1 > 0$. Then there exists a linear continuous mapping

$$\gamma : W_r^{1,2}(D) \to L_r^2(\Gamma_1)$$

such that

$$\gamma v = v|_{\Gamma_1} \quad \forall v \in C^1(\overline{D}).$$

For the proof see [**14**, Sec. 1].

Due to this trace theorem we can define the space

$$V = \{v \in W_r^{1,2}(D) \mid \gamma v = 0 \text{ on } \Gamma_u\}.$$

LEMMA 3.2 (Generalized Friedrichs' inequality) Let $v \in W_r^{1,2}(D)$ be such that $\gamma v = 0$ on Γ_1, where $\Gamma_1 \subset \partial D \backslash \Gamma_\sigma$ is measurable and meas $\Gamma_1 > 0$. Then

$$|v|_{1,r,D} \geq C\|v\|_{1,r,D}, \tag{3.1}$$

where the constant $C > 0$ is independent of v.

For the proof see [**7**, p. 510].

Further let

$$\mathcal{H} = L_r^2(D) \times L_r^2(D),$$

$$\|q\|_{0,r,D} = \left(\int_D (q_r^2 + q_z^2) r \, dr \, dz \right)^{1/2}, \quad q = (q_r, q_z)^T \in \mathcal{H},$$

and introduce the bilinear form on $\mathcal{H} \times \mathcal{H}$

$$(q, p)_{\mathcal{H}} = \int_D (a_r^{-1} q_r p_r + a_z^{-1} q_z p_z) r \, dr \, dz. \tag{3.2}$$

For the induced norm

$$\|q\|_{\mathcal{H}} = (q, q)_{\mathcal{H}}^{1/2}$$

we have

$$C_1\|q\|_{0,r,D} \leq \|q\|_{\mathcal{H}} \leq C_2\|q\|_{0,r,D}, \tag{3.3}$$

where C_1, C_2 are positive constants. The bilinear form (3.2) is a scalar product and the space $\mathcal{H}$, equipped with this product, is a Hilbert space. Finally, we set

$$\mathcal{H}_1 = \{q \in \mathcal{H} \mid \exists v \in V : q = \text{cograd}\, v\},$$
$$\mathcal{H}_2 = \{q \in \mathcal{H} \mid b(q, v) = 0 \quad \forall v \in V\},$$

where

$$b(q, v) = \int_D \left(q_r \frac{\partial v}{\partial r} + q_z \frac{\partial v}{\partial z} \right) r \, dr \, dz. \tag{3.4}$$

LEMMA 3.3   The spaces $\mathcal{H}_1$ and $\mathcal{H}_2$ are closed subspaces of $\mathcal{H}$ and the following orthogonal decomposition holds

$$\mathcal{H} = \mathcal{H}_1 \oplus \mathcal{H}_2$$

with respect to the scalar product (3.2).

For the proof see [7, p. 511].

Introducing the linear form

$$L(v) = \int_D fvr\,dr\,dz + \int_{\Gamma_g} g\gamma vr\,ds, \quad v \in V,$$

we define

$$Q_{f,g} = \{q \in \mathcal{H} \mid b(q,v) = L(v) \quad \forall v \in V\}.$$

The *primal variational formulation* of the problem (2.6) consists of finding $u \in \bar{u} + V$ which minimizes the functional of potential energy

$$\mathcal{I}(v) = \tfrac{1}{2}\|\operatorname{cograd} u\|_{\mathcal{H}}^2 - L(v) \tag{3.5}$$

over the set $\bar{u} + V$.

The *dual variational formulation* of the problem (2.6) consists of finding $p \in Q_{f,g}$ which minimizes the functional of complementary energy

$$\mathcal{J}(q) = \tfrac{1}{2}\|q\|_{\mathcal{H}}^2 - b(q,\bar{u}) \tag{3.6}$$

over the set $Q_{f,g}$.

The following *principle of minimum of complementary energy* is proved in [7, p. 512].

THEOREM 3.4   If u and p is the solution of the primal and dual formulation, respectively, then

$$p = \operatorname{cograd} u. \tag{3.7}$$

If $\bar{u} = 0$ then we can prove like in [11] (or [7, p. 514]) that

$$\mathcal{I}(u) + \mathcal{J}(p) = 0.$$

This is the reason why $\mathcal{J}$ is called the functional of complementary energy.

4   A POSTERIORI ERROR ESTIMATES

First we recall the *primal finite element method* for the problem (2.6). It consists of finding $u_h \in \bar{u} + V_h$ which minimizes the functional (3.5) over the set $\bar{u} + V_h$, where V_h is an arbitrary nonempty finite element subspace of V.

Second we introduce how to derive the dual finite element method. The set $Q_{f,g}$ is obviously an affine manifold. Thus if we find some element $\bar{p} \in Q_{f,g}$ then

$$Q_{f,g} = \bar{p} + \mathcal{H}_2,$$

i.e., any element $q \in Q_{f,g}$ can be written in the form

$$q = \bar{p} + q^0, \quad q^0 \in \mathcal{H}_2. \tag{4.1}$$

Substituting (4.1) into the functional $\mathcal{J}$, we easily find, by (3.7), that the functional

$$J(q^0) = \tfrac{1}{2}\|q^0\|_{\mathcal{H}}^2 + (\bar{p}, q^0)_{\mathcal{H}} - b(q^0, \bar{u})$$

attains its minimum over the subspace $\mathcal{H}_2$ at the point p^0 if and only if $p^0 = \operatorname{cograd} u - \bar{p}$.

Thus the task of minimizing $\mathcal{J}$ over $Q_{f,g}$ can be split into two independent problems:

(i) find some (fixed) $\bar{p} \in Q_{f,g}$, and

(ii) minimize the functional J over the subspace $\mathcal{H}_2$, i.e., find $p^0 \in \mathcal{H}_2$ such that

$$J(p^0) = \min_{q^0 \in \mathcal{H}_2} J(q^0).$$

A detailed discussion on solving the problem (i) is given in [**7**, Sec. VII]. Thus we shall deal with an approximation of the problem (ii) under the assumption that some solution of (i) is known. For the purpose of approximations, we shall construct nonempty finite-dimensional subspaces $Q_h \subset \mathcal{H}_2$ in terms of finite elements (see [**7**]).

The *dual finite element method* consists of finding $p_h = \bar{p} + p_h^0$, where the approximation $p_h^0 \in Q_h$ of p^0 is defined via the Ritz method, i.e.,

$$J(p_h^0) = \min_{q_h \in Q_h} J(q_h^0).$$

THEOREM 4.1   (A posteriori error estimates) Let u and p be the solutions of the primal and dual variational formulation of the problem (2.6), respectively (cf. (3.5) and (3.6)). Let u_h and p_h be the corresponding finite element solutions. Then there exist constants $C_1, C_2 > 0$ independent of V_h and Q_h such that

$$C_1\|u - u_h\|_{1,r,D} \le \|\operatorname{cograd} u_h - p_h\|_{\mathcal{H}}, \tag{4.2}$$

$$C_2\|p - p_h\|_{0,r,D} \le \|\operatorname{cograd} u_h - p_h\|_{\mathcal{H}}. \tag{4.3}$$

Proof: By (4.1) and (3.7), for any $q \in Q_{f,g}$

$$\operatorname{cograd} u - q \in \mathcal{H}_2, \tag{4.4}$$

and clearly $\operatorname{cograd} u_h - \operatorname{cograd} u \in \mathcal{H}_1$. According to Lemma 3.3, the spaces $\mathcal{H}_1$ and $\mathcal{H}_2$ are $(\cdot, \cdot)_{\mathcal{H}}$-orthogonal and thus

$$\| \operatorname{cograd} u_h - q\|_{\mathcal{H}}^2 = \| \operatorname{cograd} u_h - \operatorname{cograd} u + \operatorname{cograd} u - q\|_{\mathcal{H}}^2 \tag{4.5}$$

$$= \| \operatorname{cograd}(u_h - u)\|_{\mathcal{H}}^2 + \| \operatorname{cograd} u - q\|_{\mathcal{H}}^2.$$

Setting $q = p_h$, we see, by (3.7) and (3.3), that (4.3) holds.

From (4.5), (3.2), (2.5) and the generalized Friedrichs' inequality (3.1), we obtain

$$\| \operatorname{cograd} u_h - p_h\|_{\mathcal{H}}^2 \ge \| \operatorname{cograd}(u - u_h)\|_{\mathcal{H}}^2 \ge C\|u - u_h\|_{1,r,D}^2.$$

Hence, (4.2) is valid.   □

REMARK 4.2   If we apply the primal and dual finite element method to solve the problem (2.6), then the right-hand side of inequalities (4.2) and (4.3) is known, and it can be used to estimate the errors. Notice that $C_2 = 1$ in the case of the Laplace operator. The spaces V_h and Q_h in Theorem 4.1 may correspond to different triangulations.

The simultaneous use of the primal and dual finite element method enables us to get two-sided bounds of energy [**7**, p. 514] (see also [**11**], [**15**]). We can also apply the hypercircle method firstly proposed in [**17**].

THEOREM 4.3   (The hypercircle method) Under the assumptions of Theorem 4.1 we have

$$\| \operatorname{cograd} u - \tfrac{1}{2}(\operatorname{cograd} u_h + p_h)\|_{\mathcal{H}} = \tfrac{1}{2}\| \operatorname{cograd} u_h - p_h\|_{\mathcal{H}}. \qquad (4.6)$$

Proof: By Lemma 3.3, the vectors $\operatorname{cograd}(u - u_h) \in \mathcal{H}_1$ and $p - p_h \in \mathcal{H}_2$ are $(\cdot, \cdot)_{\mathcal{H}}$-orthogonal. Therefore, from (3.7) we get

$$\|2 \operatorname{cograd} u - \operatorname{cograd} u_h - p_h\|_{\mathcal{H}}^2 = \| \operatorname{cograd} u - \operatorname{cograd} u_h + \operatorname{cograd} u - p_h\|_{\mathcal{H}}^2$$
$$= \| \operatorname{cograd}(u_h - u)\|_{\mathcal{H}}^2 + \| \operatorname{cograd} u - p_h\|_{\mathcal{H}}^2$$
$$= \| \operatorname{cograd} u_h - \operatorname{cograd} u + \operatorname{cograd} u - p_h\|_{\mathcal{H}}^2 = \| \operatorname{cograd} u_h - p_h\|_{\mathcal{H}}^2.$$

□

REMARK 4.4   The cogradient of the true solution u lies on the hypercircle with center $\tfrac{1}{2}(\operatorname{cograd} u_h + p_h)$ and radius $\tfrac{1}{2}\| \operatorname{cograd} u_h - p_h\|_{\mathcal{H}}$. If the average $\tfrac{1}{2}(\operatorname{cograd} u_h + p_h)$ is considered as an approximate solution, then the norm of the error is exactly equal to the radius of the hypercircle (compare (4.6)).

REFERENCES

1. M. Ainsworth and A. Craig, *A posteriori error estimators in the finite element method*, Numer. Math. **60** (1992), 429–463.
2. I. Babuška and A. Miller, *A feedback finite element method with a posteriori error estimation. Part I*, Comput. Methods Appl. Mech. Engrg. **61** (1987), 1–40.
3. R. E. Bank and A. Weiser, *Some a posteriori error estimator for elliptic partial differential equations*, Math. Comp. **44** (1985), 283–301.
4. Z. Chen, L^p-*posteriori error analysis of mixed methods for linear and quasilinear elliptic problems*, in "Modeling, Mesh Generation, and Adaptive Numerical Methods for Partial Differential Equations," I. Babuška et al (eds.), Springer, Berlin, 1995, pp. 187–199.
5. R. Durán, M. A. Muschietti and R. Rodríguez, *On the asymptotic exactness of error estimators for linear triangular finite elements*, Numer. Math. **59** (1991), 107–127.
6. J. Haslinger and I. Hlaváček, *Convergence of a finite element method based on the dual variational formulation*, Apl. Mat. **21** (1976), 43–65.

7. I. Hlaváček and M. Křížek, *Dual finite element analysis of three-dimensional axisymmetric elliptic problems, Parts I–II*, Numer. Methods Partial Differential Equations **9** (1993), 507–526, 527–500.

8. I. Hlaváček and R. Mäkinen, *On the numerical solution of axisymmetric domain optimization problems by dual finite element method*, Numer. Methods Partial Differential Equations **10** (1994), 637–650.

9. C. Johnson, "Numerical solutions of partial differential equations by the finite element method," Cambridge University Press, Cambridge, 1988.

10. M. Křížek and P. Neittaanmäki, "Finite element approximation of variational problems and applications," Pitman Monographs and Surveys in Pure and Applied Mathematics 50, Longman Scientific & Technical, Harlow (Copubl. John Wiley & Sons, New York), 1990.

11. M. Křížek and P. Neittaanmäki, "Mathematical and numerical modelling in electrical engineering: theory and applications," Kluwer, Amsterdam, 1996.

12. Q. Lin and N. Yan, *A rectangle test for singular solution with irregular meshes*, in "Proc. of Systems Sci. & Systems Engrg.," Great Wall (H.K.) Culture Publ. Co., 1991, pp. 236–237.

13. Q. Lin and Q. Zhu, *Asymptotic expansion for the derivative of finite elements*, J. Comput. Math. **2** (1984), 361–363.

14. B. Mercier and G. Raugel, *Résolution d'un problème aux limites dans un ouvert axisymétrique par éléments finis en r, z et séries de Fourier en Θ*, RAIRO Anal. Numér. **16** (1982), 405–461.

15. J. Nečas and I. Hlaváček, "Mathematical theory of elastic and elasto-plastic bodies: an introduction," Elsevier, Amsterdam, 1981.

16. R. Stenberg, *Postprocessing schemes for some mixed finite elements*, RAIRO Modèl. Math. Anal. Numér. **25** (1991), 151–168.

17. J. L. Synge, "The hypercircle in mathematical physics," Cambridge University Press, Cambridge, 1957.

18. L. B. Wahlbin, *Local behavior in finite element methods*, in "Handbook of Numerical Analysis II," P. G. Ciarlet and J. L. Lions (eds.), North-Holland, Amsterdam, 1991, pp. 353–522.

19. J. Weisz, *On a-posteriori error estimate of approximate solutions to a mildly nonlinear nonpotential elliptic boundary value problem*, Math. Nachr. **153** (1991), 231–236.

20. O. C. Zienkiewicz, J. Z. Zhu, *The superconvergent patch recovery and a posteriori error estimates. Part 1*, Internat. J. Numer. Methods Engrg. **33** (1992), 1331–1364.

Hierarchical basis error estimators for Raviart–Thomas discretizations of arbitrary order

R. H. W. HOPPE and B. WOHLMUTH Mathematical Institute, University of Augsburg, D–86135 Augsburg, Germany, e-mail: hoppe@math.uni-augsburg.de, wohlmuth@math.uni-augsburg.de

Abstract We consider mixed finite element discretizations of linear second order elliptic boundary value problems with respect to an adaptively generated hierarchy of possibly highly nonuniform simplicial triangulations. In particular, we present a hierarchical basis error estimator for Raviart–Thomas mixed finite element discretizations of order l on simplicial triangulations. The introduction of the error estimator is based on the principle of defect correction in higher order ansatz spaces. By means of appropriate localization and decoupling techniques of the flux ansatz space, we obtain an easily computable, efficient and reliable a posteriori error estimator for the flux error and for flux and primal error components.

1 INTRODUCTION

We consider the following boundary value problem for a linear second order elliptic differential operator

$$Lu := -\operatorname{div}(a\nabla u) + bu = f \quad \text{in } \Omega,$$
$$u = g \quad \text{on } \Gamma := \partial\Omega \tag{1}$$

where Ω stands for a bounded, polygonal domain in the Euclidean space $\mathbb{R}^2$, $f \in L^2(\Omega)$ and $g \in H^{1/2}(\Gamma)$. The matrix-valued function $a = (a_{ij})_{i,j=1}^2$, $a_{ij} \in L^\infty(\Omega)$, $1 \le i,j \le 2$ is assumed to be symmetric and uniformly positive definite, and $b \in L^\infty(\Omega)$, $0 \le b \le \bar{b}$. All subsequent results can be easily applied to more general boundary conditions.

Introducing the auxiliary variable $\mathbf{j} := -a\nabla u$, the original problem (1) is transformed into a first order system. The natural ansatz space for the flux $\mathbf{j}$ is the Hilbert space

$$H(\operatorname{div};\Omega) := \left\{ \mathbf{q} \in (L^2(\Omega))^2 \mid \operatorname{div}\mathbf{q} \in L^2(\Omega) \right\}$$

associated with the inner product $(\mathbf{p}, \mathbf{q})_{\mathrm{div}} := (\mathbf{p}, \mathbf{q})_0 + (\mathrm{div}\,\mathbf{p}, \mathrm{div}\,\mathbf{q})_0$ and the norm $\|\mathbf{q}\|_{\mathrm{div}} := (\mathbf{q}, \mathbf{q})_{\mathrm{div}}^{1/2}$.

Then, the variational formulation leads to the following saddle point problem:
Find $(\mathbf{j}, u) \in H(\mathrm{div}; \Omega) \times L^2(\Omega)$ such that

$$
\begin{aligned}
a(\mathbf{j}, \mathbf{q}) + b(\mathbf{q}, u) &= -(g, \mathbf{q} \cdot \mathbf{n})_{0;\Gamma}, \quad \mathbf{q} \in H(\mathrm{div}; \Omega), \\
b(\mathbf{j}, v) - c(u, v) &= -(f, v)_{0;\Omega}, \qquad v \in L^2(\Omega),
\end{aligned}
\tag{2}
$$

where the bilinear forms $a\,(\cdot, \cdot)$, $b\,(\cdot, \cdot)$ and $c\,(\cdot, \cdot)$ are given by

$$
a\,(\mathbf{p}, \mathbf{q}) := \int_\Omega a^{-1}\mathbf{p} \cdot \mathbf{q}\, dx, \quad \mathbf{p}, \mathbf{q} \in H(\mathrm{div}; \Omega),
$$

$$
b\,(\mathbf{q}, v) := -\int_\Omega \mathrm{div}\,\mathbf{q}\, v\, dx, \quad \mathbf{q} \in H(\mathrm{div}; \Omega),\ v \in L^2(\Omega),
$$

$$
c\,(u, v) := \int_\Omega b\,u\,v\, dx, \qquad u, v \in L^2(\Omega).
$$

The existence and uniqueness of the solution of system (2) are guaranteed (cf., e.g., [**7**, §II, Thm. 1.2]).

For the discretization, we consider Raviart–Thomas mixed finite elements with respect to a simplicial triangulation $\mathcal{T}_h$ of Ω. The sets of vertices and edges are denoted by $\mathcal{P}_h$ and $\mathcal{E}_h$, respectively. Furthermore, $P_k(D)$, $D \subseteq \Omega$, $k \geq 0$, stands for the set of polynomials of degree $\leq k$ on D. In order to discretize the flux $\mathbf{j} \in H(\mathrm{div}; \Omega)$, we choose the Raviart–Thomas ansatz space of order l, $l \geq 0$,

$$
RT_l(\Omega; \mathcal{T}_h) := \{\mathbf{q}_h \in H(\mathrm{div}; \Omega)|\quad \mathbf{q}_h|_T \in RT_l(T),\ T \in \mathcal{T}_h\},
$$

where $RT_l(T)$, $T \in \mathcal{T}_h$, stands for the Raviart–Thomas element

$$
RT_l(T) := P_l(T)^2 + P_l(T)\mathbf{x}, \quad \mathbf{x} = (x_1, x_2)^T.
$$

The degrees of freedom of $RT_l(T)$ are given by the following moments

$$
\int_{e_i} \mathbf{n} \cdot \mathbf{q}_h p\, d\sigma, \quad p \in P_l(e_i), \qquad \int_T \mathbf{q}_h \cdot \mathbf{p}\, dx, \quad \mathbf{p} \in (P_{l-1}(T))^2,
$$

where $\mathbf{n}$ is the outer normal on ∂T and e_i, $1 \leq i \leq 3$, are the edges of T. Note that the dimension is $\dim RT_l(T) = (l+1)(l+3)$. The discrete ansatz space for the primal variable $u \in L^2(\Omega)$ associated with $RT_l(\Omega; \mathcal{T}_k)$ is given by piecewise polynomials of degree less or equal than l:

$$
W_l(\Omega; \mathcal{T}_h) := \{v_h \in L^2(\Omega)|\quad v_h|_T \in P_l(T),\ T \in \mathcal{T}_h\}.
$$

Thus, the natural requirement $\mathrm{div}\,RT_l(\Omega; \mathcal{T}_h) = W_l(\Omega; \mathcal{T}_h)$ is guaranteed.

In particular, we consider a hierarchy $(\mathcal{T}_k)_{k=0}^j$ of simplicial triangulations of Ω generated by the well-known refinement process due to Bank et al [5]. Then, there exist constants $0 < \kappa_0 \le \kappa_1$ such that

$$\kappa_0 h_{e_i}^2 \le |T| \le \kappa_1 h_{e_i}^2, \quad 1 \le i \le 3, \ T \in \mathcal{T}_k, \tag{3}$$

where $|T|$ denotes the area of T.

We refer to $(\mathbf{j}, u) \in H(\mathrm{div}; \Omega) \times L^2(\Omega)$ as the unique solution of the mixed variational problem (2) and to $(\mathbf{j}_{RT_l}, u_{RT_l}) \in RT_l(\Omega; \mathcal{T}_k) \times W_l(\Omega; \mathcal{T}_k)$ as the Raviart–Thomas finite element solution satisfying

$$
\begin{aligned}
a(\mathbf{j}_{RT_l}, \mathbf{q}) + b(\mathbf{q}, u_{RT_l}) &= -(g, \mathbf{q} \cdot \mathbf{n})_{0;\Gamma}, \quad \mathbf{q} \in RT_l(\Omega; \mathcal{T}_k), \\
b(\mathbf{j}_{RT_l}, v) - c(u_{RT_l}, v) &= -(f, v)_{0;\Omega}, \qquad v \in W_l(\Omega; \mathcal{T}_k).
\end{aligned}
\tag{4}
$$

Note that the Babuška–Brezzi condition holds true and that system (4) admits a unique solution (cf., e.g., [7, §II, Prop. 2.11]). Further, we denote by $(\tilde{\mathbf{j}}_{RT_l}, \tilde{u}_{RT_l}) \in RT_l(\Omega; \mathcal{T}_k) \times W_l(\Omega; \mathcal{T}_k)$ an available approximation obtained by means of an appropriate iterative solution process. It is assumed that $(\tilde{\mathbf{j}}_{RT_l}, \tilde{u}_{RT_l})$ satisfies the second equation of the discrete saddle point problem (4) exactly. This can be achieved, for instance, by using the algorithm proposed by Ewing and Wang [11] for a vanishing Helmholtz term $b \equiv 0$ which was later generalized in [12] and [23] to the case of a nonvanishing b.

Reliable and efficient a posteriori error estimators are an indispensable tool for efficient adaptive algorithms. We refer to the pioneering work done by Babuška and Rheinboldt [3], [4], Eriksson, Johnson and Hansbo [10], [16], [17] and the recent survey articles by Bornemann et al [6] and Verfürth [22], [21], [20] (cf. also [16], [19], [24]).

Denoting the total error to be estimated by ϵ, an estimator η is said to be efficient, if there exists a constant $\gamma > 0$ independent of the refinement level such that $\gamma\eta \le \epsilon$, whereas η is called reliable, if there exists another constant $\Gamma \ge \gamma$ independent of the refinement level such that $\epsilon \le \Gamma\eta$.

In the following, we present a hierarchical error estimator which is based on a defect correction in an appropriate higher order ansatz space and a hierarchical splitting as well as some localization techniques (cf., e.g., [8], [9], [14], [18], [22]). We propose an estimator for the $H(\mathrm{div}; \Omega)$-norm of the error in the flux variable as well as an estimator for the combined error in the flux and in the primal variable.

2   THE HIERARCHICAL BASIS ERROR ESTIMATOR

In case of standard conforming finite element discretizations, the hierarchical basis error estimator has been investigated by Deuflhard, Leinen, Yserentant [8]. Recently, this concept has been generalized by Achchab et al [1] for the mixed setting, but no easily accessible local error estimator is proposed. By means of appropriate localization and decoupling techniques of the flux ansatz space, we obtain an easily computable, efficient and reliable a posteriori error estimator for the flux error and for both error components. In the lowest order case, the hierarchical error estimator has been investigated in [12], [13], [15], [23]. Here, we generalize the main ideas to higher order ansatz spaces.

We use the higher order Raviart–Thomas ansatz space $RT_{l+1}(\Omega; \mathcal{T}_k)$ for the flux and the ansatz space $W_{l+1}(\Omega; \mathcal{T}_k)$ for the primal variable to obtain a better approximation of (2). If the solution $(\mathbf{j}, u)$ is smooth enough, the errors $\|\|\mathbf{j} - \mathbf{j}_{RT_{l+1}}\|\|_{\text{div}}$ and $\|\|\mathbf{j} - \mathbf{j}_{RT_{l+1}}\|\|_{\text{div}} + \|u - u_{RT_{l+1}}\|_0$ are of order $O(h^{l+2})$ whereas the errors $\|\|\mathbf{j} - \mathbf{j}_{RT_l}\|\|_{\text{div}}$ and $\|\|\mathbf{j} - \mathbf{j}_{RT_l}\|\|_{\text{div}} + \|u - u_{RT_l}\|_0$ are of order $O(h^{l+1})$, (see, e.g., [7]), where $\|\|\mathbf{q}\|\|^2_{\text{div}} := \int_\Omega a^{-1}\mathbf{q}\mathbf{q}\,dx + \int_\Omega \text{div}\,\mathbf{q}\,\text{div}\,\mathbf{q}\,dx$. To some extent, these a priori estimates justify the following saturation assumptions

$$\|\|\mathbf{j} - \mathbf{j}_{RT_{l+1}}\|\|_{\text{div}} \le \beta_k \|\|\mathbf{j} - \mathbf{j}_{RT_l}\|\|_{\text{div}}, \qquad\qquad \beta_k > 0, \text{(5)}$$

$$\|\|\mathbf{j} - \mathbf{j}_{RT_{l+1}}\|\|^2_{\text{div}} + \|u - u_{RT_{l+1}}\|^2_0 \le \hat\beta_k \left(\|\|\mathbf{j} - \mathbf{j}_{RT_l}\|\|^2_{\text{div}} + \|u - u_{RT_l}\|^2_0 \right), \quad \hat\beta_k > 0 \text{ (6)}$$

with $\beta_k \le \beta_\infty < 1$ and $\hat\beta_k \le \hat\beta_\infty < 1$. The saturation assumption (5) yields

$$
\begin{aligned}
(1 + \beta_\infty)^{-1} \left(\|\|\mathbf{j}_{RT_{l+1}} - \tilde{\mathbf{j}}_{RT_l}\|\|_{\text{div}} - \beta_\infty \|\|\mathbf{j}_{RT_l} - \tilde{\mathbf{j}}_{RT_l}\|\|_{\text{div}} \right) &\le \|\|\mathbf{j} - \tilde{\mathbf{j}}_{RT_l}\|\|_{\text{div}}, \\
(1 - \beta_\infty)^{-1} \left(\|\|\mathbf{j}_{RT_{l+1}} - \tilde{\mathbf{j}}_{RT_l}\|\|_{\text{div}} + \beta_\infty \|\|\mathbf{j}_{RT_l} - \tilde{\mathbf{j}}_{RT_l}\|\|_{\text{div}} \right) &\ge \|\|\mathbf{j} - \tilde{\mathbf{j}}_{RT_l}\|\|_{\text{div}}.
\end{aligned}
\tag{7}
$$

On the other hand, $(e_{\mathbf{j}}, e_u) := (\mathbf{j}_{RT_{l+1}} - \tilde{\mathbf{j}}_{RT_l}, u_{RT_{l+1}} - \tilde{u}_{RT_l})$ satisfies the following discrete defect problem

$$
\begin{aligned}
a(e_{\mathbf{j}}, \mathbf{q}) + b(\mathbf{q}, e_u) &= r(\mathbf{q}) - (g, \mathbf{q} \cdot \mathbf{n})_{0;\Gamma}, \quad \mathbf{q} \in RT_{l+1}(\Omega; \mathcal{T}_k), \\
b(e_{\mathbf{j}}, v) - c(e_u, v) &= -(f - \Pi_l f, v)_{0;\Omega}, \quad v \in W_{l+1}(\Omega; \mathcal{T}_k),
\end{aligned}
\tag{8}
$$

where the residual $r(\cdot)$ is defined as $r(\mathbf{q}) := -a(\tilde{\mathbf{j}}_{RT_l}, \mathbf{q}) - b(\mathbf{q}, \tilde{u}_{RT_l})$ and Π_l stands for the orthogonal L^2-projection onto $W_l(\Omega; \mathcal{T}_k)$. Unfortunately, (8) requires the solution of a global saddle point problem and, therefore, the approximation $(e_{\mathbf{j}}, e_u)$ is not suited for an easily computable error estimator.

In a first step, we consider a hierarchical two-level splitting of the mixed ansatz spaces $RT_{l+1}(\Omega; \mathcal{T}_k)$ and $W_{l+1}(\Omega; \mathcal{T}_k)$ which is already used in [1]. In terms of Π_l and the interpolation operator $\rho_l^{RT} : RT_{l+1}(\Omega; \mathcal{T}_k) \longrightarrow RT_l(\Omega; \mathcal{T}_k)$ given locally by

$$
\int_{e_i} \mathbf{n} \cdot (\rho_l^{RT}\mathbf{q})p\,d\sigma = \int_{e_i} \mathbf{n} \cdot \mathbf{q}p\,d\sigma, \quad p \in P_l(e_i),
$$

$$
\int_T (\rho_l^{RT}\mathbf{q}) \cdot \mathbf{p}\,dx = \int_T \mathbf{q} \cdot \mathbf{p}\,dx, \quad \mathbf{p} \in (P_{l-1}(T))^2,
$$

we obtain a hierarchical splitting of $RT_{l+1}(\Omega; \mathcal{T}_k)$ and $W_{l+1}(\Omega; \mathcal{T}_k)$ according to

$$W_{l+1}(\Omega; \mathcal{T}_k) = W_l(\Omega; \mathcal{T}_k) \oplus \widehat{W}_{l+1}(\Omega; \mathcal{T}_k),$$

where $\widehat{W}_{l+1}(\Omega; \mathcal{T}_k) := (\text{Id} - \Pi_l)\,W_{l+1}(\Omega; \mathcal{T}_k)$ and

$$RT_{l+1}(\Omega; \mathcal{T}_k) = RT_l(\Omega; \mathcal{T}_k) \oplus \widehat{RT}_{l+1}(\Omega; \mathcal{T}_k)$$

where $\widehat{RT}_{l+1}(\Omega; \mathcal{T}_k) := (\text{Id} - \rho_l^{RT})\,RT_{l+1}(\Omega; \mathcal{T}_k)$. In [1], it is proved that the coupling between $\widehat{RT}_{l+1}(\Omega; \mathcal{T}_k)$ and $RT_l(\Omega; \mathcal{T}_k)$ can be neglected and that it is sufficient to consider the discrete defect problem restricted to $\widehat{RT}_{l+1}(\Omega; \mathcal{T}_k) \times \widehat{W}_{l+1}(\Omega; \mathcal{T}_k)$. But no iteration error is taken into account and more importantly, no further localization is given. Therefore in [1], the resulting defect problem again has a global structure.

In this paper, the main idea is to investigate a further decomposition of $\widehat{RT}_{l+1}(\Omega; \mathcal{T}_k)$ into a divergence-free subspace denoted by $\widehat{RT}^0_{l+1}(\Omega; \mathcal{T}_k)$ and the remaining part $\widehat{RT}^1_{l+1}(\Omega; \mathcal{T}_k)$

$$\widehat{RT}_{l+1}(\Omega; \mathcal{T}_k) = \widehat{RT}^0_{l+1}(\Omega; \mathcal{T}_k) \oplus \widehat{RT}^1_{l+1}(\Omega; \mathcal{T}_k),$$

where

$$\widehat{RT}^0_{l+1}(\Omega; \mathcal{T}_k) := \mathbf{curl}\, \hat{S}_{l+2}(\Omega; \mathcal{T}_k),$$

$$\widehat{RT}^1_{l+1}(\Omega; \mathcal{T}_k) := \{\mathbf{q} \in \widehat{RT}_{l+1}(\Omega; \mathcal{T}_k) \mid \mathbf{q} \cdot \mathbf{n}_e|_e = 0,\ e \in \mathcal{E}_k,$$

$$\int_\Omega \mathbf{q} \cdot \mathbf{q}^0\, dx = 0,\ \mathbf{q}^0 \in \widehat{RT}^0_{l+1}(\Omega; \mathcal{T}_k) \text{ with } \mathbf{q}^0 \cdot \mathbf{n}_e|_e = 0,\ e \in \mathcal{E}_k\}.$$

Here, the hierarchical surplus $\hat{S}_{l+2}(\Omega; \mathcal{T}_k)$ is given by the following two-level splitting

$$S_{l+2}(\Omega; \mathcal{T}_k) = S_{l+1}(\Omega; \mathcal{T}_k) \oplus \hat{S}_{l+2}(\Omega; \mathcal{T}_k),$$

where $S_{l+1}(\Omega; \mathcal{T}_k)$ and $S_{l+2}(\Omega; \mathcal{T}_k)$ refer to the conforming ansatz spaces associated with the standard $P(l+1)$ and $P(l+2)$ approximations. Furthermore, it is easy to see that $\widehat{RT}^1_{l+1}(\Omega; \mathcal{T}_k)$ can be written as the direct sum of local $(l+2)$-dimensional subspaces $\widehat{RT}^1_{l+1}(T)$ which correspond to "interior" degrees of freedom of $RT_{l+1}(T)$:

$$\widehat{RT}^1_{l+1}(\Omega; \mathcal{T}_k) = \bigoplus_{T \in \mathcal{T}_k} \widehat{RT}^1_{l+1}(T), \quad \widehat{RT}^1_{l+1}(T) := \left\{\mathbf{q} \in \widehat{RT}^1_{l+1}(\Omega; \mathcal{T}_k) \mid \mathbf{q}|_{\Omega \backslash T} = 0\right\}.$$

Note that the $\widehat{RT}^1_{l+1}(T)$ can be identified with an appropriate subspace of the hierarchical surplus of the Brezzi–Douglas–Marini ansatz space of order $l+2$.

Due to the special structure of $W_{l+1}(\Omega; \mathcal{T}_k)$, the subspace $\widehat{W}_{l+1}(\Omega; \mathcal{T}_k)$ can be decomposed into the direct sum of local $(l+2)$-dimensional subspaces:

$$\widehat{W}_{l+1}(\Omega; \mathcal{T}_k) = \bigoplus_{T \in \mathcal{T}_k} \widehat{W}_{l+1}(T), \quad \widehat{W}_{l+1}(T) := \left\{v \in \widehat{W}_{l+1}(\Omega; \mathcal{T}_k) \mid v|_{\Omega \backslash T} = 0\right\}.$$

Obviously, $\operatorname{div} \widehat{RT}^1_{l+1}(T) = \widehat{W}_{l+1}(T)$.

Now, we replace the original bilinear form $a(\cdot, \cdot)$ by a spectrally equivalent bilinear form $\tilde{a}(\cdot, \cdot)$. According to the splitting of $RT_{l+1}(\Omega; \mathcal{T}_k)$, the vector fields $\mathbf{q}, \mathbf{p} \in RT_{l+1}(\Omega; \mathcal{T}_k)$ can be uniquely decomposed as follows

$$\mathbf{q} = \mathbf{q}^0 + \hat{\mathbf{q}}^0 + \hat{\mathbf{q}}^1, \quad \mathbf{p} = \mathbf{p}^0 + \hat{\mathbf{p}}^0 + \hat{\mathbf{p}}^1,$$

where $\mathbf{q}^0, \mathbf{p}^0 \in RT_l(\Omega; \mathcal{T}_k)$, $\hat{\mathbf{q}}^0, \hat{\mathbf{p}}^0 \in \widehat{RT}^0_{l+1}(\Omega; \mathcal{T}_k)$ and $\hat{\mathbf{q}}^1, \hat{\mathbf{p}}^1 \in \widehat{RT}^1_{l+1}(\Omega; \mathcal{T}_k)$. Then, the bilinear form $\tilde{a}(\cdot, \cdot)$ is given by means of

$$\tilde{a}(\mathbf{q}, \mathbf{p}) := a(\mathbf{q}^0, \mathbf{p}^0) + a(\hat{\mathbf{q}}^0, \hat{\mathbf{p}}^0) + a(\hat{\mathbf{q}}^1, \hat{\mathbf{p}}^1).$$

It is easy to see that the decomposition of the ansatz space of the flux is orthogonal with respect to $\tilde{a}(\cdot, \cdot)$.

LEMMA 2.1 There exist constants $0 < c_{RT} \leq C_{RT}$ independent of the refinement level such that

$$c_{RT}^2 \tilde{a}|_T \left(\mathbf{q}, \mathbf{q}\right) \leq a|_T \left(\mathbf{q}, \mathbf{q}\right) \leq C_{RT}^2 \tilde{a}|_T \left(\mathbf{q}, \mathbf{q}\right), \quad \mathbf{q} \in RT_{l+1}(\Omega; \mathcal{T}_k), \ T \in \mathcal{T}_k. \quad (9)$$

Proof: Since the upper bound holds true with $C_{RT}^2 = 3$, only the lower bound remains to be proved. Due to the affine equivalence of the Raviart–Thomas elements, it is easy to see that there exist constants $0 < c_{l+1} \leq C_{l+1}$ independent of the refinement level such that for each $T \in \mathcal{T}_k$, $\mathbf{q} \in RT_{l+1}(\Omega; \mathcal{T}_k)$

$$\alpha_T c_{l+1} \left(\sum_{i=1}^{3} \sum_{j=1}^{l+2} \left(\int_{e_i} \mathbf{n} \cdot \mathbf{q} \varphi_{i,j} \, d\sigma \right)^2 + \sum_{j=1}^{(l+1)(l+2)} \left(\int_{T} \mathbf{q} \cdot \varphi_j \, dx \right)^2 \right) \leq a|_T \left(\mathbf{q}, \mathbf{q}\right),$$

$$a|_T \left(\mathbf{q}, \mathbf{q}\right) \leq \alpha_T C_{l+1} \left(\sum_{i=1}^{3} \sum_{j=1}^{l+2} \left(\int_{e_i} \mathbf{n} \cdot \mathbf{q} \varphi_{i,j} \, d\sigma \right)^2 + \sum_{j=1}^{(l+1)(l+2)} \left(\int_{T} \mathbf{q} \cdot \varphi_j \, dx \right)^2 \right)$$
$$(10)$$

hold true where $\alpha_T := 0.5(\alpha_T^0 + \alpha_T^1)$ and α_T^0 and α_T^1 is the smallest resp. the largest eigenvalue of a^{-1} restricted to T. Here, $\varphi_{i,j} \in P_{j-1}(e_i)$, $1 \leq j \leq l+2$, $1 \leq i \leq 3$, with

$$\int_{e_i} \varphi_{i,j_1} \varphi_{i,j_2} \, d\sigma = \delta_{j_1,j_2} h_{e_i},$$

h_{e_i} denoting the length of the edge e_i. Therefore, the set $\{\varphi_{i,j}\}_{1 \leq j \leq l+1}$ is a hierarchical orthonormal basis of $P_{l+1}(e_i)$, $1 \leq i \leq 3$. Furthermore, the set $\{\varphi_j\}_{1 \leq j \leq (i+1)(i+2)}$ stands for an orthonormal basis of $(P_i(T))^2$, $1 \leq i \leq l$,

$$\int_{T} \varphi_{j_1} \cdot \varphi_{j_2} \, dx = \delta_{j_1,j_2}$$

and $\{\varphi_j\}_{1 \leq j \leq (l+1)(l+2)}$ gives a hierarchical orthonormal basis of $(P_l(T))^2$. Then, we obtain

$$a|_T \left(\mathbf{q}, \mathbf{q}\right) \geq \alpha_T c_{l+1} \left(\sum_{i=1}^{3} \left(\sum_{j=1}^{l+1} \left(\int_{e_i} \mathbf{n} \cdot (\rho_l^{RT} \mathbf{q}) \varphi_{i,j} \, d\sigma \right)^2 \right. \right.$$

$$\left. + \left(\int_{e_i} \mathbf{n} \cdot (\mathbf{q} - \rho_l^{RT} \mathbf{q}) \varphi_{i,l+2} \, d\sigma \right)^2 \right)$$

$$+ \sum_{j=1}^{l(l+1)} \left(\int_{T} (\rho_l^{RT} \mathbf{q}) \cdot \varphi_j \, dx \right)^2 + \sum_{j=l(l+1)+1}^{(l+1)(l+2)} \left(\int_{T} \mathbf{q} \cdot \varphi_j \, dx \right)^2 \right)$$

$$\geq \frac{c_{l+1}}{2C_l} a|_T \left(\rho_l^{RT} \mathbf{q}, \rho_l^{RT} \mathbf{q}\right)$$

$$+ \frac{c_{l+1}}{2C_{l+1}} \min\left(1, \frac{c_{l+1}}{2C_l}\right) a|_T \left(\mathbf{q} - \rho_l^{RT} \mathbf{q}, \mathbf{q} - \rho_l^{RT} \mathbf{q}\right).$$

In a second step, we have to consider $a|_T \left(\mathbf{q} - \rho_l^{RT} \mathbf{q}, \mathbf{q} - \rho_l^{RT} \mathbf{q} \right)$ in more detail. We investigate the hierarchical surplus $\widetilde{RT}_{l+1}^0 (\Omega; \mathcal{T}_k)$ and remark that there are two different types of vector fields. Let $\{\Phi_i\}_{i \in \mathcal{I}_k}$, $\mathcal{I}_k := \{1, \ldots, \dim \hat{S}_{l+2}(\Omega; \mathcal{T}_k)\}$, be the set of nodal basis functions of $\hat{S}_{l+2}(\Omega; \mathcal{T}_k)$. Then, the first type of nodal basis functions is associated with the edges of the triangulation and the second one is living in the interior of the elements,

$$\text{span}\{\Phi_i\}_{i \in \mathcal{I}_k} = \text{span}\{\Phi_e\}_{e \in \mathcal{E}_k} + \bigoplus_{T \in \mathcal{T}_k} \text{span}\{\Phi_{T;i}\}_{i=1}^l. \tag{11}$$

Altogether, we obtain $(\text{card}\,\mathcal{E}_k + l\,\text{card}\,\mathcal{T}_k)$ degrees of freedom for $\hat{S}_{l+2}(\Omega; \mathcal{T}_k)$ resp. $\widetilde{RT}_{l+1}^0 (\Omega; \mathcal{T}_k)$. Let $\mathbf{q}^B \in \widetilde{RT}_{l+1}^B (\Omega; \mathcal{T}_k) := \text{span}\{\mathbf{curl}\,\Phi_e\}_{e \in \mathcal{E}_k}$. Then, it can be shown that

$$\alpha_T \hat{c}_{l+1} \sum_{i=1}^3 \left(\int_{e_i} \mathbf{n} \cdot \mathbf{q}^B \varphi_{i,l+2}\, d\sigma \right)^2 \leq a|_T \left(\mathbf{q}^B, \mathbf{q}^B \right)$$

$$\leq \alpha_T \hat{C}_{l+1} \sum_{i=1}^3 \left(\int_{e_i} \mathbf{n} \cdot \mathbf{q}^B \varphi_{i,l+2}\, d\sigma \right)^2 \tag{12}$$

holds true with some constants $0 < \hat{c}_{l+1} \leq \hat{C}_{l+1}$. Relations (10) and (12) yield the following equivalence

$$\tilde{c}_{l+1} \left(a|_T \left(\mathbf{q}^B, \mathbf{q}^B \right) + a|_T \left(\mathbf{q}^I, \mathbf{q}^I \right) \right) \leq a|_T \left(\mathbf{q}^B + \mathbf{q}^I, \mathbf{q}^B + \mathbf{q}^I \right)$$

$$\leq \tilde{C}_{l+1} \left(a|_T \left(\mathbf{q}^B, \mathbf{q}^B \right) + a|_T \left(\mathbf{q}^I, \mathbf{q}^I \right) \right),$$

where $\mathbf{q}^I \in \widetilde{RT}_{l+1}^I (\Omega; \mathcal{T}_k) := \{\mathbf{q}^I \in \widetilde{RT}_{l+1}(\Omega; \mathcal{T}_k) \mid \mathbf{n}_e \cdot \mathbf{q}^I|_e = 0,\ e \in \mathcal{E}_k\}$ and the constants $0 < \tilde{c}_{l+1} \leq \tilde{C}_{l+1}$ are independent of the refinement level. In case $l = 0$, there is no non-trivial divergence free vector field in $\widetilde{RT}_{l+1}^I (\Omega; \mathcal{T}_k)$.

Now, $\mathbf{q} - \rho_l^{RT} \mathbf{q} \in \widetilde{RT}_{l+1}(\Omega; \mathcal{T}_k)$ can be uniquely decomposed as

$$\mathbf{q} - \rho_l^{RT} \mathbf{q} = \mathbf{q}^B + \mathbf{q}^I, \quad \mathbf{q}^B \in \widetilde{RT}_{l+1}^B (\Omega; \mathcal{T}_k),\ \mathbf{q}^I \in \widetilde{RT}_{l+1}^I (\Omega; \mathcal{T}_k).$$

If we remind $\mathbf{q} - \rho_l^{RT} \mathbf{q} = \hat{\mathbf{q}}^0 + \hat{\mathbf{q}}^1$, we finally end up with

$$a|_T \left(\mathbf{q} - \rho_l^{RT} \mathbf{q}, \mathbf{q} - \rho_l^{RT} \mathbf{q} \right)$$

$$\geq \tilde{c}_{l+1} \left(a|_T \left(\mathbf{q}^B, \mathbf{q}^B \right) + a|_T \left(\hat{\mathbf{q}}^1 + \hat{\mathbf{q}}^0 - \mathbf{q}^B, \hat{\mathbf{q}}^1 + \hat{\mathbf{q}}^0 - \mathbf{q}^B \right) \right)$$

$$= \tilde{c}_{l+1} \left(a|_T \left(\mathbf{q}^B, \mathbf{q}^B \right) + a|_T \left(\hat{\mathbf{q}}^1, \hat{\mathbf{q}}^1 \right) + a|_T \left(\hat{\mathbf{q}}^0 - \mathbf{q}^B, \hat{\mathbf{q}}^0 - \mathbf{q}^B \right) \right)$$

$$\geq \tilde{c}_{l+1} \left(\tilde{C}_{l+1}^{-1} a|_T \left(\hat{\mathbf{q}}^0, \hat{\mathbf{q}}^0 \right) + a|_T \left(\hat{\mathbf{q}}^1, \hat{\mathbf{q}}^1 \right) \right).$$

Taking into account the definition of $\tilde{a}(\cdot, \cdot)$ and $\rho_l^{RT} \mathbf{q} = \mathbf{q}_0$, the theorem is proved. $\qquad\square$

Now, we consider a modified discrete variational problem which is obtained from (8), if we replace the bilinear form $a(\cdot, \cdot)$ by $\tilde{a}(\cdot, \cdot)$:

Find $(\tilde{e}_{\mathbf{j}}, \tilde{e}_u) \in RT_{l+1}(\Omega; \mathcal{T}_k) \times W_{l+1}(\Omega; \mathcal{T}_k)$ such that

$$\tilde{a}(\tilde{e}_{\mathbf{j}}, \mathbf{q}) + b(\mathbf{q}, \tilde{e}_u) = r(\mathbf{q}) - (g, \mathbf{q} \cdot \mathbf{n})_{0;\Gamma}, \quad \mathbf{q} \in RT_{l+1}(\Omega; \mathcal{T}_k),$$

$$b(\tilde{e}_{\mathbf{j}}, v) - c(\tilde{e}_u, v) = -(f - \Pi_l f, v)_{0;\Omega}, \quad v \in W_{l+1}(\Omega; \mathcal{T}_k). \tag{13}$$

The following theorem states lower and upper bounds for the total error in the flux.

THEOREM 2.2   The solutions $(e_{\mathbf{j}}, e_u)$ and $(\tilde{e}_{\mathbf{j}}, \tilde{e}_u) \in RT_{l+1}(\Omega; \mathcal{T}_k) \times W_{l+1}(\Omega; \mathcal{T}_k)$ of the discrete variational problems (8) and (13) are equivalent in the sense that there exist constants $0 < c_{\mathbf{j};RT} \leq C_{\mathbf{j};RT}$ and $c_{d;RT}, c_{u;RT}, C_{d;RT}, C_{u;RT} > 0$ independent of $\mathcal{T}_k$, $k \geq 1$, such that

$$c_{\mathbf{j};RT}^2 \tilde{a}\,(\tilde{e}_{\mathbf{j}}, \tilde{e}_{\mathbf{j}}) \leq a\,(e_{\mathbf{j}}, e_{\mathbf{j}}) \leq C_{\mathbf{j};RT}^2 \tilde{a}\,(\tilde{e}_{\mathbf{j}}, \tilde{e}_{\mathbf{j}}) \,, \tag{14a}$$

$$\left.\begin{array}{l} \| \operatorname{div} \tilde{e}_{\mathbf{j}} \|_0 \leq \| \operatorname{div} e_{\mathbf{j}} \|_0 + c_{d;RT}\, a\,(e_{\mathbf{j}}, e_{\mathbf{j}})^{1/2} \,, \\[2mm] \| \operatorname{div} e_{\mathbf{j}} \|_0 \leq \| \operatorname{div} \tilde{e}_{\mathbf{j}} \|_0 + C_{d;RT}\, \tilde{a}\,(\tilde{e}_{\mathbf{j}}, \tilde{e}_{\mathbf{j}})^{1/2} \,, \end{array}\right\} \tag{14b}$$

$$\left.\begin{array}{l} \|\tilde{e}_u\|_0 \leq \|e_u\|_0 + c_{u;RT}\, a\,(e_{\mathbf{j}}, e_{\mathbf{j}})^{1/2} \,, \\[2mm] \|e_u\|_0 \leq \|\tilde{e}_u\|_0 + C_{u;RT}\, \tilde{a}\,(\tilde{e}_{\mathbf{j}}, \tilde{e}_{\mathbf{j}})^{1/2} \,. \end{array}\right\} \tag{14c}$$

Proof: Based on (8) and (13), we compare the solutions $e_{\mathbf{j}}$ and $\tilde{e}_{\mathbf{j}}$ and obtain

$$a\,(e_{\mathbf{j}} - \tilde{e}_{\mathbf{j}}, e_{\mathbf{j}} - \tilde{e}_{\mathbf{j}}) + c\,(e_u - \tilde{e}_u, e_u - \tilde{e}_u) = \tilde{a}\,(\tilde{e}_{\mathbf{j}}, e_{\mathbf{j}} - \tilde{e}_{\mathbf{j}}) - a\,(\tilde{e}_{\mathbf{j}}, e_{\mathbf{j}} - \tilde{e}_{\mathbf{j}})$$

as well as

$$\tilde{a}\,(e_{\mathbf{j}} - \tilde{e}_{\mathbf{j}}, e_{\mathbf{j}} - \tilde{e}_{\mathbf{j}}) + c\,(e_u - \tilde{e}_u, e_u - \tilde{e}_u) = \tilde{a}\,(e_{\mathbf{j}}, e_{\mathbf{j}} - \tilde{e}_{\mathbf{j}}) - a\,(e_{\mathbf{j}}, e_{\mathbf{j}} - \tilde{e}_{\mathbf{j}}) \,.$$

In view of (9) and the fact that $c\,(\cdot, \cdot)$ is positive semidefinite, by straightforward calculations we get

$$a\,(e_{\mathbf{j}} - \tilde{e}_{\mathbf{j}}, e_{\mathbf{j}} - \tilde{e}_{\mathbf{j}})^{1/2} \leq \left(C_{RT} + \frac{1}{c_{RT}}\right) \tilde{a}\,(\tilde{e}_{\mathbf{j}}, \tilde{e}_{\mathbf{j}})^{1/2} \,,$$

$$\tilde{a}\,(e_{\mathbf{j}} - \tilde{e}_{\mathbf{j}}, e_{\mathbf{j}} - \tilde{e}_{\mathbf{j}})^{1/2} \leq \left(C_{RT} + \frac{1}{c_{RT}}\right) a\,(e_{\mathbf{j}}, e_{\mathbf{j}})^{1/2} \,.$$

The triangle inequality applied to $a\,(e_{\mathbf{j}}, e_{\mathbf{j}})^{1/2}$ and $\tilde{a}\,(\tilde{e}_{\mathbf{j}}, \tilde{e}_{\mathbf{j}})^{1/2}$ proves (14a) with $C_{\mathbf{j};RT} := 2C_{RT} + c_{RT}^{-1}$ and $c_{\mathbf{j};RT} := \left(C_{RT} + 2c_{RT}^{-1}\right)^{-1}$.
   Recalling $\operatorname{div} RT_{l+1}(\Omega; \mathcal{T}_k) = W_{l+1}(\Omega; \mathcal{T}_k)$ and

$$b\,(\tilde{e}_{\mathbf{j}} - e_{\mathbf{j}}, v) = c\,(\tilde{e}_u - e_u, v) \,, \quad v \in W_{l+1}(\Omega; \mathcal{T}_k),$$

(14b) follows from

$$\| \operatorname{div} (\tilde{e}_{\mathbf{j}} - e_{\mathbf{j}}) \|_{0;\Omega} = \sup_{\substack{v \in W_{l+1}(\Omega; \mathcal{T}_k) \\ v \neq 0}} \frac{b\,(\tilde{e}_{\mathbf{j}} - e_{\mathbf{j}}, v)}{\|v\|_{0;\Omega}} \leq \bar{b}^{1/2} c\,(e_u - \tilde{e}_u, e_u - \tilde{e}_u)^{1/2}$$

$$= \bar{b}^{1/2} \left(\tilde{a}\,(e_{\mathbf{j}} - \tilde{e}_{\mathbf{j}}, \tilde{e}_{\mathbf{j}}) - a\,(e_{\mathbf{j}} - \tilde{e}_{\mathbf{j}}, e_{\mathbf{j}})\right)^{1/2}$$

$$\leq \bar{b}^{1/2} \sqrt{2(C_{RT} + c_{RT}^{-1})} \tilde{a}\,(\tilde{e}_{\mathbf{j}}, \tilde{e}_{\mathbf{j}})^{1/4} a\,(e_{\mathbf{j}}, e_{\mathbf{j}})^{1/4} \,.$$

By means of the triangle inequality and (14a), we conclude with $C_{d;RT}^2 := 2\bar{b}C_{\mathbf{j};RT}(C_{RT} + c_{RT}^{-1})$ and $c_{d;RT}^2 := 2c_{\mathbf{j};RT}^{-1}\bar{b}(C_{RT} + c_{RT}^{-1})$.
   For the proof of (14c), we note that for all $\mathbf{q} \in RT_{l+1}(\Omega; \mathcal{T}_k)$ we have

$$b\,(\mathbf{q}, e_u - \tilde{e}_u) = \tilde{a}\,(\tilde{e}_{\mathbf{j}}, \mathbf{q}) - a\,(e_{\mathbf{j}}, \mathbf{q}) \,. \tag{15}$$

Observing

$$\|v\|_0 \le b_{sup} \sup_{\substack{\mathbf{q} \in RT_{l+1}(\Omega; \mathcal{T}_k) \\ \mathbf{q} \ne 0}} \frac{b(\mathbf{q}, v)}{\|\mathbf{q}\|_{\text{div}}}, \quad v \in W_{l+1}(\Omega; \mathcal{T}_k),$$

where $b_{sup} > 0$ is independent of the refinement level but depends on the order l of the ansatz space (see, e.g., [7]) and using (9), (14a) and (15), we obtain

$$\|e_u - \tilde{e}_u\|_0 \le \alpha_0^{-1/2} b_{sup} \left(1 + c_{RT}^{-1} c_{j;RT}^{-1}\right) a\left(e_{\mathbf{j}}, e_{\mathbf{j}}\right)^{1/2},$$

$$\|e_u - \tilde{e}_u\|_0 \le \alpha_0^{-1/2} b_{sup} \left(C_{j;RT} + c_{RT}^{-1}\right) \tilde{a}\left(\tilde{e}_{\mathbf{j}}, \tilde{e}_{\mathbf{j}}\right)^{1/2}.$$

$\square$

A simple consequence of Theorem 2.2 is the existence of constants $0 < c_{a^{-1}} \le C_{a^{-1}}$ such that

$$c_{a^{-1}}^2 \left(\tilde{a}\left(\tilde{e}_{\mathbf{j}}, \tilde{e}_{\mathbf{j}}\right) + \|\operatorname{div} \tilde{e}_{\mathbf{j}}\|_{0;\Omega}^2\right) \le \|e_{\mathbf{j}}\|_{\text{div}}^2 \le C_{a^{-1}}^2 \left(\tilde{a}\left(\tilde{e}_{\mathbf{j}}, \tilde{e}_{\mathbf{j}}\right) + \|\operatorname{div} \tilde{e}_{\mathbf{j}}\|_{0;\Omega}^2\right), \quad (16)$$

where $c_{a^{-1}}^{-2} := \max\left(2, 2c_{d;RT}^2 + c_{j;RT}^{-2}\right)$ and $C_{a^{-1}}^2 := \max\left(2, 2C_{d;RT}^2 + C_{j;RT}^2\right)$.

According to the hierarchical splitting of the spaces $RT_{l+1}(\Omega; \mathcal{T}_k)$ and $W_{l+1}(\Omega; \mathcal{T}_k)$, we decompose $\tilde{e}_{\mathbf{j}}$ and $\tilde{e}_u$ as follows:

$$\tilde{e}_{\mathbf{j}} = \tilde{e}_{\mathbf{j}_0} + \tilde{e}_{\mathbf{j}_1}^0 + \tilde{e}_{\mathbf{j}_1}^1, \quad \tilde{e}_{\mathbf{j}_0} \in RT_l(\Omega; \mathcal{T}_k), \; \tilde{e}_{\mathbf{j}_1}^0 \in \widehat{RT}_{l+1}^0(\Omega; \mathcal{T}_k), \; \tilde{e}_{\mathbf{j}_1}^1 \in \widehat{RT}_{l+1}^1(\Omega; \mathcal{T}_k),$$

$$\tilde{e}_u = \tilde{e}_{u_0} + \tilde{e}_{u_1}, \quad \tilde{e}_{u_0} \in W_l(\Omega; \mathcal{T}_k), \; \tilde{e}_{u_1} \in \widehat{W}_{l+1}(\Omega; \mathcal{T}_k)$$

and obtain three independent variational problems.

The first one is associated with the original ansatz space $RT_l(\Omega; \mathcal{T}_k) \times W_l(\Omega; \mathcal{T}_k)$ and gives rise to a global saddle point problem

$$\begin{aligned} a\left(\tilde{e}_{\mathbf{j}_0}, \mathbf{q}\right) + b\left(\mathbf{q}, \tilde{e}_{u_0}\right) &= r(\mathbf{q}) - (g, \mathbf{q} \cdot \mathbf{n})_{0;\Gamma}, \quad \mathbf{q} \in RT_l(\Omega; \mathcal{T}_k), \\ b\left(\tilde{e}_{\mathbf{j}_0}, v\right) - c\left(\tilde{e}_{u_0}, v\right) &= 0, \quad v \in W_l(\Omega; \mathcal{T}_k). \end{aligned} \quad (17)$$

The solution $(\tilde{e}_{\mathbf{j}_0}, \tilde{e}_{u_0})$ is zero only in case of a vanishing iteration error $(\mathring{\mathbf{j}}_{RT_l} - \mathbf{j}_{RT_l}, \tilde{u}_{RT_l} - u_{RT_l})$. An upper bound for $a\left(\tilde{e}_{\mathbf{j}_0}, \tilde{e}_{\mathbf{j}_0}\right)$ and $\|\operatorname{div} \tilde{e}_{\mathbf{j}_0}\|_{0;\Omega}^2$ can be easily established by means of the iteration error.

In contrast to (17), $\tilde{e}_{\mathbf{j}_1}^0$ is the solution of the following symmetric and positive definite variational problem

$$a\left(\tilde{e}_{\mathbf{j}_1}^0, \mathbf{q}\right) = r(\mathbf{q}) - (g, \mathbf{q} \cdot \mathbf{n})_{0;\Gamma}, \quad \mathbf{q} \in \widehat{RT}_{l+1}^0(\Omega; \mathcal{T}_k), \quad (18)$$

which can be decoupled by well-known standard techniques (cf. [6]). The bilinear form $a(\cdot, \cdot)$ applied to the discrete space $\widehat{RT}_{l+1}^0(\Omega; \mathcal{T}_k) \times \widehat{RT}_{l+1}^0(\Omega; \mathcal{T}_k)$ can be seen as a bilinear form $\hat{a}(\cdot, \cdot)$ on $\hat{S}_{l+2}(\Omega; \mathcal{T}_k) \times \hat{S}_{l+2}(\Omega; \mathcal{T}_k)$. Let $\mathbf{q}, \mathbf{p} \in \widehat{RT}_{l+1}^0(\Omega; \mathcal{T}_k)$ and $\phi, \psi \in \hat{S}_{l+2}(\Omega; \mathcal{T}_k)$ with $\mathbf{q} = \operatorname{\mathbf{curl}} \phi$ and $\mathbf{p} = \operatorname{\mathbf{curl}} \psi$, then

$$a(\mathbf{q}, \mathbf{p}) = \sum_{T \in \mathcal{T}_k} \int_T a^{-1} \operatorname{\mathbf{curl}} \phi \operatorname{\mathbf{curl}} \psi \, dx = \sum_{T \in \mathcal{T}_k} \int_T \hat{a} \nabla \phi \nabla \psi \, dx =: \hat{a}(\phi, \psi). \quad (19)$$

The matrix $\hat{a}$ is defined by $\hat{a}_{11} := (a^{-1})_{22}$, $\hat{a}_{22} := (a^{-1})_{11}$ and $\hat{a}_{12} = \hat{a}_{21} := -(a^{-1})_{12}$ and has the same eigenvalues as a^{-1}. Again, let $\{\Phi_i\}_{i \in \mathcal{I}_k}$ denote the set of nodal basis functions of $\hat{S}_{l+2}(\Omega; \mathcal{T}_k)$. Then, each element $\phi \in \hat{S}_{l+2}(\Omega; \mathcal{T}_k)$ can be uniquely written as $\phi = \sum_{i \in \mathcal{I}_k} \alpha_i \Phi_i$. Now, it is well-known that there exist constants $0 < c_{\mathrm{curl}} \leq C_{\mathrm{curl}}$ such that

$$c_{\mathrm{curl}} \sum_{i \in \mathcal{I}_k} \alpha_i^2 \hat{a}|_T(\Phi_i, \Phi_i) \leq \hat{a}|_T(\phi, \phi) \leq C_{\mathrm{curl}} \sum_{i \in \mathcal{I}_k} \alpha_i^2 \hat{a}|_T(\Phi_i, \Phi_i), \quad T \in \mathcal{T}_k. \quad (20)$$

Due to (20), the coupling between different nodal basis functions can be neglected and only scalar equations have to be solved:

$$\hat{a}(\phi_{\mathbf{j}_1}^e, \Phi_e) = r(\mathbf{curl}\,\Phi_e) - (g, \mathbf{curl}\,\Phi_e \cdot \mathbf{n})_{0;\Gamma}, \quad e \in \mathcal{E}_k,$$
$$\hat{a}(\phi_{\mathbf{j}_1}^{T;i}, \Phi_{T;i}) = r(\mathbf{curl}\,\Phi_{T;i}), \qquad\qquad T \in \mathcal{T}_k,\ 1 \leq i \leq l,$$

where Φ_e and $\Phi_{T;i}$ are defined in (11). Then, the equivalence of $a\left(\tilde{e}_{\mathbf{j}_1}^0, \tilde{e}_{\mathbf{j}_1}^0\right)$ and $\sum_{e \in \mathcal{E}_k} \hat{a}(\phi_{\mathbf{j}_1}^e, \phi_{\mathbf{j}_1}^e) + \sum_{T \in \mathcal{T}_k} \sum_{i=1}^l \hat{a}(\phi_{\mathbf{j}_1}^{T;i}, \phi_{\mathbf{j}_1}^{T;i})$ is guaranteed by (20).

Due to the special structure of $\widehat{RT}_{l+1}^1(\Omega; \mathcal{T}_k)$ and $\widehat{W}_{l+1}(\Omega; \mathcal{T}_k)$, $(\tilde{e}_{\mathbf{j}_1}^1, \tilde{e}_{u_1})$ is given as the solution of independent local subproblems associated with the elements of the triangulation. For each element we have to solve a $2(l+2) \times 2(l+2)$ saddle point problem:

$$a|_T\left(\tilde{e}_{\mathbf{j}_1}^1, \mathbf{q}\right) + b|_T\left(\mathbf{q}, \tilde{e}_{u_1}\right) = r|_T(\mathbf{q}), \quad \mathbf{q} \in \widehat{RT}_{l+1}^1(T),$$
$$b|_T\left(\tilde{e}_{\mathbf{j}_1}^1, v\right) - c|_T\left(\tilde{e}_{u_1}, v\right) = (f, v)_{0;T}, \quad v \in \widehat{W}_{l+1}(T). \quad (21)$$

These preliminary remarks lead to the local definition of an a posteriori flux-oriented error estimator η_H:

$$\eta_H^2 := \sum_{T \in \mathcal{T}_k} \eta_{H;T}^2,$$

$$\eta_{H;T}^2 := \|\tilde{e}_{\mathbf{j}_1}^1|_T\|_{\mathrm{div}}^2 + \sum_{i=1}^3 w_i \hat{a}(\phi_{\mathbf{j}_1}^{e_i}, \phi_{\mathbf{j}_1}^{e_i}) + \sum_{i=1}^l \hat{a}(\phi_{\mathbf{j}_1}^{T;i}, \phi_{\mathbf{j}_1}^{T;i}), \quad T \in \mathcal{T}_k,$$

where e_i, $1 \leq i \leq 3$, denote the edges of the triangle T and w_i is a weighting factor given by $w_i := 0.5$, if e_i is an interior edge, and $w_i := 1$ else. This error estimator provides sharp lower and upper bounds for the total error in the flux, if the iteration error is small enough.

THEOREM 2.3   Under the saturation assumption (5) with $0 < \beta_k < \beta_\infty < 1$ there exist constants $c_{\mathrm{hier}}, C_{\mathrm{hier}} > 0$ and $\gamma_{\mathrm{hier}}, \Gamma_{\mathrm{hier}} > 0$, independent of the refinement level, such that

$$(1 + \beta_\infty)^{-1}\left(c_{\mathrm{hier}}\eta_H - \gamma_{\mathrm{hier}}\|\mathbf{j}_{RT_l} - \tilde{\mathbf{j}}_{RT_l}\|_{\mathrm{div}}\right) \leq \|\mathbf{j} - \tilde{\mathbf{j}}_{RT_l}\|_{\mathrm{div}},$$
$$(1 - \beta_\infty)^{-1}\left(C_{\mathrm{hier}}\eta_H + \Gamma_{\mathrm{hier}}\|\mathbf{j}_{RT_l} - \tilde{\mathbf{j}}_{RT_l}\|_{\mathrm{div}}\right) \geq \|\mathbf{j} - \tilde{\mathbf{j}}_{RT_l}\|_{\mathrm{div}}. \quad (22)$$

Proof: It remains to show that the flux of the solution of subproblem (17) is bounded by the iteration error independently of the refinement level.   For this purpose we reconsider the residual:

$$
\begin{aligned}
r(\tilde{e}_{\mathbf{j}_0}) - (g, \tilde{e}_{\mathbf{j}_0} \cdot \mathbf{n})_{0;\Gamma} &= a\left(\tilde{e}_{\mathbf{j}_0}, \tilde{e}_{\mathbf{j}_0}\right) + c\left(\tilde{e}_{u_0}, \tilde{e}_{u_0}\right) \\
&= -a\left(\tilde{\mathbf{j}}_{RT_l}, \tilde{e}_{\mathbf{j}_0}\right) - b\left(\tilde{e}_{\mathbf{j}_0}, \tilde{u}_{RT_l}\right) \\
&= a\left(\mathbf{j}_{RT_l} - \tilde{\mathbf{j}}_{RT_l}, \tilde{e}_{\mathbf{j}_0}\right) + b\left(\tilde{e}_{\mathbf{j}_0}, u_{RT_l} - \tilde{u}_{RT_l}\right) \\
&= a\left(\mathbf{j}_{RT_l} - \tilde{\mathbf{j}}_{RT_l}, \tilde{e}_{\mathbf{j}_0}\right) + c\left(\tilde{e}_{u_0}, u_{RT_l} - \tilde{u}_{RT_l}\right).
\end{aligned}
$$

In view of the equality $c\left(u_{RT_l} - \tilde{u}_{RT_l}, u_{RT_l} - \tilde{u}_{RT_l}\right) = b\left(\mathbf{j}_{RT_l} - \tilde{\mathbf{j}}_{RT_l}, u_{RT_l} - \tilde{u}_{RT_l}\right)$ and the fact that $\|\operatorname{div}(\tilde{e}_{\mathbf{j}_0})\|_{0;\Omega}^2 = b(\tilde{e}_{\mathbf{j}_0}, \operatorname{div}(\tilde{e}_{\mathbf{j}_0}))$, we obtain

$$
a\left(\tilde{e}_{\mathbf{j}_0}, \tilde{e}_{\mathbf{j}_0}\right) + c\left(\tilde{e}_{u_0}, \tilde{e}_{u_0}\right) \leq \|\|\mathbf{j}_{RT_l} - \tilde{\mathbf{j}}_{RT_l}\|\|_{\operatorname{div}}^2,
$$
$$
\|\operatorname{div}(\tilde{e}_{\mathbf{j}_0})\|_{0;\Omega}^2 \leq \bar{b}\|\|\mathbf{j}_{RT_l} - \tilde{\mathbf{j}}_{RT_l}\|\|_{\operatorname{div}}^2.
$$

Observing (7), (16) and (20), the assertions are a direct consequence of the definition of the error estimator with constants $C_{\text{hier}} := C_{a^{-1}} \max\left(1, c_{\text{curl}}^{-1/2}\right)$ and $\Gamma_{\text{hier}} := \beta_\infty \max(1, \bar{b}^{1/2}) + C_{a^{-1}}(1 + \bar{b})^{1/2}$ for the upper bounds of the error.   The constants for the lower bounds are given by $c_{\text{hier}} := c_{a^{-1}} \min\left(1, C_{\text{curl}}^{-1/2}\right)$ and $\gamma_{\text{hier}} := \beta_\infty \max(1, \bar{b}^{1/2})$. $\qquad\square$

If we are interested in an error estimator for the error in the flux and the primal variable, we have to take into account an additional term in the definition of the error estimator $\hat{\eta}_H$

$$
\hat{\eta}_H^2 := \sum_{T \in \mathcal{T}_k} \hat{\eta}_{H;T}^2,
$$
$$
\hat{\eta}_{H;T}^2 := \eta_{H;T}^2 + \|\tilde{e}_{u_1}\|_{0;T}^2, \quad T \in \mathcal{T}_k.
$$

Then, the saturation assumption (6) and the inequalities (14c), (22) guarantee that the error estimator $\hat{\eta}_H$ does provide sharp lower and upper bounds for the total error $(\|u - \tilde{u}_{RT_l}\|_0^2 + \|\|\mathbf{j} - \tilde{\mathbf{j}}_{RT_l}\|\|_{\operatorname{div}}^2)^{1/2}$, provided the iteration error is small enough.

3   NUMERICAL RESULTS

In this section, we present some numerical results obtained by an adaptive multilevel algorithm.   In particular, we have chosen $RT_0(\Omega; \mathcal{T}_k) \times W_0(\Omega; \mathcal{T}_k)$ for the discretization of (2).   The refinement process and the performance of the a posteriori error estimator is illustrated for the following test problem $-\operatorname{div}(a\nabla u) = 1$ on the unit circle with homogeneous Dirichlet boundary conditions.   The coefficient a is discontinuous: $a = 0.1$ for $x^2 + y^2 \leq 0.09$, $a = 100$ for $0.09 < x^2 + y^2 \leq 0.49$ and $a = 1$ for $0.49 < x^2 + y^2 \leq 1$.   Starting from the coarse triangulation which consists

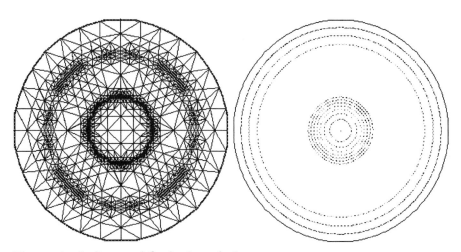

Figure 1 Isolines and final triangulation.

of four triangles obtained by rotations of the reference element $(0,0)$, $(1,0)$ and $(0,1)$ on each refinement level the discretized problem is solved by preconditioned CG-iterations. The iteration process on level $k+1$ is stopped when the estimated iteration error ε_{k+1} is less than $\varepsilon_{k+1}^2 \leq 0.01\eta_{H;k}^2 \frac{N_k}{N_{k+1}}$ where $\eta_{H;k}$ denotes the estimated error on level k and N_k and N_{k+1} stand for the number of nodes on level k and $k+1$, respectively.

Figure 1 shows the isolines of the solution and the final triangulation. The refinement process is stopped when the estimated error is less than a safety factor times the prescribed tolerance. We observe a significant adaptive refinement in the neighborhood of discontinuities of the coefficient a.

REFERENCES

1. B. Achchab, A. Agouzal, J. Baranger, and J. Maitre, *Estimateur d'erreur a posteriori hiérarchique. Application aux éléments finis mixtes*, Tech. Report 212, Equipe d'Analyse Numérique, Lyon Saint-Etienne, C.N.R.S. U.R.A. 740, 1995.

2. D. Arnold and F. Brezzi, *Mixed and nonconforming finite element methods: Implementation, post-processing and error estimates*, M^2AN Math. Modelling Numer. Anal. **19** (1985), 7–35.

3. I. Babuška and W. Rheinboldt, *Error estimates for adaptive finite element computations*, SIAM J. Numer. Anal. **15** (1978), 736–754.

4. I. Babuška and W. Rheinboldt, *A posteriori error estimates for the finite element method*, Internat. J. Numer. Methods Engrg. **12** (1978), 1597–1615.

5. R. Bank, A. Sherman, and A. Weiser, *Refinement algorithm and data structures for regular local mesh refinement*, in "Scientific Computing," R. Stepleman et al., ed., vol. 44, IMACS North-Holland, Amsterdam, 1983, pp. 3–17.

6. F. Bornemann, B. Erdmann, and R. Kornhuber, *A posteriori error estimates for elliptic problems in two and three spaces dimensions*, SIAM J. Numer. Anal.

33 (1996), 1188–1204.

7. F. Brezzi and M. Fortin, "Mixed and hybrid finite element methods," Springer, New York, 1991.

8. P. Deuflhard, P. Leinen, and H. Yserentant, *Concepts of an adaptive hierarchical finite element code*, IMPACT Comput. Sci. Engrg. **1** (1989), 3–35.

9. R. Duran and R. Rodríguez, *On the asymptotic exactness of Bank-Weiser's estimator*, Numer. Math. **62** (1992), 297–303.

10. K. Eriksson and C. Johnson, *An adaptive finite element method for linear elliptic problems*, Math. Comput. **50** (1988), 361–383.

11. R. Ewing and J. Wang, *Analysis of multilevel decomposition iterative methods for mixed finite element methods*, M^2AN Math. Modelling and Numer. Anal. **28** (1994), 377–398.

12. R. Hoppe and B. Wohlmuth, *Efficient numerical solution of mixed finite element discretizations by adaptive multilevel methods*, Appl. Math. **40** (1995), 227–248.

13. R. Hoppe and B. Wohlmuth, *Multilevel iterative solution and adaptive mesh refinement for mixed finite element discretizations*, Tech. Report 343, Math.–Nat. Fakultät, Universität Augsburg, 1995; To appear in Appl. Num. Math.

14. R. Hoppe and B. Wohlmuth, *Element-oriented and edge-oriented local error estimators for nonconforming finite element methods*, M^2AN Math. Modelling and Numer. Anal. **30** (1996), 273–263.

15. R. Hoppe and B. Wohlmuth, *A comparison of a posteriori error estimators for mixed finite elements*, Tech. Report 350, Math.–Nat. Fakultät, Universität Augsburg, 1996; Submitted to Math. Comput.

16. C. Johnson, "Numerical Solution of Partial Differential Equations by the Finite Element Method," no. 101, Cambridge University Press, Cambridge, 1987.

17. C. Johnson and P. Hansbo, *Adaptive finite element methods in computational mechanics*, Comp. Meth. Appl. Mech. Engrg. **101** (1992), 143–181.

18. T. Strouboulis and J. Oden, *A posteriori error estimation of finite element approximations in fluid dynamics*, Comp. Meth. Appl. Mech. Engrg. **78** (1990), 201–242.

19. B. Szabó and I. Babuška, "Finite Element Analysis," John Wiley & Sons, New York, 1991.

20. R. Verfürth, *A posteriori error estimates for nonlinear problems. Finite element discretizations of elliptic equations*, Math. Comp. **62** (1994), 445–475.

21. R. Verfürth, *A posteriori error estimation and adaptive mesh-refinement techniques*, J. Comp. Appl. Math. **50** (1994), 67–83.

22. R. Verfürth, "A review of a posteriori error estimation and adaptive mesh-refinement techniques," Teubner, Stuttgart, 1996.

23. B. Wohlmuth, "Adaptive Multilevel–Finite–Elemente Methoden zur Lösung elliptischer Randwertprobleme," PhD thesis, TU München, 1995.

24. O. Zienkiewicz and R. Taylor, "The Finite Element Method," vol. 1, McGraw-Hill, London, 1989.

The superconvergence of finite element methods on domains with reentrant corners

YUN-QING HUANG*   Department of Mathematics, Xiangtan University, Hunan, 411105 People's Republic of China

Abstract   There have been many studies concerned with superconvergence of finite element methods since the 1970's. This paper surveys the developments in treating the corner singularities of elliptic problems. We focus on the mesh .constructions to recover the accuracy of finite element approximations and further to obtain the superconvergence of gradient and asymptotic error expansions.

1   INTRODUCTION

Superconvergence of finite element methods and other high accuracy methods (such as extrapolation and defect correction etc.) are mainly dependent on two assumptions: uniform meshes, or with small perturbation (strongly regular), and smoothness of the solution. In general, we can construct piecewise strongly regular grid to guarantee the superconvergence but the smoothness of the solution is not satisfied in many practical situations. For example, on the domains with reentrant conners, the solution of the Poisson equation does not belong to H^2 in general. Optimal error estimates cannot be obtained by the standard finite element method. In 2-D case we can only get $\|u - u_h\|_1 \leq Ch^{\beta_M - \varepsilon}$ and $\|u - u_h\|_{L_\infty} \leq Ch^{\beta_M - \varepsilon}$ where π/β_M is the maximum angle of the domain. Schatz and Wahlbin [24], [25] proved a more delicate result

$$|(u - u_h)(x_0)| \leq C_\varepsilon h^{2\beta - \varepsilon} d^{-\beta_M},$$

where d is the distance from x_0 to the corners set on the boundary. This result shows that in the interior of the domain one can expect better approximation than on the whole domain. But the interior estimates are also not optimal though the solution is smooth enough in the interior. This is quite different for the interpolant, since finite element approximation is a global minimizer of the energy functional. This is

---

*Project supported by the NSF of China and National Education Committee.

the well-known "pollution" phenomenon. This paper surveys the investigations to overcome pollution of corner singularities and how to obtain the superconvergence.

2   SINGULARITIES OF THE SOLUTION

Let Ω be a polygonal domain in $\mathbb{R}^2$. Consider the elliptic Dirichlet boundary value problem

$$\begin{cases} Au = -D_i(a_{ij}D_ju) + a_0u = f & \text{in } \Omega, \\ u|_{\partial\Omega} = 0, \end{cases} \tag{2.1}$$

where a_{ij}, a_0, f are smooth functions, $a_0 > 0$ and $a_{ij} = a_{ji}$. Furthermore, assume that there exists a constant $\nu > 0$ such that

$$a_{ij}\xi_i\xi_j \geq \nu|\xi|^2 \quad \forall \xi \in \mathbb{R}^2.$$

Let $Q = \{Q_1, Q_2, ..., Q_l\}$ be set of all corner points of the domain. Define the weight function

$$\phi_\alpha(x) = \prod_{i=1}^{l} |x - Q_i|^{\alpha_i} \tag{2.2}$$

and the weighted Sobolev space $H_\alpha^m(\Omega)$ equipped with the norm

$$\|u\|_{m,\alpha} = \left(\|u\|_{m-1}^2 + \sum_{|k|=m} \int_\Omega \phi_\alpha^2 |D^k u|^2 dx \right)^{1/2}, \tag{2.3}$$

where $\| \cdot \|_m$ is the standard norm in the Sobolev space $H^m(\Omega)$.

In the following we set $1 - \alpha = (1 - \alpha_1, 1 - \alpha_2, ..., 1 - \alpha_l)$. For $A = -\Delta$, the solution can be represented in the local polar coordinates with origin at Q_j as (see [11], [12], [19])

$$u = \sum_{i=1}^{n} c_i r^{i\beta_j} \sin i\beta_j + w, \tag{2.4}$$

where π/β_j is the interior angle of the domain at Q_j. If β_j is an integer then a factor "$\ln r$" should be added to the singular expansion. The singularity can also be expressed by

$$|D^k u| \leq Cr^{\beta_j - |k|} \tag{2.5}$$

near the corner point Q_j. This representation does not need to know the exact forms of the singular parts. The number β_j can be replaced by an approximation $\beta_j^* \leq \beta_j$ when we construct a local refined mesh though we may get a little bit overrefined. These provide a chance to deal with equations with variable coefficients.

THEOREM 1   (Huang and Lin [17]) For problem (2.1) there exists $\beta = (\beta_1, \beta_2, ..., \beta_l)$ such that

$$|D^k u| \leq C\phi_{\beta - |k|}, \tag{2.6}$$

and the a priori estimate

$$\|u\|_{2,1-\alpha} \leq C\|f\|_{0,1-\alpha}, \quad 0 < \alpha_j \leq \min(\beta_j, 1), \tag{2.7}$$

holds, where β_j depends on the interior angle at Q_j and the eigenvalues of the coefficient matrix $(a_{ik}(Q_j))_{2\times 2}$. Moreover, $\beta_j < 1$ if the interior angle is bigger than π otherwise $\beta_j \geq 1$.

3   LOCAL REFINEMENT AND SUPERCONVERGENCE

A local refinement is introduced by I. Babuška in 1970 to deal with the corner singularities. It is based on the fact of error estimates

$$\|u - u_h\|_1 \leq C \left(\sum_\tau h_\tau^{2m} |u|_{m+1,\tau}^2 \right)^{1/2}. \tag{3.1}$$

To balance the error distribution on the domain the elements on which there are bigger norms $|u|_{m+1,\tau}$ should have smaller mesh size. The question is how to construct a triangulation T^h such that the accuracy of finite element approximation can be recovered to the optimal order. Furthermore, what kind of triangulations can preserve high accuracy properties?

3.1 σ-graded meshes and error estimates

Let T^h be a regular partition of Ω, and $\Omega_0 = \{$all elements connected with $Q\}$.

DEFINITION 1   A regular partition is called σ-graded mesh with $\sigma = (\sigma_1, \sigma_2, ..., \sigma_l)$, $\sigma_j \geq 1$, $1 \leq j \leq l$, if the following conditions hold:
   1° For any element $\tau \in \Omega_0$ which is connected with Q_j, we have

$$h_\tau = \mathrm{diam}(\tau) = O(h^{\sigma_j}). \tag{3.2}$$

  2° For any element $\tau \notin \Omega_0$ there exist constants c_0 and c_1 such that

$$c_0 h \max_\tau \phi_{1-\frac{1}{\sigma}}(x) \leq h_\tau \leq c_1 h \min_\tau \phi_{1-\frac{1}{\sigma}}(x). \tag{3.3}$$

where $1 - \frac{1}{\sigma}$ denotes the multi-index $(1 - \frac{1}{\sigma_1}, 1 - \frac{1}{\sigma_2}, ..., 1 - \frac{1}{\sigma_l})$. It is easy to see that the minimum element diameter is $O(h^{\sigma_M})$ $(\sigma_M = \max\limits_{1 \leq j \leq l} \sigma_j)$ and maximum element diameter is $O(h)$. If all $\sigma_j = 1$ then T^h becomes quasiuniform. If $\sigma_j > 1$ then T^h has local refinement near Q_j and σ_j describe the degree of refinement. So we usually call σ a local refinement index.

PROPOSITION 1   Let T^h be σ-graded mesh. Then the number of all elements $M \leq C\sigma_M h^{-2}$.
  Proof: By Definition 1 we have

$$M = \sum_{\tau \in T^h} 1 = \sum_{\tau \in T^h} h_\tau^2 \cdot h_\tau^{-2} \leq (c_0 h)^{-2} \sum_{\tau \in T^h} h_\tau^2 \frac{1}{\max_\tau \phi_{2-\frac{2}{\sigma}}(x)}$$

$$\leq (c_0 h)^{-2} \int_\Omega \phi_{\frac{2}{\sigma}-2}(x)\, dx \leq C\sigma_M h^{-2}. \tag{3.4}$$

Proposition 1 shows that the number of elements of a σ-graded mesh is proportional to that of a quasiuniform mesh. It increases linearly as the local refinement index increases.

Let

$$A(u,v) = \int_\Omega (a_{ij} D_j u D_i v + a_0 uv) \, dx. \tag{3.5}$$

The weak form and finite element approximation of (2.1) are: Find $u \in H_0^1(\Omega)$ such that

$$A(u,v) = (f,v) \quad \forall v \in H_0^1(\Omega), \tag{3.6}$$

and find $u_h \in S^h$ such that

$$A(u_h,v) = (f,v) \quad \forall v \in S^h, \tag{3.7}$$

where S^h is a piecewise polynomial conforming finite element space. We will concentrate only on the linear element.

Using σ-graded mesh we can easily obtain the error estimates.

THEOREM 2   Let Ω be a plane polygonal domain and assume that the solution of (2.1) has the the singular behaviour $|D^k u| \leq C\phi_{\beta-|k|}(x)$. If the local refinement index $\sigma_j \geq \max(1, \frac{1}{\beta_j})$, $1 \leq j \leq l$ then we have

$$\|u - u_h\|_1 \leq Ch. \tag{3.8}$$

Proof: By (3.1) and Definition 1 we find that

$$h_j = \max_{\tau \cap Q_0 = Q_j} h_\tau = O(h^{\sigma_j}),$$
$$\|u - u_I\|_{1,\Omega_0}^2 \leq C \int_{\Omega_0} \phi_{\beta-1}^2(x) \, dx \leq C \max_\tau h_j^{2\beta_j} \leq Ch^2. \tag{3.9}$$

On the element $\tau \not\subset \Omega_0$, we have

$$\|u - u_I\|_{1,\tau}^2 \leq Ch_\tau^2 \int_\tau |D^2 u|^2 \, dx \leq Ch^2 \int_\tau \phi_{1-\frac{1}{\sigma}}^2 |D^2 u|^2 \, dx. \tag{3.10}$$

Hence, by a priori estimate (Theorem 1)

$$\|u - u_h\|_1^2 \leq Ch^2 + Ch^2 \int_\Omega \phi_{1-\frac{1}{\sigma}}^2(x)|D^2 u|^2 \, dx$$
$$\leq Ch^2 + Ch^2 \int_\Omega \phi_{1-\frac{1}{\sigma}}^2(x) f^2(x) \, dx$$
$$\leq Ch^2.$$

This completes the proof.

If we increase the local refinement index we can also obtain the optimal error estimates for higher order elements. The Aubin–Nitche trick for getting L_2-norm estimates can be applied with weighted a priori estimates.

Many authors studied error estimates of the finite element methods with a local refinement. We refer to Babuška et al [1], [4], Schatz and Wahlbin [26], and Huang H.C. et al [10].

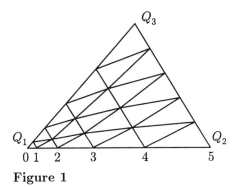

Figure 1

3.2 σ-strongly regular mesh, superconvergence

DEFINITION 2   A σ-graded mesh is called σ-strongly regular mesh if any two adjacent triangles form a convex quadrilateral and the difference of the opposite sizes of the quadrilateral satisfies

$$|\rho - \rho'| \leq Ch^2 \min_{\tau \cup \tau'} \phi_{1-\frac{2}{\sigma}}(x), \quad (\tau \cup \tau') \cap Q_0 = \emptyset. \qquad (3.11)$$

If Ω is divided into several subdomains and T^h is σ-strongly regular in every sub-domains, then we call T^h piecewise σ-strongly regular. If all the vertices of the subdomains are on the boundary of Ω, then we say T^h does not have an interior cross point, otherwise we say T^h has an interior cross point.

Now we turn to the construction of σ-strongly regular mesh. For the sake of simplicity, we only consider the construction on triangular subdomains. Assume the mesh is to be refined near the corner point Q_1.

Method 1

First divide Q_2Q_3 into N uniform intervals and then divide Q_1Q_2 and Q_1Q_3 by the points of $Q_1 + (jh_i)^\sigma \vec{e_i}$, $j = 1, 2, ..., N$, where $h_i = |Q_1Q_i|/N$ and e_i is the unit vector for $\overrightarrow{Q_1Q_i}$, $i = 1, 2$. Connect the divided points on Q_2Q_3 with Q_1Q_i ($i = 1, 2$) respectively by straight lines. Finally divide the quadrilaterals into triangles (see Fig. 1).

Method 2

Suppose the triangle $\triangle Q_1Q_2Q_3$ can be mapped to the triangle $\triangle Q_1Q_2'Q_3'$ by a radial extending transformation

$$\Psi: (r, \theta) \to (\psi(r), \theta),$$

with $\psi(r) \approx r^{\frac{1}{\sigma}}$. Construct a uniform mesh or strongly regular mesh on $\triangle Q_1Q_2'Q_3'$ and then using the points $x_j = \Psi^{-1}(y_j)$, we can obtain a σ-strongly regular mesh, where $\{y_j\}$ are grid points on $\Delta Q_1 Q_2' Q_3'$. The choice of the transformation is very flexible. For example we can do it as follows:

Construct a uniform mesh on $\Delta Q_1 Q_2 Q_3$. For any grid point P extend $Q_1 P$ such that it acrosses $Q_2 Q_3$ at P^*. Then choose P' along $\overrightarrow{Q_1 P}$ such that

$$|Q_1 P'| = \left(\frac{|Q_1 P|}{|Q_1 P^*|} \right)^{\sigma} |Q_1 P^*| = \left(\frac{|Q_1 P|}{|Q_1 P^*|} \right)^{\sigma - 1} |Q_1 P|.$$

A σ-strongly regular mesh can be obtained from the points $\{P'\}$.

We can similarly construct σ-strongly regular meshes on quadrilateral subdomains.

σ-strongly regular meshes have following geometrical properties.

PROPOSITION 2   Let T^h be σ-strongly regular mesh. Then the difference of the areas between two adjacent elements and the difference of the related angles satisfies

$$| \operatorname{meas}(\tau) - \operatorname{meas}(\tau')| \le C h^3 \min_{\tau \cup \tau'} \phi_{2 - \frac{3}{\sigma}}(x), \quad \tau \cup \tau' \not\subset \Omega_0,$$

$$|\theta - \theta'| \le C h \min_{\tau \cup \tau'} \phi_{-\frac{1}{\sigma}}(x), \quad \tau \cup \tau' \not\subset \Omega_0.$$

Proof: A straightforward computation yields the above result by Definition 2.

THEOREM 3   Suppose the solution u of (2.1) has the singularity property $|D^k u| \le C \phi_{\beta - |k|}$ and T^h is piecewise σ-strongly regular mesh with $\sigma_j > \max(\frac{2}{\beta_j}, 1)$. Then we have the superconvergence estimates

$$\|u_h - u_I\|_1 \le \begin{cases} C h^2, & \text{if } T^h \text{ does not have an interior cross point,} \\ C h^2 |\ln h|^{\frac{1}{2}}, & \text{if } T^h \text{has an interior cross point,} \end{cases}$$

where u_I is the linear interpolant of u.

The proof of the theorem is based on the element analysis and the cancellation between the adjacent elements. We do not give the details here and only give a sketch of the proof.

From [8], [9] we have on a typical element τ that

$$u - u_I = -\frac{1}{2} \sum_{i=1}^{3} h_i^2 D_i^2 u \lambda_{i+1} \lambda_{i+2} + \tilde{r} = r + \tilde{r}, \tag{3.12}$$

$$D(u - u_I) = -\frac{1}{2} \sum_{i=1}^{3} h_i^2 D_i^2 u D(\lambda_{i+1} \lambda_{i+2}) + R, \tag{3.13}$$

and estimates

$$\|\tilde{r}\|_{s,p,\tau} \le C h_\tau^{3-s} |u|_{3,p,\tau}, \quad s = 0, 1, \tag{3.14}$$

$$\|R\|_{0,p,\tau} \le C h_\tau^2 |u|_{3,p,\tau}, \tag{3.15}$$

where h_i is length of the side s_i and $\{\lambda_i\}$ are area coordinates, D_i denotes the derivative along the direction s_i.

From (3.6), (3.7) and (3.12) we have for all $v \in S^h$ that

$$A(u_h - u_I, v) = A(u - u_I, v) = A(r, v) + A(\tilde{r}, v). \tag{3.16}$$

For simplicity of the presentation we assume $A = -\Delta$. If $\tau \in \Omega_0$ then

$$\left| \int_\tau \nabla(u - u_I)\nabla v \, dx \right| \leq C h_i^{\beta_i} |v|_{1,\tau} \leq C h^2 |v|_{1,\tau}. \tag{3.17}$$

For $\tau \notin \Omega_0$, we have by (3.3)

$$\left| \int_\tau \nabla \tilde{r} \nabla v \, dx \right| \leq C h_\tau^2 |u|_{3,p,\tau} |v|_{1,\tau}$$

$$\leq C h^2 \left(\int_\tau \phi_{2-\frac{2}{\sigma}}^2 |D^3 u|^2 \, dx \right)^{\frac{1}{2}} |v|_{1,\tau}. \tag{3.18}$$

Under the assumptions of Theorem 3 the integral $\int_\tau \phi_{2-\frac{2}{\sigma}}^2 |D^3 u|^2 \, dx$ is bounded. So we need only to estimate the term $\sum_\tau \int_\tau \nabla r \nabla v \, dx$. Standard element analysis [9] gives

$$\int_\tau \nabla r \nabla v \, dx = -\frac{1}{8} \sum_{i=1}^3 h_i^2 \left\{ \cot \alpha_{i+1} \int_{s_i} \beta_0(1 - 2\lambda_{i+1}) D_i^2 u D_i v \, ds \right.$$

$$+ \frac{\sin \alpha_i}{\sin \alpha_{i+1} \sin \alpha_{i+2}} \int_{s_{i+2}} \beta_0(1 - 2\lambda_{i+1}) D_i^2 u D_{i+2} v \, ds \tag{3.19}$$

$$\left. - \frac{1}{\sin \alpha_{i+1} \sin \alpha_{i+2}} \int_\tau \beta_0(1 - 2\lambda_{i+1}) D_{i+1} D_i^2 u D_{i+2} v \, dx \right\},$$

where $\beta_0(t) = 1 - t^2$.

The derivative $D_i v$ is continuous on s_j and the line integrals on side s_i from the adjacent elements have different signs. Using the geometrical properties of the triangulation (Proposition 2), we can have

$$D_i w - D_{i'} w = O(h \phi_{-\frac{1}{\sigma}}) |\nabla w|, \tag{3.20}$$

$$D_i^2 w - D_{i'}^2 w = O(h \phi_{-\frac{1}{\sigma}}) |D^2 w|. \tag{3.21}$$

Summation over all elements $\tau \notin \Omega_0$ gives

$$\left| \int_{\Omega \backslash \Omega_0} \nabla r \nabla v \, dx \right| \leq C h^2 |v|_1 + \sum_{i=1}^M h_\tau^2 |F(D^2 u)(x_i)| \, |v(x_i)|, \tag{3.22}$$

where $\{x_i\}$ are vertices of the subdomains $\Omega_j \backslash \Omega_0$. If T^h does not have interior cross points then $\{x_i\}$ are the points nearest to Q, $v(x_i) = h_\tau D v$, so

$$|h_\tau^2 F(D^2 u)(x_i) v(x_i)| \leq C h^{2\sigma} \cdot h^{\sigma(\beta - 2)} \cdot h_\tau \cdot Dv$$

$$\leq C h^{\sigma \beta} |v|_{1,\tau} \leq C h^2 |v|_{1,\tau}, \tag{3.23}$$

If T^h has an interior cross point x_i, then $D^2 u(x_i)$ is bounded and $v(x_i)$ can only be bounded by the inverse estimate $|v(x_i)| \leq C|\ln h|^{\frac{1}{2}}|v|_1$, since there is no refinement in neighbourhood of x_i, which deteriorates the estimates.

3.3 Other refinements

Adaptive computation of finite element method usually adopt the so-called K-mesh (see Fig. 2). On every rectangle element we use bilinear trial and test functions. All nodes are divided into formal and informal groups. The nodes which are interior points of another element edges are called informal nodes. In order to keep finite element subspace S^h conforming. The values on the informal nodes are not unknowns but the linear interpolations of values on two neighbouring formal nodes. It is clear that $S^h \subset H_0^1$.

In the following we assume that the set Q^h of all elements which contain informal nodes has measure of order $O(h)$. Usually this assumption is true since local refinement often leads to blockwise uniform meshes. The informal nodes only appear on the boundary of subdomains (blocks). The area of all informal elements is:

$$O(h_1) + O(h_2) + \cdots = O(h)$$

where $h_j = c_0^{-j+1} h$ with $c_0 \geq 2$.

Since there exist informal nodes, one can not obtain the optimal order of superconvergence but there we still have the superconvergence estimate (see [14])

$$\begin{aligned}
\|u_h - u_I\|_{1,\Omega^{\delta_1}} &\leq Ch^{1.5} \left(\|u\|_{3,\Omega^{\delta_2}} + \|u\|_{2,\infty,\Omega^{\delta_2}} \right) \\
&+ C\|u - u_h\|_{0,\Omega^{\delta_2}}, \qquad \delta_1 > \delta_2,
\end{aligned} \tag{3.24}$$

where $\Omega^\delta = \{x \in \Omega \mid \mathrm{dist}(x,Q) > \delta\}$. For the subdomains which are contained in a uniform grid piece, the order of the estimate can be improved to h^2.

For the Poisson equation and special domains such as L-shaped domain, a local refinement mesh can be obtained by the rectangular elements with nodes $(jh)^\sigma$ on x and y axis (see Fig. 3). This kind of refinement can be easily implemented. The mesh is not regular if $\sigma > 1$, but optimal order superconvergence and asymptotic expansions of the error can be obtained (see Lin [21], [23]). On the other hand, the resulting system of this kind refinement has worse condition number. It was shown in [15] that

$$\mathrm{cond}(A) \geq C \max \left(\frac{h_\tau}{k_\tau} + \frac{k_\tau}{h_\tau} \right) N = O(Nh^{1-\sigma}),$$

where N is the number of unknowns.

4 COORDINATE TRANSFORMATION AND ERROR EXPANSION

In this section we construct a kind of special finite element spaces by the coordinate transformation. These finite element spaces are not piecewise polynomials. The

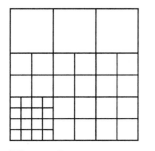

Figure 2

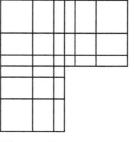

Figure 3

elements are curved and with local refinement near the corner points. The implementations and analysis can be reduced to the standard finite element framework. It is needed only to do some preprocesses for the coefficients.

4.1   Coordinate transformation and finite element space

THEOREM 4   For any $\sigma = (\sigma_1, \sigma_2, ..., \sigma_l)$, $1 \le \sigma_j < \infty$, there exists a decomposition $\Omega = \cup \Omega_i$ satisfying

1° Ω_i is a triangle or convex quadrilateral, and connected with at most one corner point of Q. Furthermore, $\{\Omega_i\}$ forms a macro quasiuniform triangulation of Ω.

2° There exists a one to one mapping $\Psi \colon \Omega \to \Omega$ such that $\Psi(\Omega_i) = \Omega_i$.

3° Ψ is continuous on $\overline{\Omega}$ and smooth on Ω_j except Q. If Ω_j is connected with the corner point $Q_i \in Q$, then by local polar coordinate with origin at Q_i, $y = \psi(x) \colon \Omega_j \to \Omega_j$ satisfy

   a) $r = O(\rho^{\sigma_i})$, $r = |x - Q_i|$, $\rho = |y - Q_i|$;

   b) $\dfrac{\partial y_\mu}{\partial x_\nu} = O\left(r^{\frac{1}{\sigma_i}-1}\right) = O(\rho^{1-\sigma_i})$, $\mu, \nu = 1, 2$;

   c) $J = \det\left(\dfrac{\partial(y_1, y_2)}{\partial(x_1, x_2)}\right) = O(r^{\frac{2}{\sigma_i}-2}) = O(\rho^{2-2\sigma_i})$.

Proof: The choice of the mapping Ψ is very flexible. We shall construct one below.

First we construct a macro triangulation $\Omega = \cup \Omega_j$ such that 1° is satisfied. On Ω_j which is not connected with Q, define

$$\Psi = I_d \quad \text{(Identity mapping)}.$$

On Ω_j which is connected with $Q_i \in Q$, let us assume that Ω_j is a triangle $\triangle Q_i AB$. For quadrilateral case we refer to Huang [13] and Huang and Lin [16]–[18]. Under local polar coordinate at origin Q_i, denote

$$x_1 = r \cos\theta, \quad x_2 = r \sin\theta. \tag{4.1}$$

Let the equation of line AB be $r = g(\theta)$. Define $y = \psi(x) \colon \Omega_j \to \Omega_j$ as follows

$$\theta' = \theta, \qquad \rho = \left(\frac{r}{g(\theta)}\right)^{\frac{1}{\sigma_i}} g(\theta), \tag{4.2}$$

$$y_1 = \rho \cos\theta', \quad y_2 = \rho \sin\theta'. \tag{4.3}$$

It is easy to see that ψ satisfies a), b), and c). We need only to verify the global continuity. Since ψ is the identity on AB, it coincides with ψ' defined on other subdomain $\Omega_{j'}$ with AB as a part of the boundary.

On side Q_iA (similarly for Q_iB) Ψ is a radial extending transformation which is determined only by the length of Q_iA and the refinement index σ_i. So it also coincides on QA with ψ' defined on $\Omega_{j'}$. This completes the proof.

Suppose T_*^h is a quasiuniform mesh which is a refinement of macro triangulation $\{\Omega_j\}$. So the boundaries $\{\partial\Omega_j\}$ consist of $\{\partial e\}$. Let S_*^h be the standard piecewise polynomial finite element space of order m. Define a triangulation T^h on Ω and finite element space S^h by

$$T^h = \psi^{-1}(T_*^h), \tag{4.4}$$

$$S^h = \{v \mid v = \chi(\psi(x)), \ \chi \in S_*^h\}. \tag{4.5}$$

From the properties of transformation Ψ we know that T^h is a σ-graded mesh, and element τ is curved in general. The functions of S^h have very complicated forms. Fortunately, in practical computation we do not involve in these difficulties. In fact, using the transformation, the bilinear form $A(u, v)$ can be written as

$$\begin{aligned}
A(u, v) &= \int_\Omega (a_{ij} D_j u D_i v + a_0 uv)\, dx = A_*(u, v) \\
&= \int_\Omega [A_1 \partial_1 u \partial_1 v + A_2(\partial_1 u \partial_2 v + \partial_2 u \partial_1 v) + A_3 \partial_2 u \partial_2 v + A_0 uv]\, dy,
\end{aligned} \tag{4.6}$$

where

$$A_1 = a_{ij} \frac{\partial y_1}{\partial x_i} \frac{\partial y_1}{\partial x_j} J^{-1}, \quad A_2 = a_{ij} \frac{\partial y_1}{\partial x_i} \frac{\partial y_2}{\partial x_j} J^{-1},$$

$$A_3 = a_{ij} \frac{\partial y_2}{\partial x_i} \frac{\partial y_2}{\partial x_j} J^{-1}, \quad A_0 = a_0(\psi^{-1}(y)) J^{-1},$$

$$\partial_1 = \frac{\partial}{\partial y_1}, \qquad\qquad \partial_2 = \frac{\partial}{\partial y_2}.$$

It is easy to see that $|A_i| \leq C$ if a_{ij}, a_0 are bounded in the subdomain Ω_j which is connected with $Q_i \in Q$. Then we have

$$|\partial^k A_i| \leq C\rho^{-|k|}, \quad i = 1, 2, 3, \quad |\partial^k A_0| \leq C\rho^{2\sigma_i - 2 - |k|}.$$

Define the finite element approximation $u_h \in S^h$ by

$$A(u_h, v) = (f, v) \quad \forall v \in S^h. \tag{4.7}$$

This is equivalent to: Find $u_h^* \in S_*^h$ such that

$$A_*(u_h^*, \chi) = (f^*, \chi) \quad \forall \chi \in S_*^h, \tag{4.8}$$

where $f^* = f(\psi^{-1}(y))/J$. The function u_h can be obtained by $u_h = u_h^*(\psi(x))$, especially the values of u_h^* at nodes of T^h are the values of u_h^* at the corresponding nodes of T_*^h. Solving (4.8) is standard. The system is SPD with the condition number $O(h^{-2})$.

4.2   Error analysis

Using the transformation, we can obtain a weighted estimate of the interpolant in S^h.

LEMMA 1   We have

$$
h \left(\int_\Omega |D(u - u_I)|^2 \, dx \right)^{\frac{1}{2}} + \left(\int_\Omega \phi_{1 - \frac{1}{\sigma}}^{-2} |u - u_I|^2 \, dx \right)^{\frac{1}{2}}
$$
$$
\leq C h^{m+1} \left(\int_\Omega \phi_{1 - \frac{1}{\sigma}}^{2m} |D^{m+1} u|^2 \, dx \right)^{\frac{1}{2}}. \tag{4.9}
$$

Proof: Transforming the left-hand side of (4.9) to y variable and using standard estimates of interpolant in S_*^h and then transforming back to x variable gives the desired result.

Note that

$$
\|D(w - w_I)\|^2 \leq C h^2 \|w\|_{2, 1 - \frac{1}{\sigma}}, \tag{4.10}
$$

and using weighted a priori estimate, we can use the Aubin–Nitche trick just like on a convex domain. Hence, all the results on a convex domain can be extended to nonconvex case by choosing a suitable refinement index σ. For the error analysis we have (see Huang and Lin [16]–[18])

THEOREM 5   Suppose the solution of (2.1) has property $|D^k u| \leq \phi_{\beta - |k|}$, and refinement index σ satisfy $\sigma_j > \max(\frac{m}{\beta_j}, 1)$. Then we have

$$
h \left(\int_\Omega |D(u - u_h)|^2 \, dx \right)^{\frac{1}{2}} + \left(\int_\Omega \phi_{1 - \frac{1}{\sigma}}^{-2} |u - u_h|^2 \, dx \right)^{\frac{1}{2}} \leq C h^{m+1}. \tag{4.11}
$$

If the refinement index satisfy $\sigma_j \geq \max(\frac{m+1}{\beta_j}, 1)$ then we have the maximum norm estimates

$$
\|u - u_h\|_{L_\infty} \leq C h^{m+1} |\ln h|, \tag{4.12}
$$
$$
\|\phi_{1 - \frac{1}{\sigma}} D(u - u_h)\|_{L_\infty} \leq C h^m |\ln h|. \tag{4.13}
$$

4.3   Asymptotic expansions

Now, we only discuss on the case of the linear element, namely $m = 1$.

DEFINITION 3 T^h is called piecewise σ-uniform if T_*^h is piecewise uniform.

Let $\mathcal{P} = \{\text{all vertexes of } \Omega_j\} \setminus Q$.

THEOREM 6 Suppose the solution of (2.1) satisfies $|D^k u| \leq \phi_{\beta - |k|}$, and T^h is piecewise σ-uniform with refinement index $\sigma_j > \max(\frac{4}{\beta_j}, 1)$, $m = 1$, then fo $\mathrm{dist}(z, \mathcal{P}) \geq \delta > 0$, we have

$$(u_h - u_I)(z) = h^2 w(z) + O(h^4 |\ln h|),$$
$$D_z(u_h - u_I)(z) = h^2 D_z w(z) + O(h^3 |\ln h|)\phi_{\frac{1}{\sigma} - 1}(z).$$

For the proof we refer to Huang and Lin [16] and Chen and Huang [9].

5   USING SINGULAR BASIS

Consider the model problem

$$-\Delta u = f \quad \text{in } \Omega,$$
$$u|_{\partial \Omega} = 0. \tag{5.1}$$

Let $r = |x - Q_j|$. As we know in the neighbourhood of Q_j, the solution can be expressed by

$$u = \sum_{i=1}^{n} k_{ij} r^{i\beta_j} \sin i\beta_j \theta + \tilde{u}. \tag{5.2}$$

If β_j is integer then there is a logarithmic term "$\ln r$" in the expansion. For simplicity we assume $\beta_j < 2$.

Let $w_j(r) \in C^k$ be a cut-off function, $0 \leq w_j(r) \leq 1$, $w_j(r) = 0$ if $r \leq r_1$ and $w_j(r) = 0$ if $r \geq r_2 > r_1$. In fact, in the interval (r_1, r_2), $w_j(r)$ can be taken as Hermite polynomial of order $2k + 1$, defined by

$$w_j(r_1) = 1, \qquad\qquad w_j(r_2) = 0,$$
$$D_r^i w_j(r_1) = D_r^i w_j(r_2) = 0, \quad i = 1, 2, ..., k.$$

Let $\psi_{ij} = w_j(r) r^{i\beta_j} \sin i\beta_j \theta$. This restricts the singularity in the neighbourhood of Q_j. In (5.2), $\tilde{u}$ has a better smoothness as n increases.

In order to improve the accuracy of finite element approximations, one can add the singular functions $\{\psi_{ij}\}$ to the standard finite element space. Let $S_\Psi^h = S^h + \mathrm{span}\{\psi_{ij}\}$. Then near the corner points the solution is approximated mainly by the singular basis $\{\psi_{ij}\}$ and the standard finite element space approximates the smooth part.

THEOREM 7 Suppose $\tilde{u} \in C^{4+\varepsilon}$ and T^h is piecewise uniform mesh without interior cross point. Then the Galerkin approximation in S_Ψ^h has the asymptotic expansion

$$(u_h - u_I)(z) = h^2 w(z) + O(h^3 |\ln h|).$$

Furthermore, if $0 < r_1 < r_2 < \delta$ then for $z \in \Omega^\delta$ we have

$$(u_h - u_I)(z) = h^2 w(z) + O(h^4 |\ln h|^2).$$

REFERENCES

1. Babuška, I., *Finite element methods for domains with corners*, Computing **6** (1970), 264–273.
2. Babuška, I. and Miller, A., "A posteriori error estimates and adaptive techniques for finite element method," Tech. Note BN-968, Inst. for Physical Science and Technology, Univ. of Maryland, June, 1981.
3. Babuška, I. and Miller, A., *A feedback finite element method with a posteriori error estimation: Part I, The finite element method and some basic properties of the a posteriori error estimator*, Comput. Meth. Appl. Mech. Engrg. **61** (1987), 1–40.
4. Babuška, I., Kellogg, R. and Pitkäranta, J., *Direct and inverse error estimates for finite elements with mesh refinement*, Numer. Math. **33** (1979), 447–471.
5. Blum, H., *On Richardson extrapolation for linear finite elements on domains with reentrant corners*, Z. Angew. Math. Mech. **67** (1987), 351–352.
6. Blum, H. and Rannacher, R., *Extrapolation techniques for reducing the pollution effect of reentrant corners in the finite element method*, Numer. Math. **52** (1988), 539–564.
7. Chen, C.M., *Superconvergence in finite element methods in domain with reentrant corners*, Natur. Sci. J. Xiangtan Univ. **12** (1990), 134–141.
8. Chen, C.M., "Finite element method and its analysis of increasing accuracy," Hunan Science and Technology Press, 1982.
9. Chen, C.M. and Huang Y.-Q., "High accuracy theory of finite element methods," Hunan Science and Technology Press, 1995.
10. E, W.N., Huang, H.C. and Han, W.M., *Error analysis of local refinements of polygonal domains*, J. Comput. Math. **5** (1987).
11. Grisvard, P., *Behaviour of the solution of an elliptic boundary value problems in a polygonal or polyhedral domain*, in "Numerical solution of PDE-III," Hubbard, B. (ed.), Academic Press, New York, 1976, pp. 207–274.
12. Grisvard, P., Wendland, W. and Whiteman, J.R. (eds.), "Singularities and constructive methods for their treatment," Springer Lecture Notes in Math. 1121, 1985.
13. Huang, Y.-Q., "Finite element methods – Extrapolation and Superconvergence," Ph.D thesis, Institute of Systems Sciences, Academia Sinica, 1987.
14. Huang, Y.-Q. and Chen, Y.P., *The superconvergence and asymptotic exact a posteriori error estimates of finite element on K-mesh*, Mathematica Numerica Sinica **16** (1994), 278–285; see also, Chinese J. Numer. Math. & Appl. **16** (1994), 66–74.
15. Huang, Y.-Q. and Chen, Y.P., *A lower bound estimate of condition number for the finite element equations on highly refined rectangle meshes*, Natur. Sci. J. Xiangtan Univ. **15** (1993), Suppl. Issue, 92–96.
16. Huang, Y.-Q. and Lin, Q., *A finite element method on polygonal domains with extrapolation*, Math. Numer. Sinica **12** (1990), 239–249. (in Chinese)
17. Huang, Y.-Q. and Lin, Q., *Elliptic boundary value problems on polygonal domains and finite element approximations*, J. Syst. Sci. Math. Sci. **12** (1992), 263–268.
18. Huang, Y.-Q. and Lin, Q., *Some estimates of Green function and its finite element approximations on polygonal domains*, J. Syst. Sci. Math. Sci. **14**

(1994), 1–8.

19. Kondrat'ev, V.A., *Boundary value problems for elliptic equations in domains with conical or angular points*, Trans. Moscow Math. Soc. **16** (1967), 227–313.

20. Křížek, M. and Neittaanmäki, P., *On superconvergence techniques*, Acta Appl. Math. **9** (1987), 175–198.

21. Lin, Q., *Fourth order eigenvalue approximation by extrapolation on domains with reentrant corners*, Numer. Math. **58** (1991), 631–640.

22. Lin, Q., *Global error expansion and superconvergence for higher order interpolation of finite elements*, J. Comp. Math. **4** (1991).

23. Lin, Q. and Yan, N.N., *A rectangle test for singular solution with irregular meshes*, in "Proceedings of Systems Science & Systems Engineering," Great Wall (H.K.) Culture Publish Co., 1991, pp. 236–237.

24. Schatz, A.H. and Wahlbin, L.B., *Interior maximum norm estimates for finite element methods*, Math. Comp. **31** (1977), 414–442.

25. Schatz, A.H. and Wahlbin, L.B., *Maximum norm estimates for finite element method on plane polygonal domains, Part I*, Math. Comp. **32** (1978), 73–109.

26. Schatz, A.H. and Wahlbin, L.B., *Maximum norm estimates for finite element method on plane polygonal domains, Part II: Refinement*, Math. Comp. **33** (1979), 465–492.

27. Strang, G. and Fix, G.J., "An analysis of the finite element method," Prentice-Hall, Englewood Cliffs, NJ, 1973.

28. Wahlbin, L.B., *On the sharpness of certain local estimates for H_0^1 projections into finite element spaces: influence of a reentrant corner*, Math. Comp. **42** (1984), 1–8.

29. Wahlbin, L.B., *Local behaviour in finite element methods*, in "Handbook of Numerical Analysis," Vol. II (Part I), Ciarlet, P.G. and Lions, J.L. (eds.), North-Holland, Amsterdam, 1991, pp. 353–522.

Error estimation for linear and nonlinear problems

ANTONIO HUERTA, PEDRO DÍEZ and JUAN JOSÉ EGOZCUE   Departamento de Matemática Aplicada III, E.T.S. de Ingenieros de Caminos, Universitat Politècnica de Catalunya, Campus Nord C–2, E–08034 Barcelona, Spain

1   INTRODUCTION

It is now widely accepted that adaptive procedures in finite element computations are necessary for any practical computations [7]. Error estimation is a key feature of an adaptive procedure. An error estimator should provide information about the global quality of the solution and the distribution of the error in the domain. A priori error estimates are the main tool for the theoretical study of the Finite Element Method but they cannot provide practical results [2]. In fact, a posteriori error estimators are employed for practical applications. These estimators may be classified into two families: flux projection (for instance, the Zienkiewicz–Zhu estimator [8]) and residual type estimators.

Residual type estimators were first introduced by Babuška and Rheinboldt [1]. The residual of the finite element solution is used as a source term in local boundary value problems associated with the error. These estimators are defined by two factors: the selection of the local interpolation space (usually a bubble space for a p-refined reference solution) and the approximation of the boundary conditions by splitting the flux jump across the edges of the elements. Indeed, flux jump can be thought as a weak residual. This kind of estimators have a solid theoretical background but their results have been found to be less robust than flux-projection estimates [2]. The biggest part of the computational cost of these estimators is taken by the flux jump computation and the flux splitting procedure.

Here, a new approach to residual type estimators is introduced, the details for linear elliptic problems can be found in [3]. The estimates are seen as projections on subspaces of locally-supported functions, these projections are computed by solving local problems. Flux jumps are not computed across the edges, therefore the main disadvantages of residual type estimators are precluded. Moreover, the estimator can be used with arbitrary meshes: structured and non-structured even if different element types are mixed. Instead of flux jumps along the element edges a multi-projection strategy is devised to account for the neighboring data.

The obtained estimate is a lower bound of the actual error. In addition, since inputs for local problems are the same as those for the standard problem, the definition and the computation of the estimator are easily integrated in a finite element code. The development and implementation details for linear problems are in [3], as well as some comparisons to the widely used Zienkiewicz–Zhu estimator [8]. This estimator which is originally developed for linear elliptic problems is further generalized for nonlinear cases where non singular positive definite tangent matrices are available. This is in fact the case for most nonlinear mechanics problems.

2 STATEMENT OF THE PROBLEM

The estimator is presented in a mechanical context in order to describe a general nonlinear problem. The unknown variable is a displacement field u having values in the domain Ω. Let us assume that the function u belongs to a function space $\mathcal{V}$.

The solution of a mechanical problem is a displacement field u in $\mathcal{V}$ verifying

$$a(u, v) = l(v) \quad \text{for all } v \in \mathcal{V}, \tag{1}$$

where the forms $a(\cdot, \cdot)$ and $l(\cdot)$ are defined by

$$a(u, v) = \int_{\Omega} \boldsymbol{\sigma}(u) : \boldsymbol{\varepsilon}(v) \, d\Omega \tag{2}$$

and

$$l(v) = \int_{\Omega} f \cdot v \, d\Omega + \int_{\Gamma_n} g_n \cdot v \, d\Gamma, \tag{3}$$

where $\boldsymbol{\sigma}(u)$ stands for the stress tensor associated with the displacements u and $\boldsymbol{\varepsilon}(v)$ stands for the strain tensor associated with the "virtual" displacements v. The body is represented by the domain Ω and it is loaded by the body forces f and the tension g_n applied in a part Γ_n of the boundary of Ω.

The form $a(u, v)$ can be seen as the internal work in the body, at a state described by u, given by a "virtual" displacement (test function) v and the form $l(v)$ is the work of external forces under the virtual displacement v.

The loads are assumed to be conservative and then $l(\cdot)$ does not depend on u and it is linear. Thus, the nonlinearity can only be introduced by $a(\cdot, \cdot)$: if geometric nonlinearity is introduced in the model, both arguments of $a(\cdot, \cdot)$ are nonlinear because $\boldsymbol{\varepsilon}(u)$ is a nonlinear function of u. The nonlinearity of the constitutive model (material nonlinearity) concerns only the first argument of $a(\cdot, \cdot)$ since it implies that $\boldsymbol{\sigma}(u)$ is a nonlinear function of $\boldsymbol{\varepsilon}(u)$.

The FE approximation of u is a function u_h belonging to a finite dimensional interpolation space $\mathcal{V}_h$ such that

$$a(u_h, v_h) = l(v_h) \quad \text{for all } v_h \in \mathcal{V}_h. \tag{4}$$

The space $\mathcal{V}_h$ is generated by a mesh of elements Ω_k, $k = 1, \ldots, M$, discretizing the domain Ω.

The space $\mathcal{V}_h$ is generated by a mesh discretizing the domain Ω. We denote by $\mathcal{B} = \{N_1, \ldots, N_n\}$ the basis of interpolation functions generating $\mathcal{V}_h$. Then, the function u_h is identified with its representation in $\mathcal{B}$, that is, the vector $\boldsymbol{u}_h$ containing the nodal values of u_h.

Internal and external discretized force vectors, $\mathbf{f}_{\text{int}}(\boldsymbol{u}_h)$ and $\mathbf{f}_{\text{ext}}$, associated with the discretization are introduced in order to express the discrete equilibrium equation (4) as a vectorial equality. We define $\mathbf{f}_{\text{int}}(\boldsymbol{u}_h)$ and $\mathbf{f}_{\text{ext}}$ from their components:

$$[\mathbf{f}_{\text{int}}(\boldsymbol{u}_h)]_i := a(u_h, N_i)$$
$$= \int_\Omega \boldsymbol{\sigma}(u_h) : \boldsymbol{\varepsilon}(N_i) \ d\Omega \ \text{ for } i = 1, \dots, n, \tag{5}$$
$$[\mathbf{f}_{\text{ext}}]_i := l(N_i).$$

Then, Eq. (4) can be rewritten as a vector equilibrium equation:

$$\mathbf{r}(\boldsymbol{u}_h) := \mathbf{f}_{\text{ext}} - \mathbf{f}_{\text{int}}(\boldsymbol{u}_h) = 0. \tag{6}$$

The goal of the error estimation is to characterize the error $e := u - u_h$ through one global measure and several local measures, defined usually elementwise, that allows to describe the spatial distribution of the error.

In the linear elastic problem, the internal force vector is a simple linear function of the displacements,

$$\mathbf{f}_{\text{int}}(\boldsymbol{u}_h) = \mathbf{K} \cdot \boldsymbol{u}_h,$$

where $[\mathbf{K}]_{ij} = a(N_i, N_j)$, and, therefore, Eq. (6) gives

$$\mathbf{K} \cdot \boldsymbol{u}_h = \mathbf{f}_{\text{ext}}.$$

The nonlinear problem is solved by dividing the total load $\mathbf{f}_{\text{ext}}$ in n_s increments $\mathbf{f}_{\text{ext}}^k$, $k = 1, \dots, n_s$. Then, an equilibrium solution is found at each step: $\boldsymbol{u}_h^k$ such that

$$\mathbf{r}^k(\boldsymbol{u}_h^k) := \mathbf{f}_{\text{ext}}^k - \mathbf{f}_{\text{int}}(\boldsymbol{u}_h^k) = 0. \tag{7}$$

Since the definition of $\mathbf{f}_{\text{int}}(\boldsymbol{u}_h)$ is nonlinear, each step needs the solution of a nonlinear system of n equations and n unknowns (the components $\boldsymbol{u}_h$).

3   THE LINEAR ESTIMATOR

This section is devoted to draw the main ideas of the linear error estimator introduced in [1]. This estimator can only be applied to linear problems. The nonlinear extension is presented in the next sections.

The goal of the error estimator is to approximate the norm of a reference solution associated with a reference finer mesh. Here, the reference mesh is build-up by the assembly of submeshes discretizing each element Ω_k of the original submesh.

The reference mesh generates the interpolation space $\mathcal{V}_{\tilde{h}}$ ($\tilde{h} \ll h$). Let us denote by $\tilde{\mathcal{B}} = \{\tilde{N}_1, \dots, \tilde{N}_{\tilde{n}}\}$ the basis of interpolation functions. The reference solution associated with the finer mesh is denoted by $u_{\tilde{h}}$ and the reference error is defined by $e_r := u_{\tilde{h}} - u_h$. Both $u_{\tilde{h}}$ and e_r are identified with the vectors of their nodal values in the reference mesh, $\boldsymbol{u}_{\tilde{h}}$ and $\hat{\mathbf{e}}_r$, respectively.

The original interpolation space $\mathcal{V}_h$ is included in the reference space $\mathcal{V}_{\tilde{h}}$ because the finer mesh is build-up by a refinement of the original mesh. Therefore, u_h can also be represented in the basis $\tilde{\mathcal{B}}$. This representation is denoted by $\tilde{\boldsymbol{u}}_h$ and, thus, $\hat{\mathbf{e}}_r = \boldsymbol{u}_{\tilde{h}} - \tilde{\boldsymbol{u}}_h$.

The reference equilibrium equation giving $\boldsymbol{u}_{\tilde{h}}$ can be written:

$$\tilde{\mathbf{r}}(\boldsymbol{u}_{\tilde{h}}) := \tilde{\mathbf{f}}_{\text{ext}} - \tilde{\mathbf{f}}_{\text{int}}(\boldsymbol{u}_{\tilde{h}}) = 0, \tag{8}$$

where $\tilde{\mathbf{f}}_{\text{ext}}$ and $\tilde{\mathbf{f}}_{\text{int}}$ are the external and internal force vectors associated with the reference discretization.

The residual $\tilde{\mathbf{r}}(\tilde{\boldsymbol{u}}_h)$ is not zero and, therefore, $\boldsymbol{u}_h$ does not satisfy the equilibrium equation (8), from the criterion given by the reference mesh. This applies also in the nonlinear case and it can be used in the generalization of the estimator. In fact, this residual is used as a source term in the error equation.

If the problem is linear, $\tilde{\mathbf{f}}_{\text{int}}(\cdot)$ linear, and it can be expressed by means of a stiffness matrix $\tilde{\mathbf{K}}$. That is,

$$\tilde{\mathbf{f}}_{\text{int}}(\boldsymbol{u}_{\tilde{h}}) = \tilde{\mathbf{K}} \cdot \boldsymbol{u}_{\tilde{h}},$$

which is obviously different from

$$\tilde{\mathbf{f}}_{\text{int}}(\tilde{\boldsymbol{u}}_h) = \tilde{\mathbf{K}} \cdot \tilde{\boldsymbol{u}}_h.$$

Thus, a linear equation for the reference error is found:

$$\tilde{\mathbf{K}} \cdot \hat{\mathbf{e}}_r = \tilde{\mathbf{f}}_{\text{ext}} - \tilde{\mathbf{K}} \cdot \tilde{\boldsymbol{u}}_h = \tilde{r}(\tilde{\boldsymbol{u}}_h). \tag{9}$$

The global solution of (8), or, alternatively, the solution of (9), is extremely expensive from a computational point of view and it must be avoided. Then, an approximation to e_r is computed by solving local problems where the unknowns are the restrictions of e_r to the elements Ω_k.

In order to relate the global equation to local problems, it is convenient to remark that Eq. (9) is the matrix form of a discrete equation coming from a continuous weak equation. In fact, Eq. (9) can be seen as the discrete version of the residual weak equation

$$a(e, v) = l(v) - a(u_h, v) \quad \text{for all } v \in \mathcal{V}, \tag{10}$$

and can also be written

$$a(e_r, v_{\tilde{h}}) = l(v_{\tilde{h}}) - a(u_h, v_{\tilde{h}}) \quad \text{for all } v_{\tilde{h}} \in \mathcal{V}_{\tilde{h}}. \tag{11}$$

Note that the projections of e_r on any subspace of $\mathcal{V}_{\tilde{h}}$ can be computed by using Eq. (11).

The linear estimator is defined from Eq. (11) by introducing subspaces of $\mathcal{V}_{\tilde{h}}$ containing locally supported functions. The projections of e (or e_r) on these subspaces can be taken as approximations of the error. Each of these subspaces is defined such that it contains locally-supported functions. Then, the approximations have local character and they can be computed by solving low cost local problems.

A first family of projections giving interior estimates is defined through elementary submeshes discretizing each element Ω_k. The space V_k (associated with the element

Ω_k) is generated by the interpolation functions $\tilde{N}_i$ of $\mathcal{V}_{\tilde{h}}$ associated with the nodes lying in the interior of Ω_k. The projection of the actual error on the local subspace V_k is denoted by ε_k and it is the local interior estimate associated with the element Ω_k. The assembly of all the local interior estimates builds up the global interior estimate ε:

$$\varepsilon = \sum_{k=1}^{m} \varepsilon_k. \tag{12}$$

Since all subspaces V_k are orthogonal the norm of ε is easily computed:

$$\|\varepsilon\|^2 = \sum_{k=1}^{m} \|\varepsilon_k\|^2. \tag{13}$$

In the linear elliptic case, since the error is measured using the energy norm $\|\cdot\|$ induced by $a(\cdot,\cdot)$, local and global interior estimates are lower bounds of the actual error ($\|\varepsilon\| \leq \|e\|$).

The global interior approximation ε is forced to vanish on the boundary of the elements. However, the actual error e is generally different of zero in these points and, therefore, the approximation can be poor. In order to preclude this restriction, we compute the projections of the error e on a new family of subspaces and we perform a complete estimation.

An enrichment of the interior estimate is defined by a second partition of the domain Ω in subdomains Λ_l, $l = 1, \ldots, m'$, called patches. A new family of subspaces U_l associated with the patches is introduced. Each subspace U_l is generated by the interpolation functions $\tilde{N}_i$ of $\mathcal{V}_{\tilde{h}}$ associated with the nodes lying in the interior of Λ_l. The projection on the space U_l is imposed to be orthogonal to the global interior approximation ε. In fact, we are projecting the error e on the space $\tilde{U}_l = U_l \cap \text{span}\{\varepsilon\}^{\perp}$ in order to obtain the patch-estimate denoted by η_l. A global patch estimate, η, is defined as being the sum of all the local patch-estimates η_l. Thus, the global complete estimate, $\varepsilon + \eta$, is also a lower bound of the actual error. The complete estimate is also defined elementwise by adding the contribution of each patch to the overlapped elements.

4 NONLINEAR GENERALIZATION

In this section, the linear error estimator described in the previous section is generalized to nonlinear problems.

Since we are dealing with a nonlinear problem, the reference residual associated with the approximate solution u_h is

$$\tilde{\mathbf{r}}(\tilde{\boldsymbol{u}}_h) = \tilde{\mathbf{f}}_{\text{ext}} - \tilde{\mathbf{f}}_{\text{int}}(\tilde{\boldsymbol{u}}_h). \tag{14}$$

But now, the simplification of Eq. (9) is not possible.

However, assuming that the reference error is small with respect to the approximate solution u_h ($\|e_r\| << \|u_h\|$), a tangent approximation can be used:

$$\begin{aligned}
\tilde{\mathbf{r}}\left(\boldsymbol{u}_{\tilde{h}}\right) &= \tilde{\mathbf{r}}\left(\tilde{\boldsymbol{u}}_h + \hat{\mathbf{e}}_r\right) \\
&\approx \tilde{\mathbf{r}}\left(\tilde{\boldsymbol{u}}_h\right) - \tilde{\mathbf{K}}_T\left(\tilde{\boldsymbol{u}}_h\right) \cdot \hat{\mathbf{e}}_r.
\end{aligned} \tag{15}$$

The tangent matrix $\widetilde{\mathbf{K}}_T$ represents the derivatives of $\tilde{\mathbf{f}}_{\text{int}}\left(\boldsymbol{u}_{\tilde{h}}\right)$ with respect to $\boldsymbol{u}_{\tilde{h}}$:

$$\widetilde{\mathbf{K}}_T = \frac{\partial \tilde{\mathbf{f}}_{\text{int}}}{\partial \boldsymbol{u}_{\tilde{h}}}.$$

Then, Eq. (8) can be written

$$\widetilde{\mathbf{K}}_T\left(\tilde{\boldsymbol{u}}_h\right) \cdot \hat{\mathbf{e}}_r \approx \tilde{\mathbf{r}}\left(\tilde{\boldsymbol{u}}_h\right) = \tilde{\mathbf{f}}_{\text{ext}} - \tilde{\mathbf{f}}_{\text{int}}\left(\tilde{\boldsymbol{u}}_h\right). \tag{16}$$

If matrix $\widetilde{\mathbf{K}}_T$ is the representation in $\widetilde{\mathcal{B}}$ of a symmetric positive definite bilinear form $a_T(\cdot, \cdot)$, then (16) is the discrete version of the integral equation

$$a_T(e, v) = l(v) - a(u_h, v) \text{ for all } v \in \mathcal{V}, \tag{17}$$

and therefore Eq. (16) can also be written

$$a_T(e_r, v_{\tilde{h}}) = l(v_{\tilde{h}}) - a(u_h, v_{\tilde{h}}) \text{ for all } v_{\tilde{h}} \in \mathcal{V}_{\tilde{h}}. \tag{18}$$

The bilinear form $a_T(\cdot, \cdot)$ is the tangent approximation of $a(\cdot, \cdot)$ around u_h and therefore it depends on the approximate solution u_h. Indeed, Eqs. (17) and (18) can be recovered by approximating

$$a(u, v) = a(u_h + e, v) \approx a(u_h, v) + a_T(e, v).$$

Since $a(\cdot, \cdot)$ is nonlinear $a_T(\cdot, \cdot)$ must depend on u_h.
In this case:
 1. The norm induced by $a_T(\cdot, \cdot)$ can be taken as an energy norm and it can be used to measure the error.
 2. Projections of e_r on any subspace of $\mathcal{V}_{\tilde{h}}$ can be computed using Eq. (18).
 3. The residual problem characterizing the reference error can be posed at each step in order to approximate $\hat{\mathbf{e}}_r^k := \boldsymbol{u}_{\tilde{h}}^k - \tilde{\boldsymbol{u}}_h^k$.

In this context the generalization of the error estimator defined in [1] is straightforward. The family of subspaces V_k generated by the elementary submeshes and the family of subspaces U_l generated by the patch-submeshes can be used in the same way.

REMARK It is important to notice that in Eqs. (16) and (18) the left hand side is associated with the tangent approximation. However, the computation of the right-hand side, i.e. the residual, requires to obtain $\tilde{\mathbf{f}}_{\text{int}}(\tilde{\boldsymbol{u}}_h)$, the vector of internal forces associated with the approximation u_h. As it is indicated in the next section, this fact makes the computation of residual in the nonlinear estimator quite different from the linear case.

REMARK The error estimation at each step is independent of the previous steps because it is based on the observed disequilibrium of the approximate solution u_h^k. At the end of step k the estimator approximates the error in the whole present solution u_h^k, and not only the error in the last increment $\Delta u_h^k := u_h^k - u_h^{k-1}$.

5   COMPUTATIONAL ASPECTS

The residual term of Eq. (16) needs the computation of $\tilde{\mathbf{f}}_{\text{int}}(u_h)$. In the linear case this can be directly computed from $\tilde{\boldsymbol{u}}_h$ (see Eq. (9)). In a nonlinear case $\tilde{\mathbf{f}}_{\text{int}}(\boldsymbol{u}_h)$ is not a linear function of $\tilde{\boldsymbol{u}}_h$ and it must be computed differently.

In the nonlinear case, the stresses cannot be computed having only the displacements u. Therefore, something else than u is needed to test equilibrium. The incremental process described in Eq. (7) is used to update the stresses after small load increments. These load increments correspond to small variations of the displacements. The constitutive relation depends on the state of the material which is represented by the history variables denoted by $\boldsymbol{\alpha}$. In the examples the employed history variable is the maximum plastic strain.

In a practical implementation, the approximate solution u_h has associated stresses $\boldsymbol{\sigma}(u_h)$ and history variables $\boldsymbol{\alpha}(u_h)$ which are known in the integration points (Gauss points) of the mesh generating $\mathcal{V}_h$.

The internal force vector $\tilde{\mathbf{f}}_{\text{int}}(u_h)$ is computed by integrating the approximate stress field $\boldsymbol{\sigma}(u_h)$ (recall Eq. (5)). Therefore, the values of $\boldsymbol{\sigma}(u_h)$ in the integration points of the reference mesh generating $\mathcal{V}_{\bar{h}}$ are needed. The computation of the components of the tangent matrix $\tilde{\mathbf{K}}_T(u_h)$ needs also the values of $\boldsymbol{\alpha}(u_h)$ in these integration points.

Obviously these values are not known. In the linear case the values of $\boldsymbol{\sigma}(u_h)$ in the integration points of the reference mesh are recovered by simple derivation of the approximate displacement field u_h or its representation over the reference mesh, $\tilde{\boldsymbol{u}}_h$. In the nonlinear case, the only available data concerning $\boldsymbol{\sigma}(u_h)$ and $\boldsymbol{\alpha}(u_h)$ are their values on the integration points of the computational mesh (the "coarse" mesh).

Then, in order to perform the error estimation in nonlinear problems the determination of $\tilde{\boldsymbol{u}}_h$ is irrelevant. In fact, the values of $\boldsymbol{\sigma}(u_h)$ and $\boldsymbol{\alpha}(u_h)$ in the integration points of the reference mesh must be computed using an interpolation technique.

The residue has two components. The first one is associated with the interior of the elements and indicates the sharpness of the stresses field. The second one, associated with the edges, indicates the discontinuity of the stresses across them (it is also called stress jump or singular residual: Kelly, 1984) Therefore, the interpolation technique must preserve both of them. An element-by-element fitting of $\boldsymbol{\sigma}(u_h)$ and $\boldsymbol{\alpha}(u_h)$ is a suitable option. It is worth remarking that the selection of the interpolation method depends essentially on the number of integration points. For instance, let us suppose that quadrilateral elements with four integration points are used.

In this case the values of any component of $\boldsymbol{\sigma}(u_h)$ or $\boldsymbol{\alpha}(u_h)$ in the four integration points correspond to four nodal values at the corner nodes of the element in a bilinear interpolation. Even if the element has other nodes, for instance if height-nodes elements are used, these are not considered. Once the nodal values are found, the values at the integration points of the reference mesh are obtained by simple interpolation.

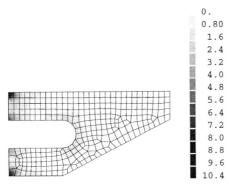

Figure 1   Console under uniform load: error distribution for the initial mesh.

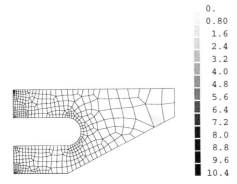

Figure 2   Console under uniform load: final mesh and associated error distribution.

6   EXAMPLES

In order to test the performance of the estimator introduced previously two examples are studied. Both examples have material and geometrical nonlinearities. The material nonlinear behavior is described by an elastoplastic law with isotropic hardening, and a large strain formulation is considered to accurately describe the continuum motion.

First an example taken from Gallimard et al [4] is reproduced. The response of a console under a uniform load is studied, see Figure 1. The goal in this analysis is to obtain a solution with a ratio: error norm with respect to solution norm, under 10%. The initial mesh with 298 elements 8-noded elements has an error 17% of the energy of the solution, see Figure 1 for the error distribution. Given this distribution and the optimality criterion presented by Li and Bettess [6], [7] an adaptive mesh refinement procedure is initiated. The adaptive procedure leads after three steps to a mesh with 473 elements that has an admissible error: 6.13%, see Figure 2 for the final error distribution.

Note that the error distribution in the final mesh is practically uniform. This is because the Li and Bettess remeshing criterion seeks for an equally distributed elementary error.

Second, a test due to Zienkiewicz and Huang [9] is reproduced. This example illustrates the formation of shear bands in a compression of a uniform plane strain specimen with a circular opening in the center.

The bilinear model employed is parameterized by the Young modulus of the elastic branch, E_e, and the Young modulus of the plastic branch E_p. A Von Mises yield criterion is employed. And the same test is carried out for different values of the ratio E_p/E_e.

If E_p/E_e is taken equal to 0.1 localization of the deformation does not appear. However, when E_p/E_e is reduced to 10^{-3}, deformation clearly localizes. Figure 3 shows for this material the initial mesh of 255 8-noded quadrilateral elements, the estimated error distribution and the computed plastic strain distribution. For this first mesh, the energy norm of the error is 4% of the energy of the solution. This is larger than an admissible error of 1% and, therefore, the error estimation is used in an adaptive procedure leading to an acceptable mesh. Again, the remeshing criterion presented in [5], [6] is used.

After three steps of the adaptive procedure (estimation and refinement) an acceptable mesh is obtained. Figure 4 presents for this last case, the discretization, the error distribution and the plastic strain distribution. The ratio error over total energy is now lower than the permissible one (i.e., 0.98% ≤ 1%) and therefore the acceptability criterion is satisfied.

The discretization clearly depicts the shear band just developed: the mesh is refined along the shear band. Thus the error estimator is able to capture the large variations in the solution.

It is important to notice that the use of a norm based on the tangent matrix precludes any case with softening: the tangent matrix must be symmetric and positive definite. On the other hand, the use of the natural (tangent) norm presents an important computational advantage in this problem. The local error norm along the plastic zones gives lower values than a local error norm based on the elastic or secant modulus because the plastic modulus E_p is clearly below the elastic or secant modulus. Therefore, the error distribution is not over-estimated and no excessive refinement is performed along the band.

The same process is carried out for a ratio E_p/E_e of 10^{-10} starting from the initial mesh of Figure 3. Figure 5 presents a zoom near the central hole of the plastic deformation in three meshes (not including the initial one due to the large errors in this zone) of the adaptive procedure. The width of the shear band clearly decreases and the maximum plastic deformation increases (over 100%) as the mesh is refined.

This results were expected, because a ratio E_p/E_e equal to 10^{-10} (much smaller than the precision of the nonlinear computation) reproduces, in fact, a virtually ideal plastic model. This non-regularized plastic problem shows mesh dependence, the plastic deformation is concentrated on the critical elements as the mesh is refined. Therefore, an increase in the number of elements and a decrease in the element size in the plastic zone does not allow to reduce the error. The two last meshes in Figure 5 are, respectively, zooms of discretizations with 775 and 857 8-noded elements with percentages of error that do not decrease under 1.5%.

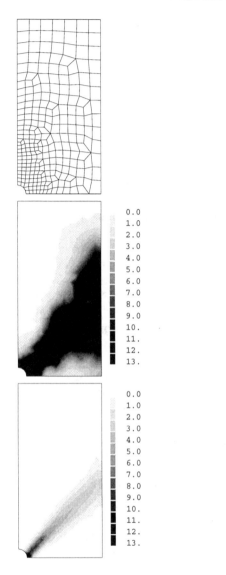 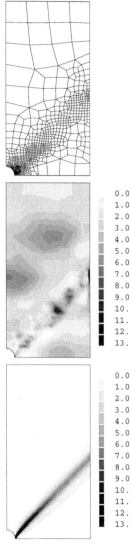

Figure 3 Initial mesh, distribution of local error and % of computed plastic strain (from top to bottom). The ratio E_p/E_e is equal to 10^{-3}.

Figure 4 Final mesh, distribution of local error and the computed % plastic strain (from top to bottom). The ratio E_p/E_e is equal to 10^{-3}.

7   CONCLUSIONS

In this work a straightforward extension of a linear estimator [3] is presented. This residual estimator does not require to compute the flux jumps across inter-element edges because it is based on a multi-projection strategy over orthogonal functional spaces. A natural energy norm is employed for measuring the error.

The generalization is based in a tangent expansion of the weak equation. There-

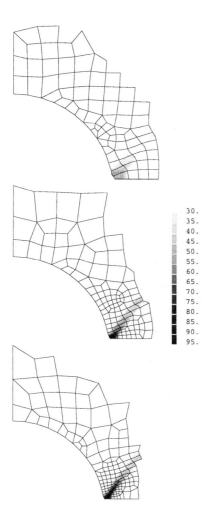

Figure 5 Plastic deformation around the hole for the successive meshes in the adaptive procedure (E_p/E_e equal to 10^{-10}).

fore, only nonlinear problems with symmetric positive definite tangent matrices are considered. In this case all the properties: convergence (Bessel's inequality) and underestimation of the global error, are also encountered in the nonlinear problems. The tangent bilinear form is used both in the residual equation and as a norm allowing to measure the error with a physical interpretation. Thus standard finite element techniques can be generalized for all the local problems that must be solved.

Finally, the estimator has been used in adaptive procedures for nonlinear analysis with excellent success. Even in problems where shear bands are formed the estimator predicts the expected results.

REFERENCES

1. Babuška, I. and C. Rheinboldt, *A-posteriori error estimates for the finite element method*, Internat. J. Numer. Methods Engrg. **12** (1978), 1597–1615.
2. Babuška, I., T. Strouboulis and C.S. Upadhyay, *A model study of a posteriori error estimators for linear elliptic problems. Error estimation in the interior of patchwise uniform grids of triangles*, Comput. Methods Appl. Mech. Engrg. **114** (1994), 307–378.
3. Díez, P., J.J. Egozcue, and A. Huerta, *A posteriori error estimation for standard finite element analysis*, Int. Report n° 75, Int. Center for Numerical Methods in Engrg., Barcelona, 1995; also submitted to Comput. Methods Appl. Mech. Engrg.
4. Gallimard, L., P. Ladevèze and J.P. Pelle, *Error estimation and adaptivity in elastoplasticity*, Internat. J. Numer. Methods Engrg. **39** (1996), 189–217.
5. Li, L.-Y., P. Bettess, W. Bull, T. Bond and I. Applegarth, *Theoretical formulations for adaptive finite element computations*, Comm. Numer. Methods Engrg. **11** (1995), 857–868.
6. Li, L.-Y. and P. Bettess, *Notes on mesh optimal criteria in adaptive finite element computations*, Comm. Numer. Methods Engrg. **11** (1995), 911–915.
7. Zhu, J.Z., E. Hinton and O.C. Zienkiewicz, *Mesh enrichment again mesh regeneration using quadrilateral elements*, Comm. Numer. Methods Engrg. **9** (1993), 547–554.
8. Zienkiewicz, O.C. and J.Z. Zhu, *A simple error estimator and adaptive procedure for practical engineering analysis*, Internat. J. Numer. Methods Engrg. **24** (1987), 337–357.
9. Zienkiewicz, O.C. and G.C. Huang, *A note on localization phenomena and adaptive finite-element analysis in forming processes*, Comm. Appl. Numer. Methods **26** (1990), 71–76.

Superconvergent recovery operators: Derivative recovery techniques

A. M. LAKHANY and J. R. WHITEMAN   BICOM, Institute of Computational Mathematics, Brunel University, Uxbridge, Middlesex UB8 3PH, The United Kingdom

Abstract   To this date many gradient recovery techniques, which involve some post-processing of the gradient of the piecewise polynomial finite element solution for linear elliptic problems on planar domains, have been published. In many cases the effect of such a post-processing is to yield a pair of piecewise polynomial, possibly continuous, functions which approximate the gradient of the weak solution to an $O(h)$ accuracy higher than the gradient of the finite element solution. In particular, for the case of piecewise linear finite element approximation, such a post-processing will result in a vector $\nabla^R u_h$, piecewise linear in both components, with the $O(h^2)$ accuracy compared to just $O(h)$ accuracy of the usual gradient of the finite element solution under sufficient regularity of the weak solution of the given boundary value problem. In the present paper we study the approximation properties of the local derivatives of the recovered gradient functions. It is shown that these derivatives are $O(h)$ approximations to the corresponding second derivatives of the weak solution when sufficient regularity of the weak solution is assumed. We also show that the Midpoint Recovered Gradient $\nabla^M u_h$, is equal to the gradient of some underlying quadratic u_h^M over each element; thereby allowing us to compute the second order derivatives of u_h^M locally over each element. These second order derivatives are $O(h)$ approximations to the corresponding second order derivatives of the weak solution in a discrete norm. We next repeat the recovery technique and obtain superconvergent second order derivatives under sufficient regularity assumptions. These recovered second order derivatives are differentiated once again to yield approximations to the third order derivatives of the weak solution over each element. We end this paper by supporting our study by means of a numerical example.

1   INTRODUCTION

The term gradient superconvergence in the finite element method was introduced approximately a quarter of century ago (cf. Strang and Fix [17] and Zlámal [23], [24]), primarily, as a natural phenomenon wherein the gradient of the finite element solution of a boundary value problem possess exceptional accuracy at some points in the domain. These points have been called stress points in the literature. In particular, for a two dimensional linear elliptic problem or system, with a regular mesh of isosceles right angled triangles over a rectangular domain, it is well known that, under conditions of sufficient regularity of the solution of the problem, at the midpoints of element sides, the tangential derivatives of a piecewise linear finite element approximation have $O(h^2)$ accuracy. If the normal derivatives appearing from the two elements sharing a common edge are averaged (or extrapolated in the case when the element side falls on the boundary of the domain), then it can be shown that the resulting derivative is superconvergent. This is the idea behind the Midpoint Recovery Operator described in Levine [11], Goodsell and Whiteman [6] and Goodsell [5]. Another common recovery operator is the Vertex Recovery Operator, based on the averaging at vertices (cf. §2, Křížek and Neittaanmäki [8], [9], Hlaváček and Křížek [7] and references therein). Superconvergence has also been obtained through Richardson Extrapolation by various authors. The technique requires asymptotic expansion of the discretization error in the finite element solution (cf. Blum *et al* [2], Lin and Xu [13], Lin and Zhu [14], Lin and Lu [12]). More recently the Patch Recovery Schemes of Zienkiewicz and Zhu (Zienkiewicz and Zhu [20], Wiberg and Abdul Wahab [19]) have gained considerable attention, especially because of their use in the "Z^2" estimator of Zienkiewicz and Zhu [21].

The structure of this paper is as follows: we shall mainly concern ourselves with the Midpoint (∇^M) and the Vertex (∇^V) Recovery Operators, described in §2. When the piecewise linear finite element method is used to solve the Poisson Problem over rectangular domain Ω triangulated using regular right angled isosceles elements $\tau_k \in T^h$, we have the following error estimate (cf. Goodsell and Whiteman [6] and Křížek and Neittaanmäki [8], [9]):

$$\left\| \nabla u - \nabla^R u_h \right\|_{L^2(\Omega)} \leq C\, h^2\, |u|_{H^3(\Omega)} \qquad R \equiv M, V. \tag{1.1}$$

We specifically note that the recovered gradient function $\nabla^R u_h$ is superior to the usual gradient of the piecewise linear finite element approximation over T^h, ∇u_h, in the following respect:

- $\nabla^R u_h$ shows better convergence than ∇u_h as seen from the superconvergence error estimate (1.1), and
- furthermore, over any element τ_k, $\nabla^R u_h \big|_{\tau_k}$ is a linear function, whereas $\nabla u_h \big|_{\tau_k}$ is only a constant.

The aforementioned properties allow us to take derivatives of the recovered gradient function locally over any element τ_k and think of these derivatives as "good approximations" to the corresponding second derivatives of the solution of the given boundary value problem in view of (1.1). In §3 we shall prove this property for the local derivatives of the recovered gradients $\nabla^M u_h$ and $\nabla^V u_h$.

It can be shown using the formula for the midpoint recovery (cf. §2) that the recovered gradient $\nabla^M u_h$ is, in fact, the gradient of some underlying quadratic function locally over each element. Thus we straightaway have for any element τ_k

$$\nabla^M u_h|_{\tau_k} = \nabla u_h^M|_{\tau_k}, \tag{1.2}$$

which essentially means that we can compute local second order derivatives of the underlying quadratic function $u_h^M \in \Pi_2/\Pi_0$ where Π_m represents the space of all polynomials of degree less than or equal to m. This property of the Midpoint Recovered Gradient gives a convenient setting for repeating the argument of Midpoint Recovery on the computed second order derivatives of u_h^M thereby obtaining the Recovered Second Order Derivatives when sufficient regularity of the weak solution is assumed. In §4 we prove the superconvergence properties of the Recovered Second Order Derivatives. Furthermore in §5 we compute the approximations to the third order derivatives of the weak solution. Finally in §6 we support our study by numerical results.

Before leaving this section we would like to emphasize that in order to prove superconvergence results for the recovered derivatives, we have assumed high regularity for the solution of our model boundary value problem introduced in the next section. This, obviously, cannot be guaranteed in general. In view of this we emphasize upon the importance of a local analysis, such as the one appearing in Wheeler and Whiteman [18], to extend the results appearing in the present paper to subdomains of the domain on which the original problem is defined. Finally, we also mention the fact that in the analysis to follow we have ignored all errors due to cubature, which in practice can be guaranteed by adopting a sufficiently accurate cubature scheme.

2   NOTATIONS AND CLASSICAL RESULTS

Throughout this paper we use $H^{s,p}(\Omega)$ to denote the general Sobolev spaces, and the notation $H^s(\Omega)$ for the Hilbert Space $H^{s,2}(\Omega)$. We recall that the norm and the semi-norm on the Sobolev Spaces $H^{m,p}(\Omega)$ are given respectively by:

$$\|u\|_{H^{m,p}(\Omega)} = \begin{cases} \left(\sum_{0 \le |\alpha| \le m} \|D^\alpha u\|_{L^p(\Omega)}^p \right)^{1/p} & 1 \le p < \infty, \\ \max_{0 \le |\alpha| \le m} \|D^\alpha u\|_{L^\infty(\Omega)} & p = \infty, \end{cases} \tag{2.1}$$

and

$$|u|_{H^{m,p}(\Omega)} = \begin{cases} \left(\sum_{0 \le |\alpha| = m} \|D^\alpha u\|_{L^p(\Omega)}^p \right)^{1/p} & 1 \le p < \infty, \\ \max_{0 \le |\alpha| = m} \|D^\alpha u\|_{L^\infty(\Omega)} & p = \infty, \end{cases} \tag{2.2}$$

wherein the derivative $D^\alpha u$ is to be taken in the weak sense. The space $L^p(\Omega)$ is the usual Lebesgue Space of equivalence classes of p-integrable measurable functions on

Ω with the norm given by:

$$\|u\|_{L^p(\Omega)} = \begin{cases} \left(\displaystyle\int_\Omega |u|^p \, dx\right)^{1/p} & 1 \le p < \infty, \\[2ex] \operatorname{ess\,sup}_\Omega |u| & p = \infty. \end{cases} \tag{2.3}$$

We also let $H_0^s(\Omega)$ be the closure of $C_0^\infty(\Omega)$ in the norm of $H^s(\Omega)$. We now consider the model Poisson problem in two dimensions with solution $u(x)$ such that

$$\begin{aligned} -\Delta u(x) &= f(x) & x \equiv (x, y) \in \Omega, \\ u(x) &= 0 & x \equiv (x, y) \in \partial\Omega, \end{aligned} \tag{2.4}$$

defined over a rectangular domain Ω in $\mathbb{R}^2$ with boundary $\partial\Omega$. The weak form of (2.4) is that in which we seek $u \in H_0^1(\Omega)$ such that

$$a(u, v) = \langle f, v \rangle \qquad \forall v \in H_0^1(\Omega), \tag{2.5}$$

where $a(u, v) \equiv \int_\Omega \nabla u \cdot \nabla v \, dx$ and $\langle f, v \rangle \equiv \int_\Omega f v \, dx$. The bilinear form $a(\cdot, \cdot)$ is symmetric and elliptic in $H_0^1(\Omega)$. It is also continuous, as is the linear form $f(\cdot) = \langle f, \cdot \rangle$, on $H_0^1(\Omega)$. We assume that $f(x)$ has any required regularity and that $u \in H^{s,p}(\Omega) \cap H_0^1(\Omega)$ for some $s \ge 3$ and $2 \le p \le \infty$ as required later. We overlay Ω with a completely regular right-angled isosceles triangulation T^h of mesh size h ($0 < h \le h_0 < 1$), such that every element in T^h has as its set of vertices either the set $\{(p, q), (p + h, q), (p, q + h)\}$ or the set $\{(p + h, q + h), (p, q + h), (p + h, q)\}$ for some real numbers p, q. We define the finite dimensional subspace $S^h \subset H_0^1(\Omega)$ consisting of piecewise linear polynomials defined over the partition T^h. For the finite element approximation to the solution of (2.5) we seek $u_h \in S^h$ such that

$$a(u_h, v_h) = \langle f, u_h \rangle \qquad \forall v_h \in S^h. \tag{2.6}$$

We shall represent by $\widehat{\Omega}$ the image of Ω under an affine mapping which transforms the partition T^h of Ω, with mesh size h, into the partition $\hat{T}^1$ of $\widehat{\Omega}$, with mesh size unity. The modified functions and operators in $\widehat{\Omega}$ will be indicated by use of a hat ($\hat{\ }$).

Let V represent the set of all vertices in the triangulation T^h, and $V^\circ \subset V$ be the set of vertices not falling on the boundary $\partial\Omega$. Corresponding to any vertex $\nu_j \in V^\circ$ there is a six element set T_j each element of which shares this common point. We recover the gradient at such a vertex by:

$$\nabla^V u_h|_{\nu_j} \equiv \frac{1}{6} \sum_{\tau_k \in T_j} \nabla u_h|_{\tau_k}. \tag{2.7}$$

For $\nu_j \in V - V^\circ$ we need to modify the scheme given in (2.7) by taking specific weighted averages in groups of elements near and including the boundary point (cf. Hlaváček and Křížek [7]).

Now, let E represent the set of all element edges in the triangulation T^h, and $E^\circ \subset E$ be the set of edges not falling on the boundary $\partial\Omega$. Then if $m_{k,l}$ represents

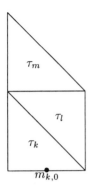

Figure 1

the midpoint of an edge in E° shared by the two elements τ_k and τ_l, we recover the gradient at such a midpoint using the formula:

$$\nabla^M u_h|_{m_{k,l}} = \frac{1}{2}\nabla u_h|_{\tau_k} + \frac{1}{2}\nabla u_h|_{\tau_l}. \qquad (2.8)$$

For midpoints of the element edges falling on the boundary $\partial\Omega$ the averaging scheme (2.8) cannot be used. Thus with τ_k, τ_l and τ_m appearing as in Fig. 1, the recovered gradient at point $m_{k,0}$ on the boundary is given by:

$$\nabla^M u_h|_{m_{k,0}} = \nabla u_h|_{\tau_k} + \frac{1}{2}\nabla u_h|_{\tau_l} - \frac{1}{2}\nabla u_h|_{\tau_m}. \qquad (2.9)$$

We emphasize that in what follows we shall implicitly assume that the mesh over Ω is a completely regular isosceles right angled triangular mesh.

3   DERIVATIVES OF THE RECOVERED GRADIENTS

We recall from our previous discussion that each component of the generic recovered gradient function $\nabla^R u_h(x,y)$ is a linear function over each element τ_k, i.e., $\nabla^R u_h|_{\tau_k} \in \Pi_1(\tau_k) \times \Pi_1(\tau_k)$. This essentially means that the recovered gradient function over each element has only a symbolic interpretation – it is not a gradient in the true sense; i.e., it is not the gradient of an underlying quadratic function. Technically, therefore, we cannot think of the further derivatives of the recovered gradient function as second order derivatives of any function – not even locally. However, we can show that the derivatives of the recovered gradient functions $\nabla^R u_h|_{\tau_k}$ are good approximations to the corresponding second order derivatives of u over τ_k. In order to quantify this statement we introduce some symbols. In the present work we shall, for the sake of convenience, denote $\partial_x \equiv D^{(1,0)}$, $\partial_y \equiv D^{(0,1)}$, $\partial_{xx} \equiv D^{(2,0)}$, $\partial_{xy} \equiv D^{(1,1)}$, $\partial_{yy} \equiv D^{(0,2)}$, etc. In view of these notations we shall represent the components of the recovered gradient functions $\nabla^R u_h$ by $\partial_x^R u_h = \mathbf{i}\cdot\nabla^R u_h$ and $\partial_y^R u_h = \mathbf{j}\cdot\nabla^R u_h$. We further introduce the tensor ∇^2 which is given by the indeterminate vector product $\nabla\nabla$ and which can be represented by the matrix

$$\nabla\nabla \equiv \begin{bmatrix} \partial_{xx} & \partial_{xy} \\ \partial_{yx} & \partial_{yy} \end{bmatrix}.$$

With these notations we now consider the convergence properties of the symbolic error norm:

$$\left\|\nabla\nabla u - \nabla\nabla^R u_h\right\|_{L^2(\Omega)}^h \equiv \sqrt{\sum_{k=1}^{N}\left\|\nabla\nabla u - \nabla\nabla^R u_h\right\|_{L^2(\tau_k)}^2}, \qquad (3.1)$$

where

$$\left\|\nabla\nabla u - \nabla\nabla^R u_h\right\|_{L^2(\tau_k)}^2 \equiv \int_{\tau_k} \left(\partial_{xx}u - \partial_x\partial_x^R u_h\right)^2 + \left(\partial_{xy}u - \partial_x\partial_y^R u_h\right)^2$$
$$+ \left(\partial_{yx}u - \partial_y\partial_x^R u_h\right)^2 + \left(\partial_{yy}u - \partial_y\partial_y^R u_h\right)^2 d\boldsymbol{x}.$$

LEMMA 3.1 If $u \in H_0^1(\Omega) \cap H^3(\Omega)$ is the weak solution of the boundary value problem (2.4) and u_h be its piecewise linear finite element approximation over T^h, then the following estimate holds:

$$\left\|\nabla\nabla u - \nabla\nabla^R u_h\right\|_{L^2(\Omega)}^h \leq Ch|u|_{H^3(\Omega)},$$

where the symbolic error norm on the left hand side is as defined in (3.1).

Proof: We first note that for $\hat{u}$ quadratic

$$\left.\widehat{\nabla}\widehat{\nabla}\hat{u}\right|_{\hat{\tau}_k} = \left.\widehat{\nabla}\widehat{\nabla}^R\hat{\Pi}\hat{u}\right|_{\hat{\tau}_k}.$$

Furthermore since $\left.\widehat{\nabla}^R\hat{\Pi}\hat{u}\right|_{\hat{\tau}_k}$ is a linear function over any $\hat{\tau}_k$, we have on making use of an inverse estimate and the local superconvergence property of $\widehat{\nabla}^R\hat{\Pi}\hat{u}$ (cf. Goodsell and Whiteman [6] and Křížek and Neittaanmäki [8])

$$\left\|\widehat{\nabla}\widehat{\nabla}\hat{u} - \widehat{\nabla}\widehat{\nabla}^R\hat{\Pi}\hat{u}\right\|_{L^2(\hat{\tau}_k)} \leq \left\|\widehat{\nabla}\widehat{\nabla}\hat{u}\right\|_{L^2(\hat{\tau}_k)} + \left\|\widehat{\nabla}\widehat{\nabla}^R\hat{\Pi}\hat{u}\right\|_{L^2(\hat{\tau}_k)}$$
$$\leq C\left\{|\hat{u}|_{H^2(\hat{\tau}_k)} + \left\|\widehat{\nabla}^R\hat{\Pi}\hat{u}\right\|_{L^2(\hat{\tau}_k)}\right\}$$
$$\leq C\left\{\|\hat{u}\|_{H^2(\hat{\tau}_k)} + \left\|\widehat{\nabla}^R\hat{\Pi}\hat{u} - \widehat{\nabla}\hat{u}\right\|_{L^2(\hat{\tau}_k)}\right\}$$
$$\leq C\|\hat{u}\|_{H^3\left(\hat{\tau}_k^{(1)}\right)},$$

where $\hat{\tau}_k^{(1)}$ denotes the union of all elements $\hat{\tau}_j$ which are involved in gradient recovery over the element $\hat{\tau}_k$.

The aforementioned facts can now be combined with standard arguments including the use of the Bramble–Hilbert Lemma (cf. Bramble and Hilbert [3], Goodsell and Whiteman [6] and Křížek and Neittaanmäki [8]) to prove that

$$\left\|\nabla\nabla u - \nabla\nabla^R\Pi u\right\|_{L^2(\Omega)}^h \leq Ch|u|_{H^3(\Omega)}.$$

Again, making use of an inverse estimate and the Oganesyan–Rukhovets Lemma (cf. Oganesyan and Rukhovets [15]) one can also show that

$$\left\|\nabla\nabla^R\Pi u - \nabla\nabla^R u_h\right\|_{L^2(\Omega)}^h \le Ch|u|_{H^3(\Omega)}.$$

The proof of Lemma 3.1 follows on combining the two previous estimates.   □

We are interested in computing the approximations to the weak second derivatives of u. To enable us to do so we introduce the symmetric tensor

$$\nabla^*\nabla^R u_h \equiv \frac{1}{2}\left(\nabla\nabla^R u_h + \left[\nabla\nabla^R u_h\right]^T\right) \tag{3.2}$$

which inherits the following approximation property of $\nabla\nabla^R u_h$:

LEMMA 3.2   Under the condition of Lemma 3.1

$$\left\|\nabla\nabla u - \nabla^*\nabla^R u_h\right\|_{L^2(\Omega)}^h \le Ch|u|_{H^3(\Omega)}.$$

Proof: The proof follows immediately from Lemma 3.1 and definition (3.2).   □

Having described the convergence properties of the derivatives of the generic recovered gradients in Lemma 3.1 and in Lemma 3.2, we now consider the special cases of the Vertex and the Midpoint Recovered Gradient functions introduced in §2.

Firstly, since $\nabla^V u_h$ is continuous the symbolic error norm in (3.2) is actually an error norm and we have the superconvergence estimate

$$\left\|(h\nabla)^j\left[\nabla u - \nabla^V u_h\right]\right\|_{L^2(\Omega)} \le Ch^2|u|_{H^3(\Omega)}, \qquad j = 0, 1 \tag{3.3}$$

or

$$\left\|\nabla u - \nabla^V u_h\right\|_{L^2(\Omega)} + h\left\|\nabla^*\left[\nabla u - \nabla^V u_h\right]\right\|_{L^2(\Omega)} \le Ch^2|u|_{H^3(\Omega)}, \tag{3.4}$$

where $\nabla^*\nabla^V$ is as defined in (3.2) and in view of this definition we have $\nabla^*\nabla u \equiv \nabla\nabla u$.

On the other hand the Midpoint Recovered Gradient $\nabla^M u_h$ has the following interesting property:

LEMMA 3.3   Let $\nabla^M u_h$ be the Midpoint Recovered Gradient, then over any element $\tau_k \in T^h$, we have

$$\nabla^M u_h\big|_{\tau_k} = \nabla u_h^M\big|_{\tau_k}$$

for some quadratic function u_h^M in the quotient space Π_2/Π_0.

Proof: The proof is very simple and straightforward and hence omitted for the sake of brevity.   □

COROLLARY 3.4   Under the conditions of Lemma 3.1 we have

$$\left\|\nabla\nabla(u - u_h^M)\right\|_{L^2(\Omega)}^h \le Ch|u|_{H^3(\Omega)}. \tag{3.5}$$

Proof: Follows immediately from the fact that

$$\nabla\nabla^M u_h\big|_{\tau_k} = \nabla\nabla u_h^M\big|_{\tau_k}$$

for some underlying quadratic u_h^M as in the argument of Lemma 3.3.   □

From Lemma 3.3 we see that the Midpoint Recovered Gradient, over any element τ_k, is in fact the gradient of some underlying quadratic – an extremely desirable property. We note, of course, that the second order derivatives obtained from the Midpoint Recovered Gradient have only local significance and we would like to give them global meaning. In order to do so we consider these second order derivatives as piecewise constant functions, and denote them by $\partial_{xx}^M u_h$, $\partial_{xy}^M u_h$ and $\partial_{yy}^M u_h$ in accordance with the symbols used for the recovered gradients. Furthermore, for the sake of simplicity, we shall also use the symbols $\partial_{xx,\tau_k}^M u_h$, etc. and $\partial_{xx,P}^M u_h$ etc. to denote these derivatives over the element τ_k and at point P respectively.

4   SUPERCONVERGENT RECOVERY OF SECOND ORDER DERIVATIVES

In this section we shall make use of the midpoint recovery technique, similar to the one employed to recover the gradient of the finite element solution, to recover the second order derivatives $\partial_{xx}^M u_h$, $\partial_{xy}^M u_h$ and $\partial_{yy}^M u_h$. Our aim is to produce (discontinuous) piecewise linear functions $\partial_{xx}^{MM} u_h(x)$, $\partial_{xy}^{MM} u_h(x)$ and $\partial_{yy}^{MM} u_h(x)$ so that the following estimate holds:

THEOREM 4.1   For $u \in H^{4,\infty}(\Omega)$ and Ω a rectangle, the following estimate holds:

$$|u - u_h|_{H^2(\Omega)}^{MM} \leq C\, h^2 \log^2\left(\frac{1}{h}\right) |u|_{H^{4,\infty}(\Omega)}, \qquad (4.1)$$

where the error semi-norm on the left hand side of (4.1) is given by:

$$\left\{|u - u_h|_{H^2(\Omega)}^{MM}\right\}^2 = \int_\Omega \left[\left(\partial_{xx}u(x) - \partial_{xx}^{MM}u_h(x)\right)^2 + \left(\partial_{xy}u(x) - \partial_{xy}^{MM}u_h(x)\right)^2\right.$$
$$\left. + \left(\partial_{yy}u(x) - \partial_{yy}^{MM}u_h(x)\right)^2\right] dx. \qquad (4.2)$$

Before attempting to prove this theorem we need to define the recovered second order derivatives. Let $T_{\partial\Omega}^h$ represent the collection of all elements sharing at least one edge with the boundary $\partial\Omega$. Let $E^{\circ\circ} \subset E$ be the set of all edges which are not being shared by the elements in $T_{\partial\Omega}^h$. Then if $m_{k,l}$ is the midpoint of an element edge in $E^{\circ\circ}$ shared commonly by elements τ_k and τ_l, we recover second order derivatives at such a midpoint using the formula:

$$\partial_{x_1 x_2, m_{k,l}}^{MM} u_h = \frac{1}{2}\left\{\partial_{x_1 x_2, \tau_k}^M u_h + \partial_{x_1 x_2, \tau_l}^M u_h\right\}, \quad (x_1, x_2) \equiv (x, x), (x, y), (y, y). \quad (4.3)$$

For edges falling in the set $E - E^{\circ\circ}$ things become complicated, primarily due to the fact that the points of exceptional accuracy of $\partial_{xx,\tau_k}^M \Pi u$ and $\partial_{yy,\tau_k}^M \Pi u$ change

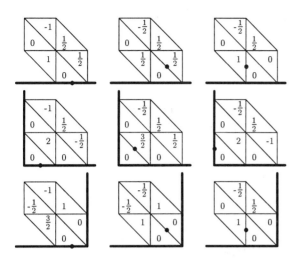

● represents midpoint under consideration

— represents boundary $\partial\Omega$

Figure 2

for any element, τ_k, adjacent to the boundary. We can, however, obtain formulæ for the recovery at the midpoint of the edges in $E - E^{\circ\circ}$ by using a local weighted averaging scheme that only involves elements not included in $T^h_{\partial\Omega}$. Such a scheme is shown in Fig. 2.

We now turn to the proof of Theorem 4.1. Let $\hat{\tau}_k \in \hat{T}^1$ be any element with element edge midpoints $\hat{m}^{(k)}_{j_1}$, $\hat{m}^{(k)}_{j_2}$ and $\hat{m}^{(k)}_{j_3}$. For any j_i we denote by $\hat{\tau}^{(1)}_{k,\hat{m}^{(k)}_{j_i}}$ the collection of all the elements involved in the recovery of the second order derivatives at the midpoint $\hat{m}^{(k)}_{j_i}, (i = 1, 2, 3)$ and by $\hat{\tau}^{(2)}_{k,\hat{m}^{(k)}_{j_i}}$ the union of elements involved in the gradient recovery over the elements included in the region $\hat{\tau}^{(1)}_{k,\hat{m}^{(k)}_{j_i}}$. We then set

$$\hat{\tau}^*_k = \bigcup_{i=1}^3 \hat{\tau}^{(1)}_{k,\hat{m}^{(k)}_{j_i}},$$

$$\hat{\tau}^{(2)}_k = \bigcup_{i=1}^3 \hat{\tau}^{(2)}_{k,\hat{m}^{(k)}_{j_i}}.$$
(4.4)

We now set ourselves to proving Lemma 4.1.

LEMMA 4.1   For any $\hat{\tau}_k \in \hat{T}^1$ and $\hat{u} \in H^4\left(\hat{\Omega}\right)$, we have:

$$\left|\hat{u} - \widehat{\Pi}\hat{u}\right|^{MM}_{H^2(\hat{\tau}_k)} \leq C\left\|\hat{u}\right\|_{H^4\left(\hat{\tau}^{(2)}_k\right)}.$$
(4.5)

Proof: Let us analyze a single integral on the left hand side of (4.5), i.e.,

$$I_1^2 = \int_{\hat{\tau}_k} \left| \hat{\partial}_{\hat{x}\hat{x}} \hat{u} - \hat{\partial}_{\hat{x}\hat{x},\hat{\tau}_k}^{MM} \widehat{\Pi} \hat{u} \right|^2 d\hat{x}\, d\hat{y}.$$

Without loss of generality we assume that element $\hat{\tau}_k$ has edge midpoints $\hat{m}_{j_1}^{(k)}(p + 1/2, q)$, $\hat{m}_{j_2}^{(k)}(p + 1/2, q + 1/2)$ and $\hat{m}_{j_3}^{(k)}(p, q + 1/2)$; then

$$I_1 \leq C \left\{ \sup_{\hat{\tau}_k} \left| \hat{\partial}_{\hat{x}\hat{x}} \hat{u} \right| + T \right\}, \qquad (4.6)$$

where

$$
\begin{aligned}
T \leq \max \Bigg\{ &\left| \hat{\partial}_{\hat{x}\hat{x},\hat{m}_{j_1}^{(k)}}^{MM} \widehat{\Pi}\hat{u} - \hat{\partial}_{\hat{x}\hat{x},\hat{m}_{j_2}^{(k)}}^{MM} \widehat{\Pi}\hat{u} + \hat{\partial}_{\hat{x}\hat{x},\hat{m}_{j_3}^{(k)}}^{MM} \widehat{\Pi}\hat{u} \right|, \\
&\left| \hat{\partial}_{\hat{x}\hat{x},\hat{m}_{j_1}^{(k)}}^{MM} \widehat{\Pi}\hat{u} + \hat{\partial}_{\hat{x}\hat{x},\hat{m}_{j_2}^{(k)}}^{MM} \widehat{\Pi}\hat{u} - \hat{\partial}_{\hat{x}\hat{x},\hat{m}_{j_3}^{(k)}}^{MM} \widehat{\Pi}\hat{u} \right|, \\
&\left| -\hat{\partial}_{\hat{x}\hat{x},\hat{m}_{j_1}^{(k)}}^{MM} \widehat{\Pi}\hat{u} + \hat{\partial}_{\hat{x}\hat{x},\hat{m}_{j_2}^{(k)}}^{MM} \widehat{\Pi}\hat{u} + \hat{\partial}_{\hat{x}\hat{x},\hat{m}_{j_3}^{(k)}}^{MM} \widehat{\Pi}\hat{u} \right| \Bigg\} \\
\leq C &\left\{ \left| \hat{\partial}_{\hat{x}\hat{x},\hat{m}_{j_1}^{(k)}}^{MM} \widehat{\Pi}\hat{u} \right| + \left| \hat{\partial}_{\hat{x}\hat{x},\hat{m}_{j_2}^{(k)}}^{MM} \widehat{\Pi}\hat{u} \right| + \left| \hat{\partial}_{\hat{x}\hat{x},\hat{m}_{j_3}^{(k)}}^{MM} \widehat{\Pi}\hat{u} \right| \right\} \\
\equiv C &\{T_1 + T_2 + T_3\} \quad \text{(say)}.
\end{aligned}
\qquad (4.7)
$$

If we denote by $\alpha_{\hat{\tau}_l,\hat{m}_{j_i}^{(k)}}$ the weights associated with various elements $\hat{\tau}_l$ involved in recovery at any midpoint $\hat{m}_{j_i}^{(k)}$, $(i = 1, 2, 3)$, we have

$$T_1 = \left| \sum_{\hat{\tau}_l \in \hat{\tau}_{k,\hat{m}_{j_1}^{(k)}}^{(1)}} \alpha_{\hat{\tau}_l,\hat{m}_{j_1}^{(k)}} \hat{\partial}_{\hat{x}\hat{x},\hat{\tau}_l}^{M} \widehat{\Pi}\hat{u} \right| \leq C \sum_{\hat{\tau}_l \in \hat{\tau}_{k,\hat{m}_{j_1}^{(k)}}^{(1)}} \left| \hat{\partial}_{\hat{x}\hat{x},\hat{\tau}_l}^{M} \widehat{\Pi}\hat{u} \right|,$$

whereupon using an inverse estimate and the superconvergence property of $\widehat{\nabla}^M \widehat{\Pi}\hat{u}$ we have

$$T_1 \leq C \sum_{\hat{\tau}_l \in \hat{\tau}_{k,\hat{m}_{j_1}^{(k)}}^{(1)}} \left\| \widehat{\nabla}^M \widehat{\Pi}\hat{u} - \widehat{\nabla}\hat{u} \right\|_{L^2(\hat{\tau}_l)} + \left\| \widehat{\nabla}\hat{u} \right\|_{L^2(\hat{\tau}_l)} \leq C \left\| u \right\|_{H^3\left(\hat{\tau}_{k,\hat{m}_{j_1}^{(k)}}^{(2)}\right)}.$$

Using this and similar estimates for the terms T_2 and T_3 in (4.7) we have the desired proof after using an embedding result to bound the first term on the right hand side of (4.6).   □

We next establish the exactness of the recovered second order derivatives for a cubic polynomial in the following lemma.

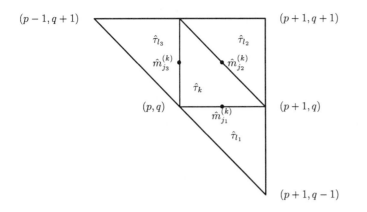

Figure 3

LEMMA 4.2 Let $\hat{u}$ be a cubic polynomial then over any $\hat{\tau}_k$ we have:

$$\hat{\partial}^{MM}_{\hat{x}\hat{x},\hat{\tau}_k}\widehat{\Pi}\hat{u} = \hat{\partial}_{\hat{x}\hat{x}}\hat{u},$$

$$\hat{\partial}^{MM}_{\hat{x}\hat{y},\hat{\tau}_k}\widehat{\Pi}\hat{u} = \hat{\partial}_{\hat{x}\hat{y}}\hat{u}, \tag{4.8}$$

$$\hat{\partial}^{MM}_{\hat{y}\hat{y},\hat{\tau}_k}\widehat{\Pi}\hat{u} = \hat{\partial}_{\hat{y}\hat{y}}\hat{u}.$$

Proof: We prove the lemma only for $\hat{\tau}_k$ situated in the interior of Ω. Following the same argument we can also establish this result for elements near the boundary, making use of the scheme shown in Fig. 2.

With reference to Fig. 3, we can write the second order derivatives over various elements involved using the formula for the Midpoint Gradient Recovery (cf. §2), as follows:

$$\hat{\partial}^{M}_{\hat{x}\hat{x},\hat{\tau}_k}\widehat{\Pi}\hat{u} = \hat{u}(p-1,q+1) + \hat{u}(p+1,q+1) - 2\hat{u}(p,q+1)$$

$$\hat{\partial}^{M}_{\hat{x}\hat{y},\hat{\tau}_k}\widehat{\Pi}\hat{u} = \hat{u}(p+1,q+1) + \hat{u}(p,q) - \hat{u}(p,q+1) - \hat{u}(p+1,q)$$

$$\hat{\partial}^{M}_{\hat{y}\hat{y},\hat{\tau}_k}\widehat{\Pi}\hat{u} = \hat{u}(p+1,q-1) + \hat{u}(p+1,q+1) - 2\hat{u}(p+1,q)$$

$$\hat{\partial}^{M}_{\hat{x}\hat{x},\hat{\tau}_{l_1}}\widehat{\Pi}\hat{u} = \hat{u}(p+2,q-1) + \hat{u}(p,q-1) - 2\hat{u}(p+1,q-1)$$

$$\hat{\partial}^{M}_{\hat{x}\hat{y},\hat{\tau}_{l_1}}\widehat{\Pi}\hat{u} = \hat{u}(p+1,q) + \hat{u}(p,q-1) - \hat{u}(p,q) - \hat{u}(p+1,q-1)$$

$$\hat{\partial}^{M}_{\hat{y}\hat{y},\hat{\tau}_{l_1}}\widehat{\Pi}\hat{u} = \hat{u}(p,q-1) + \hat{u}(p,q+1) - 2\hat{u}(p,q)$$

$$\hat{\partial}^{M}_{\hat{x}\hat{x},\hat{\tau}_{l_2}}\widehat{\Pi}\hat{u} = \hat{u}(p+2,q) + \hat{u}(p,q) - 2\hat{u}(p+1,q) \tag{4.9}$$

$$\hat{\partial}^{M}_{\hat{x}\hat{y},\hat{\tau}_{l_2}}\widehat{\Pi}\hat{u} = \hat{u}(p+1,q+1) + \hat{u}(p,q) - \hat{u}(p+1,q) - \hat{u}(p,q+1)$$

$$\hat{\partial}^{M}_{\hat{y}\hat{y},\hat{\tau}_{l_2}}\widehat{\Pi}\hat{u} = \hat{u}(p,q+2) + \hat{u}(p,q) - 2\hat{u}(p,q+1)$$

$$\hat{\partial}^{M}_{\hat{x}\hat{x},\hat{\tau}_{l_3}}\widehat{\Pi}\hat{u} = \hat{u}(p-1,q) + \hat{u}(p+1,q) - 2\hat{u}(p,q)$$

$$\hat{\partial}^{M}_{\hat{x}\hat{y},\hat{\tau}_{l_3}}\widehat{\Pi}\hat{u} = \hat{u}(p-1,q) + \hat{u}(p,q+1) - \hat{u}(p,q) - \hat{u}(p-1,q+1)$$

$$\hat{\partial}^{M}_{\hat{y}\hat{y},\hat{\tau}_{l_3}}\widehat{\Pi}\hat{u} = \hat{u}(p-1,q+2) + \hat{u}(p-1,q) - 2\hat{u}(p-1,q+1)$$

Using (4.9) and formula (4.3) we write the recovered second order derivatives as follows:

$$\hat{\partial}^{MM}_{\hat{x}\hat{x},\hat{m}^{(k)}_{j_1}} \widehat{\Pi}\hat{u} = 1/2\big(\hat{u}(p-1,q+1) + \hat{u}(p+2,q-1) + \hat{u}(p+1,q+1)$$
$$+ \hat{u}(p,q-1) - 2\hat{u}(p+1,q-1) - 2\hat{u}(p,q+1)\big)$$

$$\hat{\partial}^{MM}_{\hat{x}\hat{y},\hat{m}^{(k)}_{j_1}} \widehat{\Pi}\hat{u} = 1/2\big(\hat{u}(p+1,q+1) + \hat{u}(p,q-1) - \hat{u}(p+1,q-1)$$
$$- \hat{u}(p,q+1)\big)$$

$$\hat{\partial}^{MM}_{\hat{y}\hat{y},\hat{m}^{(k)}_{j_1}} \widehat{\Pi}\hat{u} = 1/2\big(\hat{u}(p+1,q-1) + \hat{u}(p+1,q+1) + \hat{u}(p,q-1)$$
$$+ \hat{u}(p,q+1) - 2\hat{u}(p+1,q) - 2\hat{u}(p,q)\big)$$

$$\hat{\partial}^{MM}_{\hat{x}\hat{x},\hat{m}^{(k)}_{j_2}} \widehat{\Pi}\hat{u} = 1/2\big(\hat{u}(p-1,q+1) + \hat{u}(p+2,q) + \hat{u}(p+1,q+1)$$
$$+ \hat{u}(p,q) - 2\hat{u}(p+1,q) - 2\hat{u}(p,q+1)\big)$$

$$\hat{\partial}^{MM}_{\hat{x}\hat{y},\hat{m}^{(k)}_{j_2}} \widehat{\Pi}\hat{u} = \hat{u}(p+1,q+1) + \hat{u}(p,q) - \hat{u}(p+1,q) - \hat{u}(p,q+1) \qquad (4.10)$$

$$\hat{\partial}^{MM}_{\hat{y}\hat{y},\hat{m}^{(k)}_{j_2}} \widehat{\Pi}\hat{u} = 1/2\big(\hat{u}(p+1,q-1) + \hat{u}(p+1,q+1) + \hat{u}(p,q+2)$$
$$+ \hat{u}(p,q) - 2\hat{u}(p+1,q) - 2\hat{u}(p,q+1)\big)$$

$$\hat{\partial}^{MM}_{\hat{x}\hat{x},\hat{m}^{(k)}_{j_3}} \widehat{\Pi}\hat{u} = 1/2\big(\hat{u}(p-1,q+1) + \hat{u}(p-1,q) + \hat{u}(p+1,q+1)$$
$$+ \hat{u}(p+1,q) - 2\hat{u}(p,q+1) - 2\hat{u}(p,q)\big)$$

$$\hat{\partial}^{MM}_{\hat{x}\hat{y},\hat{m}^{(k)}_{j_3}} \widehat{\Pi}\hat{u} = 1/2\big(\hat{u}(p-1,q) + \hat{u}(p+1,q+1) - \hat{u}(p-1,q+1)$$
$$- \hat{u}(p+1,q)\big)$$

$$\hat{\partial}^{MM}_{\hat{y}\hat{y},\hat{m}^{(k)}_{j_3}} \widehat{\Pi}\hat{u} = 1/2\big(\hat{u}(p-1,q+2) + \hat{u}(p-1,q) + \hat{u}(p+1,q-1)$$
$$+ \hat{u}(p+1,q+1) - 2\hat{u}(p-1,q+1) - 2\hat{u}(p+1,q)\big).$$

Since the weights involved in recovery at any midpoint add to one, and since the second order derivatives in (4.9) are exact for all quadratics, we only need to prove the exactness of the recovered derivatives in (4.10) for the cubic base

$$\{\hat{x}^3, \hat{x}^2\hat{y}, \hat{x}\hat{y}^2, \hat{y}^3\}.$$

Case I $\hat{u} = \hat{x}^3$

$$\hat{\partial}^{MM}_{\hat{x}\hat{x},\hat{m}^{(k)}_{j_1}} \widehat{\Pi}\hat{u} = 3(2p+1) = \hat{D}^{(2,0)}\hat{u}(p+1/2,q)$$

$$\hat{\partial}^{MM}_{\hat{x}\hat{x},\hat{m}^{(k)}_{j_2}} \widehat{\Pi}\hat{u} = 3(2p+1) = \hat{D}^{(2,0)}\hat{u}(p+1/2,q+1/2)$$

$$\hat{\partial}^{MM}_{\hat{x}\hat{x},\hat{m}^{(k)}_{j_3}} \widehat{\Pi}\hat{u} = 6p \qquad = \hat{D}^{(2,0)}\hat{u}(p,q+1/2)$$

Case II $\quad \hat{u} = \hat{x}^2 \hat{y}$

$$\hat{\partial}^{MM}_{\hat{x}\hat{x}, \hat{m}^{(k)}_{j_1}} \widehat{\Pi} \hat{u} = 2q \qquad = \hat{D}^{(2,0)} \hat{u}(p + 1/2, q)$$

$$\hat{\partial}^{MM}_{\hat{x}\hat{x}, \hat{m}^{(k)}_{j_2}} \widehat{\Pi} \hat{u} = 2q + 1 = \hat{D}^{(2,0)} \hat{u}(p + 1/2, q + 1/2)$$

$$\hat{\partial}^{MM}_{\hat{x}\hat{x}, \hat{m}^{(k)}_{j_3}} \widehat{\Pi} \hat{u} = 2q + 1 = \hat{D}^{(2,0)} \hat{u}(p, q + 1/2)$$

$$\hat{\partial}^{MM}_{\hat{x}\hat{y}, \hat{m}^{(k)}_{j_1}} \widehat{\Pi} \hat{u} = 2p + 1 = \hat{D}^{(1,1)} \hat{u}(p + 1/2, q)$$

$$\hat{\partial}^{MM}_{\hat{x}\hat{y}, \hat{m}^{(k)}_{j_2}} \widehat{\Pi} \hat{u} = 2p + 1 = \hat{D}^{(1,1)} \hat{u}(p + 1/2, q + 1/2)$$

$$\hat{\partial}^{MM}_{\hat{x}\hat{y}, \hat{m}^{(k)}_{j_3}} \widehat{\Pi} \hat{u} = 2p \qquad = \hat{D}^{(1,1)} \hat{u}(p, q + 1/2),$$

where we have only considered non-zero entries and for the reason of symmetry have not considered cases $\hat{u} = \hat{x}\hat{y}^2$ and $\hat{u} = \hat{y}^3$. Furthermore, since p and q are arbitrary, it is clear that a similar result can also be established on the element diagonally opposite to $\hat{\tau}_k$ ($\hat{\tau}_{l_2}$ in Fig. 3) situated in the interior of the domain $\hat{\Omega}$. Since a unique linear function interpolates three values, we have the desired result.

$\square$

Proof of Theorem 4.1: Taking into account definition (4.2) and using the triangle inequality we have

$$\left\{ |u - u_h|^{MM}_{H^2(\Omega)} \right\}^2 \leq \left\{ |u - \Pi u|^{MM}_{H^2(\Omega)} \right\}^2 + \left\| \partial^{MM}_{xx} (\Pi u - u_h) \right\|^2_{L^2(\Omega)}$$

$$+ \left\| \partial^{MM}_{xy} (\Pi u - u_h) \right\|^2_{L^2(\Omega)} + \left\| \partial^{MM}_{yy} (\Pi u - u_h) \right\|^2_{L^2(\Omega)} \qquad (4.11)$$

$$= T_A + T_{B1} + T_{B2} + T_{B3} \quad \text{(say)}.$$

Using the properties of the affine transformation $\Omega \leftrightarrow \hat{\Omega}$ we have from Lemma 4.1, Lemma 4.2 and the Bramble–Hilbert Lemma

$$T_A = \sum_{k=1}^{n} \left\{ |u - \Pi u|^{MM}_{H^2(\tau_k)} \right\}^2 \leq \sum_{k=1}^{n} C \, h^4 |u|^2_{H^4\left(\tau_k^{(2)}\right)}$$

so that with an appropriate (blow up) constant we have

$$T_A \leq C \, h^4 |u|^2_{H^4(\Omega)}. \qquad (4.12)$$

Next making use of formula (4.3) or Fig. 2 whichever be the case, we have

$$T_{B1} = \sum_{k=1}^{n} \left\| \partial^{MM}_{xx} (\Pi u - u_h) \right\|^2_{L^2(\tau_k)} \leq C \sum_{k=1}^{n} \left\| \partial^{M}_{xx} (\Pi u - u_h) \right\|^2_{L^2\left(\tau_k^{(*)}\right)} \qquad (4.13)$$

$$\leq C \left\| \partial^{M}_{xx} (\Pi u - u_h) \right\|^2_{L^2(\Omega)}.$$

In what follows we shall prove the following:

$$\left\| \partial^{M}_{xx} (\Pi u - u_h) \right\|_{L^2(\Omega)} \leq C \, h^2 \log^2 \left(\frac{1}{h} \right) |u|_{H^{4,\infty}(\Omega)}. \qquad (4.14)$$

Assuming this for the time being we have

$$T_{B1} \le C\, h^4 \log^4 \left(\frac{1}{h}\right) |u|^2_{H^{4,\infty}(\Omega)}.$$

With similar estimates for the terms T_{B2} and T_{B3} we have the required proof of Theorem 4.1.   □

COROLLARY 4.2   For $u \in H^4(\Omega)$ and Ω a rectangle, the following estimate holds:

$$|u - u_h|^{MM}_{H^2(\Omega)} \le C\, h^{3/2-\delta} |u|_{H^4(\Omega)}, \qquad \delta > 0. \tag{4.15}$$

Proof: The estimate for the term T_A in (4.12) carries over to the present case whereas we modify the estimate for the term T_{B1} in (4.13) by first using an inverse estimate, then using the local boundedness property of the operator ∇^M (cf. Goodsell and Whiteman [6]) and finally the Oganesyan–Rukhovets Lemma to obtain the following:

$$T_{B1} \le C\, h^{-2} \left\|\nabla^M(\Pi u) - \nabla^M u_h\right\|^2_{L^2(\Omega)} \le C\, h^2 |u|^2_{H^3(\Omega)}. \tag{4.16}$$

with similar estimates for the terms T_{B2} and T_{B3}. Alternatively, if we assume the higher regularity $u \in H^{5+\varepsilon}(\Omega)$ ($\varepsilon > 0$) we also have from the proof of Theorem 4.1 by means of an embedding result the following

$$T_{B1} \le C\, h^4 \log^4 \left(\frac{1}{h}\right) |u|^2_{H^{5+\varepsilon}(\Omega)}. \tag{4.17}$$

Using estimates (4.16) and (4.17) along with the interpolation result

$$\left(H^3(\Omega)\,;H^{5+\varepsilon}(\Omega)\right)_{\left[\frac{1}{2+\varepsilon}\right]} = H^4(\Omega)$$

we have the following estimate for T_{B1}

$$T_{B1} \le C\, h^{3-\delta_1} \log^{\frac{4}{2+\varepsilon}} \left(\frac{1}{h}\right) |u|^2_{H^4(\Omega)},$$

where $\delta_1 = \frac{\varepsilon}{2+\varepsilon} > 0$ can be chosen as small as desired. Since $h^{-\varepsilon}$, $\varepsilon > 0$ dominates $\log(1/h)$ we end up with the estimate:

$$T_{B1} \le C\, h^{3-2\delta} |u|^2_{H^4(\Omega)},$$

where again $\delta > 0$ can be chosen as small as we desire. With similar estimates for the terms T_{B2} and T_{B3} we have the desired proof.   □

We close this section by proving (4.14). Our approach is identical to the one used by Lin and Xu [13] in proving their asymptotic expansion for the discretization error in $\nabla u_h - \nabla \Pi u$. However for the sake of completeness we reproduce some of the technical details.

For any $z \in \Omega$ the Green's function G^z and the approximate Green's function G_h^z satisfy

$$a\left(G^z, \phi\right) = \phi(z) \qquad \forall \phi \in H^{1,\infty}\left(\Omega\right), \tag{4.18}$$

and

$$a\left(G_h^z, \chi\right) = \chi(z) \qquad \forall \chi \in S^h\left(\Omega\right). \tag{4.19}$$

respectively, where $a\left(\cdot, \cdot\right)$ is the bilinear form associated with the boundary value problem (2.4). Thus letting χ equal $u_h - \Pi u$ in (4.19) we have from Lin and Xu [13]

$$(u_h - \Pi u)(z) = a\left(u - \Pi u, G_h^z\right) = h^2 \int_\Omega \tilde{\nabla}^4 u\, G_h^z\, dx + O(h^3) \int_\Omega \left|\nabla G_h^z\right| \left|\nabla^4 u\right| dx, \tag{4.20}$$

where $z \in \tau_k$ with vertices $\nu_{j_1}^{(k)}, \nu_{j_2}^{(k)}$ and $\nu_{j_3}^{(k)}$ and $\tilde{\nabla}^4 u$ is some combination of the fourth order derivatives of u. We consider the associated problem

$$\begin{aligned} -\Delta w &= \tilde{\nabla}^4 u \qquad \text{in } \Omega, \\ w &= 0 \qquad \text{on } \partial\Omega. \end{aligned} \tag{4.21}$$

Then since Ω is convex we have $w \in H^{2,q}\left(\Omega\right)$ for any $2 \leq q < \infty$ (cf. Lin and Xu [13]). Let $\lambda_{j_1}^{(k)}, \lambda_{j_2}^{(k)}$ and $\lambda_{j_3}^{(k)}$ be the barycentric coordinates of point z with respect to τ_k, then using the fact that $\sum_{i=1}^3 \nabla \lambda_{j_i}^{(k)} = 0$ we have on letting $\chi \equiv u_h - \Pi u$

$$\begin{aligned} \nabla \chi(z) &= \sum_{i=1}^3 \chi\big|_{\nu_{j_i}^{(k)}} \nabla \lambda_{j_i}^{(k)} \\ &= \left\{\chi\big|_{\nu_{j_1}^{(k)}} - \chi\big|_{\nu_{j_3}^{(k)}}\right\} \nabla \lambda_{j_1}^{(k)} + \left\{\chi\big|_{\nu_{j_2}^{(k)}} - \chi\big|_{\nu_{j_3}^{(k)}}\right\} \nabla \lambda_{j_2}^{(k)} \\ &= a\left(\chi, G_h^{\nu_{j_1}^{(k)}} - G_h^{\nu_{j_3}^{(k)}}\right) \nabla \lambda_{j_1}^{(k)} + a\left(\chi, G_h^{\nu_{j_2}^{(k)}} - G_h^{\nu_{j_3}^{(k)}}\right) \nabla \lambda_{j_2}^{(k)}. \end{aligned}$$

Hence from (4.20) we have

$$\begin{aligned} \nabla \chi(z) &= h^2 \nabla \lambda_{j_1}^{(k)} \int_\Omega \tilde{\nabla}^4 u (G_h^{\nu_{j_1}^{(k)}} - G_h^{\nu_{j_3}^{(k)}})\, dx \\ &\quad + h^2 \nabla \lambda_{j_2}^{(k)} \int_\Omega \tilde{\nabla}^4 u (G_h^{\nu_{j_2}^{(k)}} - G_h^{\nu_{j_3}^{(k)}})\, dx \\ &\quad + O(h^3) \nabla \lambda_{j_1}^{(k)} \int_\Omega \left|\nabla (G_h^{\nu_{j_1}^{(k)}} - G_h^{\nu_{j_3}^{(k)}})\right| \left|\nabla^4 u\right| dx \\ &\quad + O(h^3) \nabla \lambda_{j_2}^{(k)} \int_\Omega \left|\nabla (G_h^{\nu_{j_2}^{(k)}} - G_h^{\nu_{j_3}^{(k)}})\right| \left|\nabla^4 u\right| dx \\ &= T_1 + T_2 + T_3 + T_4 \quad \text{(say)}. \end{aligned} \tag{4.22}$$

Now using a result on the discrete Green's function (cf. Frehse and Rannacher [4]) we have

$$\begin{aligned} T_3 &= O(h^2) \int_\Omega \left|\nabla (G_h^{\nu_{j_1}^{(k)}} - G_h^{\nu_{j_3}^{(k)}})\right| \left|\nabla^4 u\right| dx \\ &= O(h^2) \left\|\nabla^4 u\right\|_{L^\infty(\Omega)} \int_\Omega \left|\nabla (G_h^{\nu_{j_1}^{(k)}} - G_h^{\nu_{j_3}^{(k)}})\right| dx \\ &= O\left(h^3 \log\left(\frac{1}{h}\right)\right) \left\|\nabla^4 u\right\|_{L^\infty(\Omega)}. \end{aligned} \tag{4.23}$$

T_4 can be bounded similarly. We next bound T_1 as follows:

$$T_1 = h^2 \int_\Omega \tilde{\nabla}^4 u (G_h^{\nu_{j_1}^{(k)}} - G_h^{\nu_{j_3}^{(k)}}) \nabla \lambda_{j_1}^{(k)} \, d\boldsymbol{x}$$

$$= h^2 \int_\Omega \tilde{\nabla}^4 u \left\{ (G_h^{\nu_{j_1}^{(k)}} - G^{\nu_{j_1}^{(k)}}) - (G_h^{\nu_{j_3}^{(k)}} - G^{\nu_{j_3}^{(k)}}) \right\} \nabla \lambda_{j_1}^{(k)} \, d\boldsymbol{x}$$

$$+ h^2 \int_\Omega \tilde{\nabla}^4 u (G^{\nu_{j_1}^{(k)}} - G^{\nu_{j_3}^{(k)}}) \nabla \lambda_{j_1}^{(k)} \, d\boldsymbol{x} = T_5 + T_6 \quad \text{(say)},$$

where using the approximability properties of discrete Green's function (Zhu and Lin [**22**]) we have

$$T_5 = h^2 \int_\Omega \tilde{\nabla}^4 u \left(G_h^{\nu_{j_1}^{(k)}} - G^{\nu_{j_1}^{(k)}} \right) \nabla \lambda_{j_1}^{(k)} \, d\boldsymbol{x} - h^2 \int_\Omega \tilde{\nabla}^4 u \left(G_h^{\nu_{j_3}^{(k)}} - G^{\nu_{j_3}^{(k)}} \right) \nabla \lambda_{j_1}^{(k)} \, d\boldsymbol{x}$$

$$\leq C h \left\| \nabla^4 u \right\|_{L^\infty(\Omega)} \int_\Omega \left| G_h^{\nu_{j_1}^{(k)}} - G^{\nu_{j_1}^{(k)}} \right| + \left| G_h^{\nu_{j_3}^{(k)}} - G^{\nu_{j_3}^{(k)}} \right| \, d\boldsymbol{x}$$

$$\leq C h^3 \log^2 \left(\frac{1}{h} \right) \left\| \nabla^4 u \right\|_{L^\infty(\Omega)}. \tag{4.24}$$

Combining these results and retaining only the dominant term we have

$$\nabla \chi(\boldsymbol{z}) = h^2 \int_\Omega \tilde{\nabla}^4 u \left(G^{\nu_{j_1}^{(k)}} - G^{\nu_{j_3}^{(k)}} \right) \nabla \lambda_{j_1}^{(k)} \, d\boldsymbol{x} + h^2 \int_\Omega \tilde{\nabla}^4 u \cdot$$

$$\cdot \left(G^{\nu_{j_2}^{(k)}} - G^{\nu_{j_3}^{(k)}} \right) \nabla \lambda_{j_2}^{(k)} \, d\boldsymbol{x} + O \left(h^3 \log^2 \left(\frac{1}{h} \right) \right) \left\| \nabla^4 u \right\|_{L^\infty(\Omega)}$$

$$= h^2 \int_\Omega -\Delta w \left(G^{\nu_{j_1}^{(k)}} - G^{\nu_{j_3}^{(k)}} \right) \nabla \lambda_{j_1}^{(k)} \, d\boldsymbol{x} + h^2 \int_\Omega -\Delta w \cdot$$

$$\cdot \left(G^{\nu_{j_2}^{(k)}} - G^{\nu_{j_3}^{(k)}} \right) \nabla \lambda_{j_2}^{(k)} \, d\boldsymbol{x} + O \left(h^3 \log^2 \left(\frac{1}{h} \right) \right) \left\| \nabla^4 u \right\|_{L^\infty(\Omega)}$$

$$= h^2 \int_\Omega \nabla w \cdot \nabla \left(G^{\nu_{j_1}^{(k)}} - G^{\nu_{j_3}^{(k)}} \right) \nabla \lambda_{j_1}^{(k)} + \nabla w \cdot \tag{4.25}$$

$$\cdot \nabla \left(G^{\nu_{j_2}^{(k)}} - G^{\nu_{j_3}^{(k)}} \right) \nabla \lambda_{j_2}^{(k)} \, d\boldsymbol{x} + O \left(h^3 \log^2 \left(\frac{1}{h} \right) \right) \left\| \nabla^4 u \right\|_{L^\infty(\Omega)}$$

$$= h^2 \left\{ \left(w|_{\nu_{j_1}^{(k)}} - w|_{\nu_{j_3}^{(k)}} \right) \nabla \lambda_{j_1}^{(k)} + \left(w|_{\nu_{j_1}^{(k)}} - w|_{\nu_{j_3}^{(k)}} \right) \nabla \lambda_{j_2}^{(k)} \right\}$$

$$+ O \left(h^3 \log^2 \left(\frac{1}{h} \right) \right) \left\| \nabla^4 u \right\|_{L^\infty(\Omega)}$$

$$= h^2 \nabla (\Pi w)(\boldsymbol{z}) + O \left(h^3 \log^2 \left(\frac{1}{h} \right) \right) \left\| \nabla^4 u \right\|_{L^\infty(\Omega)},$$

which gives

$$\partial_{xx}^M (u_h - \Pi u)(\boldsymbol{z}) = h^2 \partial_{xx}^M (\Pi w)(\boldsymbol{z}) + O \left(h^2 \log^2 \left(\frac{1}{h} \right) \right). \tag{4.26}$$

Integrating over τ_k we get

$$\int_{\tau_k} \left| \partial xx^M (u_h - \Pi u) \right|^2 dz$$

$$\leq C h^4 \int_{\tau_k} \left| \partial xx^M \Pi w \right|^2 dz + O\left(h^6 \log^4 \left(\frac{1}{h}\right) \right) \left\| \nabla^4 u \right\|^2_{L^\infty(\Omega)}$$

$$\leq C h^4 \int_{\tau_k} \left| \partial xx^M \Pi w - D^{(2,0)} w \right|^2 dz + C h^4 \int_{\tau_k} \left| D^{(2,0)} w \right|^2 dz$$

$$+ O\left(h^6 \log^4 \left(\frac{1}{h}\right) \right) \left\| \nabla^4 u \right\|^2_{L^\infty(\Omega)}$$

$$\leq C h^4 |w|^2_{H^2\left(\tau_k^{(1)}\right)} + O\left(h^6 \log^4 \left(\frac{1}{h}\right) \right) \left\| \nabla^4 u \right\|^2_{L^\infty(\Omega)}.$$

Thus summing over all the elements τ_k we have

$$\int_\Omega \left| \partial_{xx}^M (u_h - \Pi u) \right|^2 dz$$

$$\leq C h^4 |w|_{H^2(\Omega)} + O\left(h^4 \log^4 \left(\frac{1}{h}\right) \right) \left\| \nabla^4 u \right\|^2_{L^\infty(\Omega)}$$

$$\leq C h^4 \| -\Delta w \|_{L^2(\Omega)} + O\left(h^4 \log^4 \left(\frac{1}{h}\right) \right) \left\| \nabla^4 u \right\|^2_{L^\infty(\Omega)}$$

$$\leq C h^4 \log^4 \left(\frac{1}{h}\right) \left\| \nabla^4 u \right\|^2_{L^\infty(\Omega)}.$$

With similar contributions for the terms $\partial_{xy}^M(u_h - \Pi u)$ and $\partial_{yy}^M(u_h - \Pi u)$ we have the desired proof. $\qquad\qquad\square$

5   ESTIMATION OF THIRD ORDER DERIVATIVES

Having done the superconvergent recovery of second order derivatives, we can go one step further and obtain approximate third order derivatives in a manner similar to §3.1 of the present chapter. Thus we define :

$$\begin{aligned}
\partial_{xxx}^* u_h &\equiv \partial_x \partial_{xx}^{MM} u_h \\
\partial_{xxy}^* u_h &\equiv 1/2(\partial_x \partial_{xy}^{MM} u_h + \partial_y \partial_{xx}^{MM} u_h) \\
\partial_{xyx}^* u_h &\equiv \partial_{xxy}^* u_h \\
\partial_{xyy}^* u_h &\equiv 1/2(\partial_x \partial_{yy}^{MM} u_h + \partial_y \partial_{xy}^{MM} u_h) \\
\partial_{yxx}^* u_h &\equiv \partial_{xxy}^* u_h \\
\partial_{yxy}^* u_h &\equiv \partial_{xyy}^* u_h \\
\partial_{yyx}^* u_h &\equiv \partial_{xyy}^* u_h \\
\partial_{yyy}^* u_h &\equiv \partial_y \partial_{yy}^{MM} u_h.
\end{aligned} \qquad (5.1)$$

The symbolic derivatives on the left-hand side of (5.1) can be thought of as components of a tensor of order three which we symbolically represent as $\nabla^*(\nabla\nabla)^M u_h$.

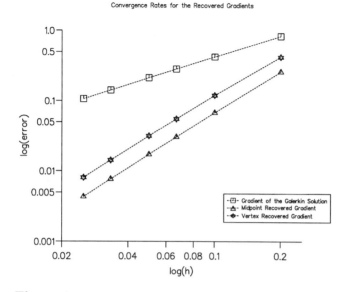

Figure 4.a

Following the ideas of the proof of Lemma 3.1 and Lemma 3.2, we close this section with the following assertion valid for $u \in H_0^1(\Omega) \cap H^4(\Omega)$:

$$\left\| \nabla\nabla\nabla u - \nabla^*(\nabla\nabla)^M u_h \right\|_{L^2(\Omega)}^h \leq Ch^{3/2-\delta} |u|_{H^4(\Omega)} \qquad \delta > 0.$$

6   NUMERICAL RESULTS

To demonstrate the superconvergent error estimates derived in the previous sections we take the following specific example:

$$
\begin{aligned}
-\Delta u &= \left(2 + \pi^2(1 - x^2)\right)\sin(\pi y) && \text{in } \Omega \equiv (-1, -1) \times (1, 1), \\
u &= 0 && \text{on } \partial\Omega,
\end{aligned}
\qquad (6.1)
$$

with known solution $u = (1 - x^2)\sin(\pi y)$. The convergence rates for the Midpoint and Vertex Recovered Gradients introduced earlier in §2 appear in Fig. 4a. From the figure it is evident that not only are the recovered gradients more accurate than the usual gradient of the Galerkin Approximation but they are also superconvergent. The convergence rates in the discrete norm $\left\| \nabla\nabla u - \nabla^*\nabla^R u_h \right\|_{L^2(\Omega)}^h$, defined in the opening section of this chapter for the computed derivatives of the recovered gradients, appear in Fig. 4b. The convergence of the Recovered Second Ordered Derivatives in the norm defined in Theorem 3.1 also appears in Fig. 4b. The recovered second order derivatives are not only more accurate but are also superconvergent. Because of the smoothness of the solution of the boundary value problem (6.1) an almost $O(h^2)$ convergence is obtained. Finally convergence rates for the estimated third order derivatives are produced in Fig 4c.

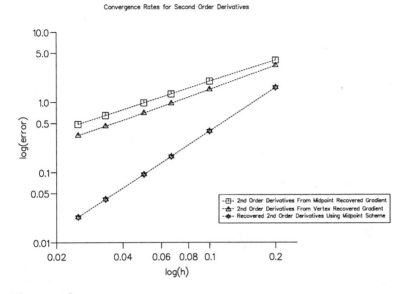

Figure 4.b

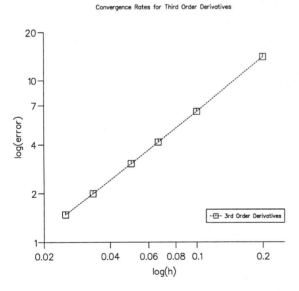

Figure 4.c

Acknowledgement   We take this opportunity to thank Prof. Manil Suri, University of Maryland, for his continuous interest and assistance in this research and we also thank the Aga Khan Foundation, Geneva for funding the first author's work in BICOM.

REFERENCES

1. Adams, R., "Sobolev spaces," Academic Press, 1975.
2. Blum, H. , Lin, Q. and Rannacher, R., *Asymptotic error expansion and Richardson extrapolation for linear finite elements*, Numer. Math. **49** (1986), 11–37.
3. Bramble, J.H. and Hilbert, S.R., *Estimation of linear functional on Sobolev spaces with application to Fourier transforms and spline interpolation*, SIAM J. Numer. Anal. **7** (1970), 112–124.
4. Frehse, J. and Rannacher, R., *Eine L^1 Fehlerabschätzung der diskrete Grundlösungen in der Methode der finiten Element*, Bonn. Math. Schrift. **89** (1976), 92–114.
5. Goodsell, G., "PhD Thesis," Dept. of Math., Brunel U., 1988.
6. Goodsell, G., and Whiteman, J.R., *A unified treatment of superconvergent recovered gradient functions for piecewise linear finite element approximations*, Tech. Rep., BICOM 87/1, Brunel Inst. of Comp. Math..
7. Hlaváček, I. and Křížek, M., *On a superconvergent finite element scheme for elliptic systems, Parts I–III*, Appl. Math. **32** (1987), 131–154, 200–213, 276–289.
8. Křížek, M. and Neittaanmäki, P., *Superconvergence phenomenon in the finite element method arising from averaging gradients*, Numer. Math. **45** (1984), 105–116.
9. Křížek, M. and Neittaanmäki, P., *On a global superconvergence of the gradient of linear triangular elements*, J. Comp. Appl. Math. **18** (1987), 221–233.
10. Křížek, M. and Neittaanmäki, P., *On superconvergence techniques*, Acta Appl. Math. **9** (1987), 175–198.
11. Levine, N.D., *Superconvergent recovery of gradient from piecewise linear finite element approximation*, IMA J. Numer. Anal. **5** (1985), 407–427.
12. Lin, Q. and Lu, T., *Asymptotic expansion for finite element eigenvalue and finite element solution*, Bonn. Math. Schrift. **158** (1984), 1–10.
13. Lin, Q. and Xu, J., *Linear finite elements with high accuracy*, J. Comp. Math. **2** (1985), 115–133.
14. Lin, Q. and Zhu, Q., *Asymptotic expansion for the derivative of finite elements*, J. Comp. Math. **2** (1984), 361–363.
15. Oganesyan, L.A. and Rukhovets, L.A., *Study of the rate of convergence of variational difference scheme for second order elliptic equations in two dimensional field with smooth boundary*, Zh. Vychisl. Math. Fiz. **9** (1969), 1102–1120.
16. Rodríguez, R., *Some remarks on Zienkiewicz–Zhu estimator*, Numer. Methods Partial Differential Equations (to appear).
17. Strang, G. and Fix, G.J., "An analysis of the finite element method," Prentice Hall, New Jersey, 1973.
18. Wheeler, M.F. and Whiteman, J.R., *Superconvergent recovery of gradients on subdomains from piecewise linear finite element approximations*, Numer. Methods Partial Differential Equations **3** (1987), 65–82.
19. Wiberg, N.E. and Abdul Wahab, F., *Patch recovery based on superconvergent derivatives and equilibrium*, Internat. J. Numer. Methods Engrg. **36** (1993), 2703–2724.
20. Zienkiewicz, O.C. and Zhu, J.Z., *Superconvergent patch recovery and a posteriori error estimation, Part I: The recovery technique*, Internat. J. Numer. Methods Engrg. **33** (1992), 1331–1364.

21. Zienkiewicz, O.C. and Zhu, J.Z., *Superconvergent patch recovery and a posteriori error estimation, Part II: Error estimates and adaptivity*, Internat. J. Numer. Methods Engrg. **33** (1992), 1365–1382.
22. Zhu, Q. and Lin, Q., "Superconvergent theory of finite elements," Hunan Scientific Publication House, 1989.
23. Zlámal, M., *Some superconvergence results in the finite element method*, in "Mathematical Aspects of Finite Element Methods," Proc. Conf. Rome, Lecture Notes in Mathematics 606, de Boor (ed.), Springer, Berlin, 1977, pp. 353–362.
24. Zlámal, M., *Superconvergence and reduced integration in the finite element method*, Math. Comp. **32** (1978), 663–685.

Superclose FE-theory becomes a table of integrals

QUN LIN   Institute of Systems Science, Academia Sinica, Beijing 100080, People's Republic of China, e-mail: qlin@bamboo.iss.ac.cn

1   METHODOLOGY: FROM INFINITY TO ONE

We face an implicit and complex problem:

      (1) the solution error between a PDE and the associated FEE (finite element equation) of any reasonable type.

Such a problem is resolved into several explicit and pure problems:

      (2) some integral forms of the interpolation error

but nothing about the PDE or FEE itself.

The integral forms in (2) are defined on a complex domain composed of several (convex) quadrilateral subdomains. They are now resolved into

      (3) the integral forms defined on a quadrilateral

which is then transformed into a rectangle. So we have a focus on

      (4) rectangle integrals

or, essentially,

      (5) one variable calculus.

They are now simple enough and may have expansions with sharp estimates (see the last section).

These sharp estimates can be listed in a table of integrals. The sharp FE-error analysis or the "infinite system" (1) therefore becomes a table of integrals, or, essentially, the one variable calculus in (5).

Thus, infinity returns one!

2 NOTATION AND TABLE OF INTEGRALS (BILINEAR ELEMENT)

Let Ω be a domain of integration with the boundary $\partial\Omega$ and

$$T^h \subset \Omega : \quad \text{a FE-mesh,}$$
$$V^h : \quad \text{a FE-space over } T^h,$$
$$v \in V^h : \quad \text{any test function,}$$
$$v \in V_0^h : \quad v \in V^h \text{ and } v = 0 \text{ on } \partial\Omega,$$
$$u : \quad \text{a given function (the exact solution of a PDE),}$$
$$u^I \in V^h : \quad \text{an interpolation of } u,$$
$$w = u - u^I : \quad \text{the interpolation error.}$$

Let us forget PDEs or FEEs but focus on the integral forms or functionals of w, such as

$$\frac{\int_\Omega w_x v_x}{\|v\|_1}, \quad \frac{\int_\Omega w_x v_x}{\|v\|_0}, \quad \frac{\int_\Omega w_x v_y}{\|v\|_1}, \quad \frac{\int_\Omega w_x v_y}{\|v\|_0}, \quad \frac{\int_\Omega w_x v}{\|v\|_0},$$

where $v \in V^h$ and

$$\|v\|_i = \left(\int_\Omega |D^i v|^2 \right)^{1/2}.$$

A natural question arises: can we have the identities or expansions for these integrals?

Let us start from the simplest case where T^h is a

(6) rectangular mesh

and V^h is the bilinear FE-space.

In Table 1, AU is the abbreviation for "almost uniformity" and the star $*$ means that the boundary condition $v = 0$ on $\partial\Omega$ can be replaced with $\dfrac{\partial v}{\partial n} = 0$ on $\partial\Omega$. Thus, a non-convergent integral becomes convergent, or a slowly convergent integral becomes fastly convergent, when the FE-mesh is carefully chosen. In contrast, the careless choice of the mesh may lead to slow convergence even non-convergence.

Once we have sharp estimates for integrals on the rectangular mesh (6), they might be transported to the integrals on

(7) deformed rectangular mesh (see Fig. 1)

because the two meshes (6) and (7) are equivalent under a bilinear transformation and the two corresponding integrals have only a difference in the variable coefficients (caused by the Jacobian).

Furthermore, a general polygonal domain is a composite of several convex quadrilateral subdomains and each of them can be covered with a deformed rectangular mesh. This is called

(8) piecewise deformed rectangular mesh

It is possible to have sharp estimates even for more general meshes, like

(9) mostly almost uniform rectangular mesh (see Fig. 2),

(10) local refinement rectangular mesh (see Fig. 3).

In Table 2 is shown a table of integrals for meshes (7)–(10). Here α denotes a variable coefficient.

Going back to PDEs with the bilinear FE-solution u^h, we have:

Table 1 Table of Integrals for Mesh (6)

integral	general mesh	(6)		(6) & AU	
		$v \in V^h$	$v \in V_0^h$	$v \in V^h$	$v \in V_0^h$
$\dfrac{\int w_x v_x}{\|v\|_1}$	$O(h)$	$O(h^2)$	$O(h^2)$	$O(h^2)$	$O(h^2)$
$\dfrac{\int w_x v_x}{\|v\|_0}$	$O(1)$	$O(h^{3/2})$	$O(h^2)^*$	$O(h^{3/2})$	$O(h^2)^*$
$\dfrac{\int w_x v_y}{\|v\|_1}$	$O(h)$	$O(h^{3/2})$	$O(h^2)$	$O(h^2|\ln h|^{1/2})$	$O(h^2)$
$\dfrac{\int w_x v_y}{\|v\|_0}$	$O(1)$	$O(h^{1/2})$	$O(h)$	$O(h)$	$O(h^2)$
$\dfrac{\int w_x v}{\|v\|_0}$	$O(h)$	$O(h)$	$O(h)$	$O(h^{3/2})$	$O(h^2)$

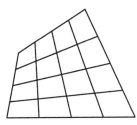

Figure 1

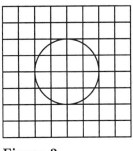

Figure 2

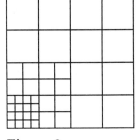

Figure 3

CONCLUSION 1   For the second order elliptic equation, we get the supercloseness:

$$\|u^h - u^I\|_1 = \begin{cases} O(h^2|\ln h|^{1/2}), & T^h \in (8) \text{ \& AU}, \\ O(h^{3/2}), & T^h \in (8)\text{--}(10), \end{cases}$$

one or a half order higher than standard estimates.

Table 2   Table of Integrals for Meshes (7)–(10).

integral	$v \in V^h$			$v \in V_0^h$
	(8) & AU	(9)	(8), (10)	(7) & AU
$\dfrac{\int \alpha w_x v_x}{\|v\|_1}$	$O(h^2\|\ln h\|^{1/2})$	$O(h^{3/2})$	$O(h^{3/2})$	$O(h^2)$
$\dfrac{\int \alpha w_x v_y}{\|v\|_1}$	$O(h^2\|\ln h\|^{1/2})$	$O(h^{3/2})$	$O(h^{3/2})$	$O(h^2)$
$\dfrac{\int \alpha w_x v}{\|v\|_0}$	$O(h^{3/2})$	$O(h^{3/2})$	$O(h)$	$O(h^2)$

CONCLUSION 2   For a first order hyperbolic equation, we get sharp estimates:

$$\|u^h - u^I\|_0 = \begin{cases} O(h^2), & T^h \in (7) \text{ \& AU,} \\ O(h^{3/2}), & T^h \in (8) \text{ \& AU or (9),} \end{cases}$$

one or a half order higher than standard estimates.

CONCLUSION 3   For the biharmonic equation with Herrmann–Miyoshi FE-solution (denoted by u^h, $\tilde{v}^h$), we get the superconvergence: when $T^h \in (6)$ & AU,

$$\|\tilde{v}^h - \partial_i \partial_j u\|_0 + \|u^h - u\|_0 + |u^h - u^I|_1 = O(h^2)$$

one or two order higher than the standard estimates, especially a non-convergent FE-solution becomes convergent.

Sometime one adopts

(11) different independent meshes

(without mesh matching) and different FE-spaces in different subdomains, e.g., a piecewise deformed rectangular mesh (8) in a subdomain with the bilinear FE-space, an independently arbitrary regular mesh in another subdomain with a quadratic FE-space and a bigger mesh on the interface with a quadratic FE-space. Properly modifying the functions of the first two FE-spaces on their elements adjoined with the interface we get, for the second order elliptic equation, still a supercloseness:

$$\|u^h - u^I\|_1 = O(h^{3/2}).$$

3   TABLE OF INTEGRALS (ADINI ELEMENT)

Among the high degree elements, the Adini element is the most attractive because it has high accuracy and less degrees of freedom and, especially, contains both of the solution and the derivatives at element vertices (see Table 3).

CONCLUSION 4   For a second order elliptic equation, we have a natural super-convergence: when $T^h \in (8)$ & AU or (9) and z is any vertex,

$$\nabla(u^h - u)(z) = O(h^{7/2})$$

Table 3   Table of Integrals.

integral	(8) & AU, (9)	(6)
$\dfrac{\int \alpha w_x v_x}{\|v\|_1}$	$O(h^{3.5})$	
$\dfrac{\int \alpha w_x v_y}{\|v\|_1}$	$O(h^{3.5})$	
$\dfrac{\int w_{xx} v_{xx}}{\|v\|_2}$		$O(h^2)$
$\dfrac{\int w_{xy} v_{xy}}{\|v\|_2}$		$O(h^2)$

Table 4   Table of Integrals.

integral	$v \in V^h$		$v \in V_0^h$
	(9)	(8) & AU	(6) & AU
$\dfrac{\int \alpha w_{xx} v_{xx}}{\|v\|_2}$	$O(h^{5/2})$	$O(h^3)$	$O(h^4)$
$\dfrac{\int \alpha w_{xy} v_{xy}}{\|v\|_2}$	$O(h^{5/2})$	$O(h^3)$	$O(h^4)$
$\dfrac{\int \alpha w_{xy} v_{xx}}{\|v\|_2}$	$O(h^{5/2})$	$O(h^3)$	

a half order higher than the standard estimate.

CONCLUSION 5   For the biharmonic equation, together with a similar integral expansion for the nonconforming term, we have a sharp estimate: when $T^h \in (6)$,

$$\|u^h - u\|_2 = O(h^2).$$

4   TABLE OF INTEGRALS (HERMITE BICUBIC ELEMENT)

The Hermite bicubic element is a conforming element for the fourth order elliptic equation, but not very expensive (see Table 4).

CONCLUSION 6   For the biharmonic equation, we have the supercloseness:

$$\|u^h - u^I\|_2 = \begin{cases} O(h^{5/2}), & T^h \in (9), \\ O(h^3), & T^h \in (8) \text{ \& AU or } (6), \\ O(h^4), & T^h \in (6) \text{ \& AU.} \end{cases}$$

Table 5 Table of Integrals (ECHL).

integral	general mesh	(8) (9)	(6) $\vec{n} \times \vec{E} = 0$ on $\partial\Omega$	(6) & AU $\vec{n} \times \vec{E} = 0$ on $\partial\Omega$ & $\psi = 0$ on $\partial\Omega$	
$\dfrac{(\vec{E} - \vec{E}^I, \vec{\nabla} \times \psi)}{\|\psi\|_0}$	$O(1)$	$O(h^{1/2})$	$O(h)$	$O(h^{3/2})$	$O(h^2)$

Table 6 Table of Integrals (MECEL).

integral	(6), $\vec{n} \times \vec{E} = 0$ on $\partial\Omega$
$\dfrac{(\vec{E} - \vec{E}^I, \vec{\nabla} \times \psi)}{\|\psi\|_0}$	$O(h^2)$

Table 7 Table of Integrals (NEDELECS).

integral	(6)
$\nabla \times (\vec{E} - \vec{E}^I), \psi)$	0

5 TABLE OF INTEGRALS (MIXED ELEMENTS)

One uses a pair of FE-spaces in the mixed methods, e.g., the ECHL element for solving the Maxwell equation (see Table 5). Let

$$\vec{V}^h : \quad \text{piecewise constant space,}$$
$$W^h : \quad \text{piecewise linear space,}$$
$$\vec{E}^I = \frac{1}{|e|} \int_e \vec{E} \in \vec{V}^h, \quad \psi \in W^h.$$

Similar situations for MECEL element and Nedelecs element when $\vec{E}^I$ are defined properly in Tables 6 and 7.

CONCLUSION 7   For the Maxwell equation, we have sharp estimates:

$$\|\vec{E}^h - \vec{E}\|_0 = \begin{cases} O(h^{1/2}), & T^h \in (8) \text{ or } (9), \\ O(h), & T^h \in (6), \\ O(h^2), & T^h \in (6) \text{ \& AU,} \end{cases}$$

for ECHL element, and

$$\|\vec{E}^h - \vec{E}\|_0 = O(h^2), \quad T^h \in (6)$$

Table 8   Table of Integrals for Mesh (6).

integral	B-R space	$Q_2 - P_1$ space	P_1 space*
$\dfrac{(\nabla(\vec{u} - \vec{u}^I), \nabla\vec{\psi})}{\|\vec{\psi}\|_1}$	$O(h^2)$	$O(h^3)$	$O(h^2)$
$\dfrac{(P - P^I, \operatorname{div}\vec{\psi})}{\|\vec{\psi}\|_1}$	$O(h^2)$	$O(h^3)^\star$	$O(h^2)$
$\dfrac{(\varphi, \operatorname{div}(\vec{u} - \vec{u}^I))}{\|\varphi\|_0}$	0	$O(h^3)$	$O(h^{3/2})$

for MECEL element and Nedelecs element.

Therefore, the non-convergent FE-solution becomes convergent and the low convergent FE-solution becomes fast convergent when the mesh is of care choice.

We list in Table 8 some tables of integrals for Bernadi–Raugel element, $Q_2 - P_1$-element and P_1-element for solving the Stokes equation.

In Table 8 we need the condition AU in the place*.

CONCLUSION 8   For the Stokes equation, we have superconvergence:

$$\|\vec{u}^h - \vec{u}^I\|_1 + \|p^h - p^I\|_0 = \begin{cases} O(h^2), & T^h \in (6), \\ O(h^3), & T^h \in (6) \ \& \ \text{AU}, \\ O(h^{3/2}), & T^h \in (6) \ \& \ \text{AU}, \end{cases}$$

for Bernadi–Raugel element, $Q_2 - P_1$ element and P_1-element, respectively.

Similar conclusions hold for other mixed elements, e.g., Raviart–Thomas element and BDFM element for the Laplace equation.

6   TABLE OF INTEGRALS (HERMITE TRIANGULAR ELEMENT)

By our methodology, we get the sharp estimates for various FEs with a rectangular mesh or it's deformation. There is a technical complication to get the sharp estimates even for a special triangular mesh like

(12) isosceles right-angled triangular mesh

because the triangle is really a two variable calculus and this is outside our methodology but has been done by Schatz, Sloan and Wahlbin (see Wahlbin's book). But our integral expansion argument may work for the Hermite triangular FE as in Table 9.

CONCLUSION 9   For the Laplace equation, we have a natural superconvergence: when $T^h \in (12)$ and z is any vertex,

$$\nabla(u^h - u)(z) = O(h^{7/2})$$

Table 9 Table of Integrals for Mesh (11).

integral	(11)
$\dfrac{\int w_x v_x}{\|v\|_1}$	$O(h^{7/2})$

a half order higher than the standard estimate.

REMARK   An interesting example is the Carey triangular element. By our integral expansion argument, without the restriction on the mesh, we can find for the Laplace equation that the linear interpolation of Carey FE-solution is nothing but the linear FE-solution.

7   POSTPROCESSING FOR FE-SOLUTION

So far we get is mainly the supercloseness to the solution interpolation u^I rather than to the solution u. By using a higher interpolation process I^{2h} (defined on a coarse mesh of mesh size $2h$) we get the superconvergence to the solution itself:

$$\|I^{2h}u^h - u\|_1 \approx \|u^h - u^I\|_1.$$

Without such a postprocess, the FE-error will be worse with the "good" meshes (6)–(10) for the elliptic problems. This phenomenon is observed from the Pythagorean theorem:

$$|u^h - u|_1^2 = |u^I - u|_1^2 - |u^h - u^I|_1^2,$$

which holds for the Poisson equation.

We can also use such interpolation postprocessing to integral equations instead of using the popular iterated postprocessing (Lin and Zhang).

8   ADD ON METHODOLOGY

We have pointed out at the beginning section how the original problem (1) can be reduced into some rectangle integrals in (4). We now explain how to get the integral expansions by the same methodology. Resolve the test function $v \in V^h$ in the rectangle integrals by the Taylor expansion, and calculate then the integrals for each component. Integration by parts and using one variable error functions

$$E(x) = \frac{1}{2}\left[(x - x_e)^2 - h_e^2\right], \quad F(y) = \frac{1}{2}\left[(y - y_e)^2 - k_e^2\right]$$

on each element $e = [x_e - h_e, x_e + h_e] \times [y_e - k_e, y_e + k_e]$ when $\Omega \subset \mathbb{R}^2$, we can get the integral expansions of sharp estimates. They are essentially one variable calculus or the integration by parts!

See the following references starting from 1990.

REFERENCES

1. Q. Lin, "An integral identity and interpolated postprocess in superconvergence," Research Report 90–7, Inst. of Sys. Sci., Academia Sinica, 1990.

2. Q. Lin, *A rectangular test for finite element analysis*, in "Proc. of Systems Science and Systems Engineering," Great Wall (H.K.) Culture Publish Co., 1991, pp. 213–216.

3. Q. Lin, *A new observation in FEM*, in "Proc. of Systems Science and Systems Engineering," Great Wall (H.K.) Culture Publish Co., 1991, pp. 389–391.

4. Q. Lin, J. Li and A. Zhou, *A rectangle test for Ciarlet–Raviart and Herrmann–Miyoshi schemes*, in "Proc. of Systems Science and Systems Engineering," Great Wall (H.K.) Culture Publish Co., 1991, pp. 230–233.

5. Q. Lin, J. Li and A. Zhou, *A rectangle test for Stokes equations*, in "Proc. of Systems Science and Systems Engineering," Great Wall (H.K.) Culture Publish Co., 1991, pp. 240–241.

6. Q. Lin and N. Yan, *A rectangle test for singular solution with irregular meshes*, in "Proc. of Systems Science and Systems Engineering," Great Wall (H.K.) Culture Publish Co., 1991, pp. 236–237.

7. Q. Lin and N. Yan, *A rectangle test in $\mathbb{R}^3$*, in "Proc. of Systems Science and Systems Engineering," Great Wall (H.K.) Culture Publish Co., 1991, pp. 242–246.

8. Q. Lin, N. Yan and A. Zhou, *A rectangle test for interpolated finite elements*, in "Proc. of Systems Science and Systems Engineering," Great Wall (H.K.) Culture Publish Co., 1991, pp. 217–229.

9. Q. Lin and A. Zhou, *A rectangle test for first order hyperbolic equation*, in "Proc. of Systems Science and Systems Engineering," Great Wall (H.K.) Culture Publish Co., 1991, pp. 234–235.

10. Q. Lin, *Interpolated finite elements and global superconvergence*, in "Proc. of 2nd Conf. on Numer. Meth. for PDE," 1991, Tianjing, World Scientific Publishing Company, 1992, pp. 91–95.

11. Q. Lin, *Global error expansion and superconvergence for higher order interpolation of finite elements*, J. Comput. Math., Supplementary Issue (1992), 286–289.

12. Q. Lin and P. Luo, *Global superconvergence of Hermite bicubic element for the biharmonic equation*, Beijing Math. **1** (1995), 52–64.

13. Q. Lin and P. Luo, *Error expansions and extrapolations for Adini nonconforming finite element*, Beijing Math. **1** (1995), 65–83.

14. Q. Lin, N. Yan and A. Zhou, *Reasonal coupling for different elements on different domains*, Beijing Math. **1** (1995), 19–26.

15. Q. Lin, N. Yan and J. Zhou, *Integral-identity argument and natural superconvergence for cubic elements of Hermite type*, Beijing Math. **1** (1995), 37–41.

16. Q. Lin and J. Wang, *Assessing finite element meshes by computer testifying*, Beijing Math. **1** (1995), 37–41.

17. Q. Lin and J. Wang, *The implementation of the finite element method with high accuracy and its numerical analysis*, Beijing Math. **1** (1995), 42–51.

18. Q. Lin and P. Luo, *High accuracy analysis for a nonconforming membrane element*, J. Math. Study **28:3** (1995), 1–5.

19. A. Schatz, I. Sloan and L. Wahlbin, *Superconvergence in finite element methods*

and meshes that are symmetric with respect to a point, SIAM J. Numer. Anal. **33** (1996), 505–521.

20. L. Wahlbin, "Superconvergence in Galerkin Finite Element Methods," Springer Lecture Notes in Mathematics 1605, 1995.

Adaptive finite element methods for systems of reaction-diffusion equations

ROBERT SANDBOGE   Department of Mathematics, Chalmers University of Technology, S–412 96 Göteborg, Sweden

Abstract   We consider an adaptive finite element method with quantitative error control for systems of reaction-diffusion equations. The adaptive algorithm is based on an a posteriori error estimate involving residuals of the computed solution and certain strong stability factors, which are estimated by solving an associated linearized dual problem numerically. To take into account that the stability properties in space and time are different, separate stability factors are used for the space discretization and the time discretization.

1 INTRODUCTION

We consider systems of reaction-diffusion equations of the form

$$u_t - \mathcal{E}\Delta u = f(u) \quad \text{in } \Omega \times I,$$
$$\frac{\partial u}{\partial \mathbf{n}} = 0 \qquad \text{in } \partial\Omega \times I, \tag{1}$$
$$u(\cdot, 0) = u_0 \qquad \text{in } \Omega,$$

where Ω is a bounded region in $\mathbb{R}^2$, $I = (0, T)$ is a time interval, $u \in \mathbb{R}^q$ is the solution vector, $\dfrac{\partial u}{\partial \mathbf{n}}$ is the outward normal derivative of u on the boundary $\partial\Omega$, $u_0 \in \mathbb{R}^q$ with $q \geq 1$ is a given initial value, $u_t = \dfrac{\partial u}{\partial t}$, $\mathcal{E} = \operatorname{diag}(\varepsilon_1, \varepsilon_2, \ldots, \varepsilon_q)$ is a given diagonal matrix of diffusion coefficients and $f \in \mathbb{R}^q$ is a given Lipschitz continuous vector valued function representing the reactive sources.

Systems of reaction-diffusion equations typically model chemical reactions between many species, combined with diffusion and heat conduction. Solutions of such problems often have various local features in space such as travelling reaction fronts or fast transient phenomena in time, or combinations thereof. Reaction-diffusion systems also model phenomena in physics and biology with complex features such as metastability, phase transition, oscillations, finite time blow up and pattern formation.

In order to solve problems of this type, it is necessary to use adaptive methods, where the computational mesh automatically resolves the local features of the solution both in space and time. More precisely, it is desirable to use adaptive methods with reliable and efficient quantitative error control.

In this article we shall present an adaptive method for (1) with the following global error control: $\max_{0 \le t \le T} \|e(t)\| \le TOL$, here e denotes the error and TOL is a given tolerance, T is the final time and $\| \cdot \|$ is the L_2-norm in space. The adaptive method is based on an a posteriori error estimate, where the error is estimated in terms of the residuals of the calculated solution, and certain strong stability factors. The a posteriori estimate indicates where the mesh should be refined in space, and it is also used to determine the timesteps.

We will give evidence that the presented adaptive method is reliable in the sense that the real error is less than the given tolerance. This requires accurate computations of the strong stability factors. Further the adaptive method is efficient in the sense that the number of degrees of freedom in the computations are close to optimal, this is achieved by (i) using a sharp a posteriori error estimate, which again demands that the stability factors are of the right size, (ii) using an efficient underlying Galerkin method to solve the equations numerically.

Following [1]-[6], we will use the Galerkin method using piecewise linear basis functions which are continuous in space and discontinuous in time, the cG(1)dG(1)-method, in the adaptive algorithm, which includes numerical computations of the stability factors. The stability factors reflects certain properties of the solution which indicates that more amount of computation work is necessary if the factors are big. To solve problems where the stability factors are large, it is important that the error accumulation in time is sufficiently small, this makes the cG(1)dG(1)-method suitable. Combining the cG(1)dG(1)-method with strong stability and Galerkin orthogonality makes long time integration without error accumulation possible for parabolic problems. To carry out this, it is natural to use L_∞ in time, in the a posteriori estimate, whereas L_2 can be used in space.

2   THE FINITE ELEMENT METHOD

We introduce some notation and define the finite element spaces. Let $0 = t_0 < t_1 < \ldots < t_n < \ldots < t_N = T$, be a sequence of discrete time steps with the corresponding time interval: $I_n := [t_n, t_{n+1})$. For each I_n we associate a triangulation $\mathcal{T}_n = \{\mathcal{K}\}$ of Ω into triangles $\mathcal{K}$ of diameter $h_\mathcal{K}$ and a corresponding mesh function $h_n(x)$ with the following properties

$$c_1 h_\mathcal{K}^2 \le \int_\mathcal{K} dx \qquad \forall \mathcal{K} \in \mathcal{T}_n,$$

$$c_2 h_\mathcal{K} \le h_n(x) \le h_\mathcal{K} \quad \forall x \in \mathcal{K}, \, \forall \mathcal{K} \in \mathcal{T}_n, \qquad (2)$$

where c_1 and c_2 are positive constants independent of n. Below, we shall seek a finite element solution U in a finite dimensional space V consisting of piecewise linear functions in x and t, continuous in space and discontinuous in time at the discrete time levels n, defined as follows:

$$W_n := \{v \in [H^1(\Omega)]^q : v_i|_{\mathcal{K}} \text{ is linear } \forall \mathcal{K} \in \mathcal{T}_n\},$$
$$V_n := \{v \in [H^1(\Omega \times I_n)]^q :$$
$$v_i(x,t) = w_0(x) + tw_1(x), \ w_0, w_1 \in W_n, \ (x,t) \in \Omega \times I_n\},$$

and $V := \prod_{n=0}^{N-1} V_n$. Because the functions in V_n may be discontinuous in time, we introduce the notation

$$v_n := v_n^+ = \lim_{s \to 0+} v(t_n + s), \quad v_n^- = \lim_{s \to 0-} v(t_n + s)$$

and $[v]_n = v_n^+ - v_n^-$. The inner product in $[L_2(\Omega)]^q$ is denoted by $(\cdot, \cdot)$, with the corresponding norm $\|\cdot\| = \|\cdot\|_{L_2(\Omega)}$, further $\|\cdot\|_{I_n} = \sup_{t \in t_n} \|\cdot\|$.

The cG(1)dG(1)-method for (1) is to find $U \in V$ such that

$$B(U; U, v) = (u_0, v_0^+) \quad \forall v \in V, \tag{3}$$

where

$$B(z, w, v) = \sum_{n=0}^{N-1} \int_{I_n} \{(w_t, v) + (\mathcal{E}\nabla w, \nabla v)\} \, dt + (w_0^+, v_0^+)$$
$$+ \sum_{n=1}^{N-1} ([w]_n, v_n^+) - \int_I (f(z), v) \, dt,$$

and $I = (0, t_N)$ with $t_N = T$ a given final time. By choosing v to vanish outside I_n, (3) reduces to an equation which successively can be solved on each space-time slab $\Omega \times I_n$, in a time stepping scheme, which takes the following form

$$\int_{I_n} \{(U_t, v) + (\mathcal{E} \cdot \nabla U, \nabla v)\} \, dt + (U_n, v_n)$$
$$= (U_n^-, v_n) + \int_{I_n} (f(U), v) \, dt \quad \forall v \in V_n, \ n = 0, 1, \dots, N-1. \tag{4}$$

We shall solve the discrete equation (4) by using Newton's method where we solve the corresponding linear problems successively for $U|_{\Omega \times I_n}$.

3   THE A POSTERIORI ERROR ESTIMATE

We now define the stability factors S_j that occur in the a posteriori error estimate. We introduce the following linearized dual problem associated to (1) and (3):

$$-\varphi_t - \mathcal{E}\Delta\varphi = J^t(u, U)\varphi \quad \text{in } \Omega \times I,$$
$$\frac{\partial\varphi}{\partial n} = 0 \qquad \text{on } \partial\Omega \times I, \tag{5}$$
$$\varphi(\cdot, t_N) = e_N^- \qquad \text{in } \Omega,$$

where

$$J(u,U) = \int_0^1 \mathcal{J}(su + (1-s)U)\,ds, \quad \mathcal{J}_{ij}(u) = \left(\frac{\partial f_i(u)}{\partial u_j} \right),$$

and J^t denotes the transpose of J. We assume that there are stability factors s_j such that

$$\|\varphi\|_I + \left(\int_I (\mathcal{E}\nabla\varphi, \nabla\varphi)\,dt \right)^{\frac{1}{2}} \le s_0\|e_N^-\|,$$

$$\left(\int_I (t_N - t)\|\mathcal{E}\Delta\varphi\|^2\,dt \right)^{\frac{1}{2}} \le s_1\|e_N^-\|,$$

$$\left(\int_I (t_N - t)\|\varphi_t\|^2\,dt \right)^{\frac{1}{2}} \le s_2\|e_N^-\|, \tag{6}$$

$$\left\| \int_t^{t_N} \mathcal{E}\Delta\varphi(s)\,ds \right\| \le s_3\|e_N^-\|, \quad t \in I_N,$$

where φ is the solution of (5). We define

$$S_1 = \max(s_1, s_3), \quad S_2 = \max(s_0, s_2).$$

We now state the a posteriori error estimate for (3) (a proof is given in [7]).

THEOREM 3.1   Let u be the solution of (1) and U the solution of (4), assume that $\Omega \subset \mathbb{R}^2$ is a bounded convex domain and that the estimates (6) for the dual solution φ holds. Then there are constants C_i such that

$$\|u - U\|_{\bar{I}_N} \le S_1 L_N \max_{n \le N-1} \left\{ C_1\|h_n^2 f(U)\|_{I_n} + C_2\|h_n^2 \mathcal{D}(U)\|_{I_n} \right\}$$

$$+ S_2 L_N \max_{n \le N-1} \left\{ C_3 \left\| k_n^2 \frac{d}{dt} f(U) \right\|_{I_n} + C_4 \left\| k_n^2 \frac{d}{dt} \mathcal{E}\Delta_h U \right\|_{I_n} + C_5 \| [U_n] \| \right\}$$

$$+ s_3 \left\{ C_6 \left\| h_{N-1}^2 k_{N-1} \mathcal{D}(U_t) \right\|_{I_{N-1}} + C_7 \left\| h_{N-1}^2 k_{N-1} \frac{d}{dt} f(U) \right\|_{I_{N-1}} \right\}$$

$$+ s_1 L_N \max_{n \le N-1} C_8 \left\| \frac{h_n^2}{k_n} [U_n] \right\|^*, \tag{7}$$

where $\bar{I}_N = (0, t_N)$, $L_N = \max_{n \le N-1} \left(1 + \log\frac{t_n}{k_n} \right)$, $h(x,t) = h_{\mathcal{K}}$ on $\mathcal{K} \times I_n, \mathcal{K} \in \mathcal{T}_n$, $k(t) = k_n$ on I_n and $\mathcal{D}(U)(x,t) = \max_{y \in \partial\mathcal{K}} \frac{1}{h_{\mathcal{K}}} \left| \mathcal{E}\left[\frac{\partial U(y,t)}{\partial n_{\mathcal{K}}} \right] \right|$, $x \in \mathcal{K}$. The $*$ indicates that the corresponding norm is not present if W_{n-1} is a subspace of W_n.

4   THE ADAPTIVE ALGORITHM

In order to achieve error control of the form

$$\|e_-^N\| \le TOL, \quad N = 0, 1, 2, \dots, \tag{8}$$

where TOL is a given tolerance, we present the following adaptive algorithm.

(A) Make an initial mesh, "guess" values on the stability factors, and the initial timestep ($n = 0$).

(B) Solve the discrete equation (4) approximately, by using a Newton's method where we solve the corresponding linear problems.

(C) Solve the linearized dual problem for different "initial data", and compute approximations of the stability factors s_j, see discussion below.

(D) Check with the a posteriori estimate (7) if the error is below TOL, if not, refine the mesh and calculate a new timestep, using the a posteriori estimate and go back to (B).

(E) If the time t_n is less than t_N, compute a new timestep and a new mesh, so that the a posteriori estimate is satisfied, update the solution vectors and calculate the new t_n and go to (B).

In the algorithm above, the stability factors s_j must, in general, be computed numerically. This is done by solving the dual problem (5) with u approximated by U, since u is unknown. In principle it would be sufficient to solve the dual problem with the error e_N^- as initial data, compute the left hand sides in (6) and evaluate s_j. However, the error is not known, so another methodology must be used. We have solved the dual problem with different types of initial data and chosen the maximum of the values of s_j that were obtained. It is natural to expect that s_j is fairly independent of the initial data, so we just need to solve the dual problem for a few different types of initial data, the numerical examples indicates that this is a fact in many cases. Initial data that were chosen was for instance (functions of) the residual, data that were related to the main problem and random data. For the numerical examples given below, the residual and perturbations of the residual were the most efficient choices.

5   EXAMPLES

As a first example, we consider a Volterra–Lotka system with diffusion

$$
\begin{aligned}
u_{1,t} - \varepsilon \Delta u_1 &= u_1(1 - u_2), \\
u_{2,t} - \varepsilon \Delta u_2 &= u_2(u_1 - 1) && \text{in } \Omega \times I, \\
\frac{\partial u}{\partial n} &= 0 && \text{in } \partial\Omega \times I, \\
u(\cdot, 0) &= u_0 && \text{in } \Omega,
\end{aligned}
\tag{9}
$$

where $\Omega \subset \mathbb{R}^2$.

First, we present analytical estimates for a problem where we have linearized around the stationary solution $u_1 = u_2 = 1$. This leads to bounds on the stability factors s_0 by 1, for s_1 by $\frac{1}{2}$, and for s_2 by cT, where c is a constant. Before we state the theorem we point at the corresponding dual problem (5) for (9)

$$
\begin{aligned}
-\varphi_{1,t} - \varepsilon \Delta \varphi_1 &= -\varphi_2 \\
-\varphi_{2,t} - \varepsilon \Delta \varphi_2 &= \varphi_1 && \text{in } \Omega \times [t, T] \\
\frac{\partial \varphi}{\partial n} &= 0 && \text{on } \partial\Omega \times [t, T],
\end{aligned}
\tag{10}
$$

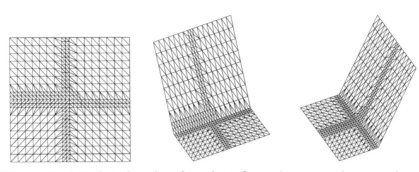

Figure 1   Initial mesh and surface plots of u_0; $\min u_{1,0} = \min u_{2,0} = 1$, $\max u_{1,0} = \max u_{2,0} = 1.25$.

and $\varphi(T)$ is initial data chosen to be e_N^- in the a posteriori error estimate.

By the following theorem (proved in [7]), we see that S_1 is bounded by a moderate constant and that S_2 is bounded by $\max(1, cT)$ for the constant solution $u_1 = u_2 = 1$. The stability factor s_3 is not included in the theorem, but can be bounded by a moderate constant by obvious estimates using equation (9) and the weak stability estimate, however, this approach won't give a sharp estimate.

THEOREM 5.1   We have the following stability estimate for the solution φ in (10)

$$\frac{1}{2}\|\varphi(t)\|^2 + \varepsilon \int_t^T \|\nabla\varphi(s)\|^2 \, ds \leq \frac{1}{2}\|\varphi(T)\|^2,$$

$$\varepsilon \int_t^T (T-s)\|\Delta\varphi(s)\|^2 \, ds \leq \frac{1}{4}\|\varphi(T)\|^2, \tag{11}$$

$$\int_t^T (T-s)\|\varphi_t(s)\|^2 \, ds + c\|\nabla\varphi(t)\|^2 \leq \frac{1+(T-t)^2}{2}\|\varphi(T)\|^2.$$

For information about strong stability of problem (9) we have done some numerical calculations which indicates that the stability factor S_1 can be bounded by 1 and S_2 by $\max(1, cT)$, where c is a constant.

In the calculations we have used $\Omega = [0,1] \times [0,1]$, $\varepsilon = 0.001$, with initial data as in Fig. 1. The strong stability factor S_1 was evaluated for 100 random polynomial initial data and for 100 random sums of Gauss-functions as initial data at time $T = 3.0$ and is plotted in Fig. 2. Calculations of the strong stability factor S_2 indicates a linear growth in time T for large T, see Fig. 3. In practice, in the calculations only a few initial data for the backward problem is necessary. For instance with the residual and with perturbated residuals as initial data we have $S_2 \sim 2.05$ at $T = 3.0$ in our calculations, as can be seen in Fig. 3, whereas if the random polynomial and Gauss-functions are used as the initial data instead, we will have $S_2 \sim 1.90$ at $T = 3.0$. In addition, the stability factor S_1 will be similar using the different approaches described above which indicates that it possible to only use a few initial data for the backward problem if the "mass" of the initial data is concentrated where the error is largest.

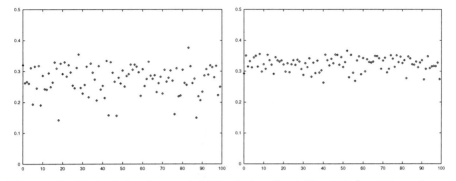

Figure 2 Plot of the strong stability factor S_1 at $T = 3.0$ for 100 random polynomial initial datas, and for 100 random sums of Gauss-functions as initial datas.

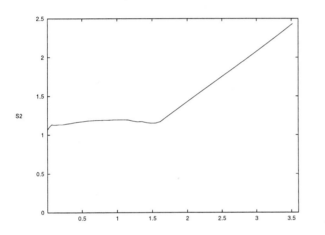

Figure 3 Plot of the strong stability factor S_2 as a function of T.

As a second example, we consider a combustion problem, which consists of a system with two equations: one for the energy, and one for the combustible material:

$$u_{1,t} - \varepsilon_\kappa \Delta u_1 = A e^{-\frac{E_a}{R u_1}} u_2,$$

$$u_{2,t} - \varepsilon_D \Delta u_2 = -B e^{-\frac{E_a}{R u_1}} u_2, \quad \text{in } \Omega \times I,$$

$$\frac{\partial u}{\partial n} = 0, \qquad\qquad \text{in } \partial\Omega \times I, \tag{12}$$

$$u(\cdot, 0) = u_0, \qquad\qquad \text{in } \Omega,$$

where u_1 is the temperature, u_2 is the density of the combustible material, $\Omega \subset \mathbb{R}^2$, and A, B, R and E_a are positive constants.

In the numerical calculations we have used the domain $\Omega = [0,1] \times [0,1]$. In the first calculation we have $\varepsilon_\kappa = 0.0001$, $\varepsilon_D = 0.001$, $A = B = 1$ and $\frac{E_a}{R} = \frac{1}{2}$. The initial mesh and data for $u_{1,0}$ is given in Fig. 4, whereas $u_{2,0} \equiv 1$. The combustion profile will be nonmonotone for $t > 0$ in this case, which can be seen in Fig. 5. The combustible material will burn very quickly with a density close to zero in the

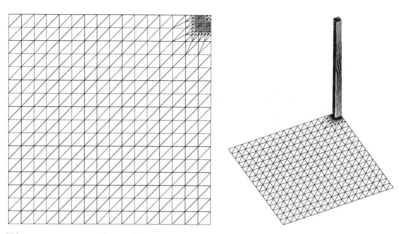

Figure 4   Initial mesh and initial data $u_{1,0}$.

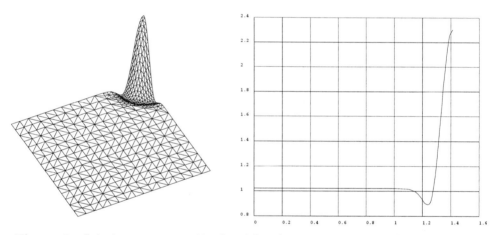

Figure 5   Solution u_1 at $t = 10$, the right plot is a crosssection from the lower (left) corner to the upper (right) corner.

entire region at $t = 15$, this can be seen reflected in the stability factor S_2 shown in Fig. 6, the stability factor S_1 is here less than 4 on the particular time interval. This example shows that, for a fix tolerance, we can use a smaller stability factor S_2 if we only are interested in the solution after a long time than if we require the solution at a time level where the combustion still is active.

In the second calculation we have used the same initial data as above except for $\dfrac{E_a}{R} = 1$, in this case the solution will be a reaction front moving slowly with corresponding stability factors S_j growing linearly in time T, see Fig. 6 for the plot of S_2, here $S_1 < S_2$.

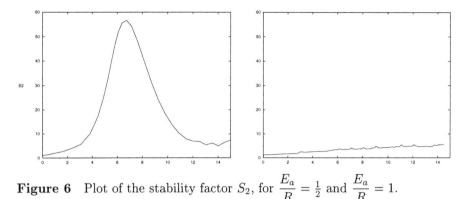

Figure 6 Plot of the stability factor S_2, for $\dfrac{E_a}{R} = \dfrac{1}{2}$ and $\dfrac{E_a}{R} = 1$.

REFERENCES

1. K. Eriksson, D. Estep, P. Hansbo, and C. Johnson, "Adaptive finite element methods," North Holland (to appear).
2. K. Eriksson and C. Johnson, *Adaptive finite element methods for parabolic problems I: A linear model problem*, SIAM J. Numer. Anal. **28** (1991), 43–77.
3. K. Eriksson and C. Johnson, *Adaptive finite element methods for parabolic problems II: Optimal error estimates in $L_\infty(L_2)$ and $L_\infty(L_\infty)$*, Preprint 1992–09, Mathematics Dept., Chalmers Univ. of Technology.
4. K. Eriksson and C. Johnson, *Adaptive finite element methods for parabolic problems IV: Nonlinear problems*, Preprint 1992–44, Mathematics Dept., Chalmers Univ. of Technology.
5. K. Eriksson and C. Johnson, *Adaptive finite element methods for parabolic problems V: Long-time integration*, Preprint 1993–04, Mathematics Dept., Chalmers Univ. of Technology.
6. K. Eriksson and C. Johnson, *Adaptive streamline diffusion finite element methods for stationary convection-diffusion problems*, Math. Comp. **60** (1993), 167–188.
7. R. Sandboge, *Adaptive finite element methods for systems of reaction-diffusion equations*.

Pointwise error estimates, superconvergence and extrapolation

ALFRED H. SCHATZ*   Department of Mathematics, Cornell University, Ithaca, NY 14853, USA, e-mail: schatz@math.cornell.edu

0   INTRODUCTION

Here, we shall roughly describe a variety of recent results concerning superconvergence and extrapolation for the finite element method for second order elliptic problems in $\mathbb{R}^N$, $N \geq 2$. The results are based on some new local pointwise estimates. An outline of the paper is as follows.

Section 1 contains preliminaries.

In Section 2 we discuss the local pointwise error estimates which are valid for a large variety of finite element spaces on irregular quasi-uniform grids.

In Section 3 we present some consequences of these estimates which are weak error expansions in the form of inequalities.

Sections 4, 5 and 6 contain some consequences of these error expansion inequalities. In particular in Section 4, conditions are given under which the L_∞ error on an element is dominated by the interpolation error on that element. Section 5 contains a new approach to Richardson extrapolation and Section 6 contains some new superconvergence results.

I would like to thank Gabriel Wittum and the people at ICA III, University Stuttgart for their kind hospitality while this paper was written.

---

*This work was supported in part by NSF Grant DMS 9403512, and ICA III, University Stuttgart, Pfaffenwaldring 27, 70569 Stuttgart.

1 PRELIMINARIES

Let Ω be a bounded domain in $\mathbb{R}^N$ with boundary $\partial\Omega$, and for convenience consider Dirichlet's Problem

$$Lu = \sum_{i,j=1}^{N} \frac{\partial}{\partial x_j}\left(a_{ij}(x)\frac{\partial u}{\partial x_i}\right) + \sum_{i=1}^{N} b_i(x)\frac{\partial u}{\partial x_i} + c(x)u = f \quad \text{in } \Omega, \qquad (1.1)$$

$$u = 0 \quad \text{on } \partial\Omega. \qquad (1.2)$$

For ease of presentation it will be assumed that the coefficients a_i, b_i and c are sufficiently smooth, say $C^\infty(\overline{\Omega})$, and furthermore that L is uniformly elliptic, i.e., there exists a constant m_{e_u} such that for all $x \in \Omega$

$$m_{e_{\ell\ell}}|\zeta|^2 \le \sum_{i,j=1}^{N} a_{ij}(x)\zeta_i\zeta_j \quad \text{for all } \zeta \in \mathbb{R}^n. \qquad (1.3)$$

Let $u \in \overset{\circ}{W}{}^1_2(\Omega)$ be a weak solution of

$$A(u,v) = (f,v) \quad \text{for all } v \in \overset{\circ}{W}{}^1_2(\Omega), \qquad (1.4)$$

where

$$A(u,v) = \int_{\Omega}\left(\sum_{i,j=1}^{N} a_{ij}\frac{\partial u}{\partial x_i}\frac{\partial v}{\partial x_j} + \sum_{i=1}^{N} b_i\frac{\partial u}{\partial x_j}v + cuv\right) dx \qquad (1.5)$$

and

$$(f,v) = \int_{\Omega} fv\,dx. \qquad (1.6)$$

Consider now a finite element approximation u_h to u. To this end let $0 < h \le 1/2$ be a parameter, $r \ge 2$ be an integer and $S^h_r(\Omega) \subset W^1_\infty(\Omega)$ be a family of finite element spaces. We can allow rather general classes of such spaces but for the purposes of this lecture they may be thought of as any one of a variety of spaces of continuous functions which, on each set τ of a quasi-uniform partition of Ω of roughly size h, contain all polynomials of degree $r - 1$. For example, $r = 2$ could correspond to piecewise linear or bilinear functions and $r = 3$ could correspond to piecewise quadratic functions, etc. Thus, they can approximate sufficiently smooth functions to order h^r in $L_\infty(\Omega)$ and h^{r-1} in $W^1_\infty(\Omega)$.

Although the local results in Sections 2 and 3 are valid up to plane boundaries, for simplicity only interior estimates will be described. For any subdomain $\Omega_0 \subset \Omega$, $S^h_r(\Omega_0)$ will denote the restriction of $S^h_r(\Omega)$ to Ω_0 and $\overset{\circ}{S}{}^h_r(\Omega_0)$ will denote the subspace of $S^h_r(\Omega_0)$ consisting of functions with support in Ω_0. Let $B_d(x)$ be the ball of radius $d > 0$ about x

$$B_d(x) = \{y : |y - x| < d\}.$$

Now for $B_d(x) \subset \Omega$, let $u - u_h$ satisfy the local equations

$$A(u - u_h, \varphi) = 0 \quad \text{for all } \varphi \in \overset{\circ}{S}{}^h_r(B_d(x)). \qquad (1.7)$$

2   SOME LOCAL ERROR ESTIMATES

As mentioned above, the results that will be described later on depend on some new local pointwise a priori estimates for the error $(u - u_h)(x)$ or $\dfrac{\partial}{\partial x_i}(u - u_h)(x)$ which will now be discussed. First some notation. For each fixed point $x \in \overline{\Omega}$, real number s, and arbitrary $y \in \mathbb{R}^N$, consider the weight function

$$\sigma_{x,h}^s(y) = \left(\frac{h}{(|x - y|^2 + h^2)^{1/2}} \right)^s. \tag{2.1}$$

Notice that if $s > 0$ and $|x - y| = 0(h)$ then $\sigma_{x,h}^s(y) = 0(1)$, and if $|x - y| = 0(1)$ then $\sigma_{x,h}^s(y) = 0(h^s)$. The following results in Theorems 2.1 and 2.2 represent a sharpening of the interior estimates given in Schatz and Wahlbin [14] and [15].

THEOREM 2.1   Suppose that $u \in L_\infty(B_d(x))$ and $u_h \in S_r^h(B_d(x))$ satisfy (1.7), where $B_d(x) \subset \Omega$. Let $0 \le s \le r - 1$, $1 \le p \le \infty$, and t be a non-negative integer. Then there exist constants M and C which depend at most $m_{e\ell\ell}$ various norms of the coefficients of $A(\cdot, \cdot)$ and constants associated with approximation and inverse properties of $S_r^h(B_d(x))$, such that if $d \ge Mh$

$$\|u - u_h\|_{L_\infty(B_h(x))} \le C \left(\ln \frac{d}{h} \right)^{\overline{s}} \left(\inf_{\chi \in S_r^h(B_d(x))} \|\sigma_{x,h}^s(u - \chi)\|_{L_\infty(B_d(x))} \right)$$
$$+ Cd^{-N/p-t}\|u - u_h\|_{W_p^{-t}(B_d(x))}. \tag{2.2}$$

Here $\overline{s} = 1$ if $s = r - 2$ and $\overline{s} = 0$ otherwise.

THEOREM 2.2   Suppose that the conditions of Theorem 2.1 hold except where $u \in W_\infty^1(B_d(x))$ and s is chosen in the range $0 \le s < r - 1$, then with M and C depending as above and $d \ge Mh$

$$\|u - u_h\|_{W_\infty^1(B_h(x))} \le C \left(\inf_{\chi \in S_r^h(B_d(x))} \|\sigma_x^s(u - \chi)\|_{W_\infty^1(B_d(x))} \right)$$
$$+ Cd^{-N/p-t-1}\|u - u_h\|_{W_p^{-t}(B_d(x))}. \tag{2.3}$$

Here, for a set G, $W_p^{-t}(G)$ denotes the dual of $\overset{\circ}{W}_q^t(G)$ with $\dfrac{1}{p} + \dfrac{1}{q} = 1$.

REMARK 2.3   As shown in Nitsche and Schatz [8] for L_2 based norms and in Schatz and Wahlbin [14], [15] in L_∞ norms, the local error is characterized by a local truncation error plus the error in an arbitrary negative norm. This latter term measures the pollution error, i.e., influences from outside of $B_d(x)$.

REMARK 2.4   When $s = 0$ and $\sigma_{x,h}^s \equiv 1$ these results were obtained in [14], [15]. The improvement in those results occurs in (2.2) and (2.3) when $s > 0$, because of the presence of the weight function $\sigma_{x,h}^s$, which implies a more local dependence of the error on u and its derivatives near x.

REMARK 2.5   Pointwise error estimates on irregular grids have been the subject of numerous investigations. We refer the reader to Natterer [5], Scott [17], Nitsche

[6], [7], Frehse and Rannacher [3], Rannacher and Scott [9], Schatz and Wahlbin [14], [15] to name a few. For a further bibliography we refer the reader to Wahlbin [18]. Global versions of Theorems 3.2 and 3.4 are contained in Schatz [10].

3   ERROR EXPANSION INEQUALITIES

Here we shall give some straightforward consequences of (2.2) and (2.3) which will be referred to as error expansion inequalities and are critical in the proofs of the superconvergence and extrapolation results to be discussed later on.

   We shall need some usual approximation assumptions that are satisfied by many commonly used finite element subspaces.

ASSUMPTION 3.1   There is a constant $C > 0$ such that if $d_1 + ch < d_2$ and $B_{d_2}(x) \subset \Omega$, then for each $u \in W_\infty^r(B_{d_2}(x))$ there exists a $\chi \in S_r^h(B_{d_2}(x))$ satisfying

$$\|u - \chi\|_{L_\infty(B_{d_1}(x))} \leq Ch^r \left(\sum_{|\gamma|=r} \|D^\gamma u\|_{L_\infty(B_{d_2}(x))} \right), \qquad (3.1)$$

and

$$\|u - \chi\|_{W_\infty^1(B_{d_1}(x))} \leq Ch^{r-1} \left(\sum_{|\gamma|=r} \|D^\gamma u\|_{L_\infty(B_{d_2}(x))} \right). \qquad (3.2)$$

THEOREM 3.2   Suppose that the conditions of Theorem 1 and (3.1) are satisfied. Let $u \in W_\infty^k(B_d(x))$ where k is an integer, $r + 1 \leq k \leq 2r - 2$ (thus $r \geq 3$).
   i) Then

$$\|u - u_h\|_{L_\infty(B_h(x))} \leq C \left(\ln \frac{d}{h} \right)^{\overline{k}} \left\{ h^r \sum_{|\gamma|=r} |D^\gamma u(x)| + \cdots \right.$$
$$\left. + h^{k-1} \left(\sum_{|\gamma|=k-1} |D^\gamma u(x)| \right) + h^k \|u\|_{W_\infty^k(B_d(x))} \right\}$$
$$+ d^{-\frac{N}{p}-1} \|u - u_h\|_{W_p^{-t}(B_d(x))}. \qquad (3.3)$$

   The constant C is independent of u, u_h, d and x. Also $\overline{k} = 1$ if $k = 2r - 2$ and $\overline{k} = 0$ otherwise.
   ii) Furthermore, if $D^\gamma u(x) = 0$ for all multi-indices $r \leq |\gamma| \leq k - 1$, then

$$\|u - u_h\|_{L_\infty(B_h(x))} \leq C \left(\left(\ln \frac{d}{h} \right)^{\overline{k}} h^k \|u\|_{W^k(B_d(x))} \right.$$
$$\left. + d^{-\frac{N}{p}-t} \|u - u_h\|_{W_p^{-t}(B_d(x))} \right). \qquad (3.4)$$

REMARK 3.3   It is the estimate (3.4) that will be used to derive our superconvergence and extrapolation results. Superconvergence of order h^k (modulo logarithms) will occur if, in addition to the above, the pollution term satisfies

$$d^{-\frac{N}{p}-t}\|u - u_h\|_{W_p^{-t}(B_d(x))} \leq Ch^k. \tag{3.5}$$

The following are analogous estimates for first derivatives of the error.

THEOREM 3.4   Suppose that the conditions of Theorem 2.2 and (3.2) are satisfied. Let $u \in W_\infty^k(B_d(x))$ where k is an integer $r + 1 \leq k \leq 2r - 1$ (thus $r \geq 2$).
   i) Then

$$\|u - u_h\|_{W_\infty^1(B_h(x))} \leq C \left(h^{r-1} \sum_{|\gamma|=r} |D^\gamma u(x)| + \cdots \right.$$

$$+ h^{k-2} \sum_{|\gamma|=k-1} |D^\gamma u(x)| + h^{k-1-\varepsilon}\|u\|_{W_\infty^k(B_d(x))} \bigg)$$

$$+ Cd^{-\frac{N}{p}-t-1}\|u - u_h\|_{W_p^{-1}(B_d(x))}. \tag{3.6}$$

ii) Furthermore if $D^\gamma u(x) = 0$ for all $r \leq |\gamma| \leq k - 1$ then

$$\|u - u_h\|_{W_\infty^1(B_h(x))} \leq C \left(h^{k-1-\varepsilon}\|u\|_{W_\infty^k(B_d(x))} \right.$$

$$+ d^{-\frac{N}{p}-t-1}\|u - u_h\|_{W_p^{-t}(B_d(x))} \bigg). \tag{3.7}$$

REMARK 3.5   A remark similar to Remark 3.3 holds for Theorem 3.4.

REMARK 3.6   The interior estimates of Theorems 2.1 through 3.4 can be easily extended up to plane boundaries where either homogeneous Dirichlet or Neumann conditions hold. Roughly, after appropriately modifying the subspaces to take care of boundary conditions the above theorems hold when distance $(x, \partial\Omega) \leq \dfrac{d}{2}$ and $B_d(x)$ is replaced by $B_d(x) \cap \overline{\Omega}$.

REMARK 3.7   We wish to again emphasize that Theorems 2.1 through 3.4 hold for a large class of finite elements defined on a general class of locally quasi-uniform (irregular) meshes. With regard to the error expansion inequalities, sharper and exact asymptotic error expansions have been derived by several authors for the Laplacian at special points and using special elements on either regular or 2 regular grids in the plane. This was mainly done by Q. Lin and coworkers. We refer the reader to, for example, Lin and Xie [2] (and the literature cited there), and to the book [18] where a comprehensive list of papers may be found. We also recommend the paper by Blum, Lin and Rannacher [1].

4   WHEN THE LOCAL INTERPOLATION ERROR IS DOMINANT

Let $\tau \subset B_h(x)$ be an element of the partition of size h. The expansion inequalities (3.5) and (3.6) suggest conditions under which, asymptotically, the errors $\|u - u_h\|_{L_\infty(\tau)}$ and $\|u - u_h\|_{W^1_\infty(\tau)}$ depend principally on the interpolation error on the element τ. This happens if we make the following reasonable assumptions.

ASSUMPTION 4.1   (Pollution is small compared to local interpolation error). Suppose that for some $d > 0$, $1 \leq p \leq \infty$, $t \geq 0$ and $\varepsilon > 0$

$$d^{-N/p-\varepsilon}\|u - u_h\|_{W^{-t}_p(B_d(x))} \leq Ch^{r+\varepsilon} \tag{4.1a}$$

holds in the case of estimating $\|u - u_h\|_{L_\infty(\tau)}$ or

$$d^{-N/p-t-1}\|u - u_h\|_{W^{-t}_p(B_d(x))} \leq Ch^{r-1+\varepsilon} \tag{4.1b}$$

when estimating $\|u - u_h\|_{W^1_\infty(\tau)}$.

ASSUMPTION 4.2   (The rth order derivatives of u in τ are not too small x compared to the derivatives of order $r + 1$). For some $\varepsilon > 0$

$$\sum_{|\gamma|+r+1} \|D^\gamma u\|_{L_\infty(B_d(x))} \leq h^{-1+\varepsilon} \left(\sum_{|\gamma|=r} \|D^\gamma u\|_{L_\infty(\tau)} \right). \tag{4.2}$$

A consequence of the error expansion inequalities is as follows (see [11]).

COROLLARY 4.3
   i) Suppose that (3.3), Assumptions 4.1 and 4.2 hold. Then

$$C_1 h^r \left(\sum_{|\gamma|=r} \|D^\gamma u\|_{L_\infty(\tau)} \right) \leq \|u - u_h\|_{L_\infty(\tau)}$$

$$\leq C_2 \left(h^r \sum_{|\gamma|=r} \|D^\gamma u\|_{L_\infty(\tau)} + h^{r+\varepsilon} \right). \tag{4.3}$$

   ii) Suppose that (3.6), Assumptions 4.1 and 4.2 hold. Then

$$C_3 h^r \left(\sum_{|\gamma|=r} \|D^\gamma u\|_{L_\infty(\tau)} \right) \leq \|u - u_h\|_{W^1_\infty(\tau)}$$

$$\leq C_4 \left(h^{r-1} \sum_{|\gamma|=r} \|D^\gamma u\|_{L_\infty(\tau)} + h^{r-1+\varepsilon} \right). \tag{4.4}$$

REMARK 4.4   Thus under these conditions the principal errors are essentially asymptotically equivalent to the interpolation error for u on τ. These estimates indicate that for smooth problems, where pollution is small compared to the local interpolation error, the local interpolation error may serve as a basis for an a posteriori error estimator. However, in many practical problems, e.g. corner problems, the pollution error may dominate. In Hoffmann, Schatz and Wittum [4] a local a posteriori error estimator will be analyzed for some smooth problems and model corner problems in the plane.

5   A NEW APPROACH TO RICHARDSON EXTRAPOLATION

In this section we shall use the expansion inequalities of Theorems 3.2 and 3.4 together with some additional ideas to give a new approach to Richardson Extrapolation. In order to do the extrapolation an additional condition will be needed that links the two subspaces that are to be used with different values of h. This condition says that if scaled near a point, one subspace is the same as the other.

DEFINITION 5.1   Let $h < h_1 = \lambda h$ with $\lambda > 1$. Let x be a fixed point in Ω and $d > 0$ so that $B_{\lambda, d}(x) \subset \Omega$. We shall say that $S_r^h(B_d(x))$ is similar to $S_r^{\lambda h}(B_{\lambda d}(x))$ (under a scaling factor λ) if the mapping

$$(T\varphi)(y) = \varphi(x + \lambda(y - x)) \tag{5.1}$$

is an isomorphism $T : S_r^{\lambda h}(B_{\lambda d}(x)) \to S_r^h(B_d(x))$.

The results of this section together with a more detailed discussion of this condition are given in Schatz [12]. As indicated in Remark 3.3 we shall need to assume some nice behavior of the pollution error. For convenience here we take this to be

$$\|u - u_h\|_{W_2^{2-r}(\Omega)} \le Ch^{2r-2}. \tag{5.2}$$

This rate of convergence can be achieved for a variety of problems (see [15] for a discussion). To simplify the results further we shall consider only special cases. Namely, it will be assumed that $A(\cdot, \cdot)$ has only quadratic terms with constant coefficients a_{ij}, i.e.,

$$A(u, v) = \int_\Omega \sum_{i,j=1}^N a_{ij} \frac{\partial u}{\partial x_i} \frac{\partial v}{\partial x_j} \, dx \tag{5.3}$$

and then restrict ourselves further to a two level extrapolation. The following results concerning extrapolation for improving the accuracy of $(u - u_h)(x)$ or $\dfrac{\partial(u - u_h)(x)}{\partial x_i}$ are easily derived from Theorems 3.2 and 3.4 when Definition 5.1 is satisfied. One uses the observation that for arbitrary scalars α and β

$$v_h(y) = \alpha u_h(y) + \beta u_{\lambda h}(x + \lambda(y - x)) \in S_r^h(B_d(x)) \tag{5.4}$$

and satisfies

$$A(v - v_h, \varphi) = 0 \quad \text{for all } \varphi \in \overset{\circ}{S}_r^h(B_d(x)). \tag{5.5}$$

Here

$$v(y) = \alpha u(y) + \beta u(x + \lambda(y - x)),$$

and when α and β are chosen (among other things) so that $D^\gamma v(x) = 0$ for all $|\gamma| = r$ we arrive at the following:

THEOREM 5.2   Suppose that Theorems 3.2 and 3.4 and Definition 5.1 hold and let $S_r^h(B_d(x))$ be similar under scaling to $S_r^{\lambda h}(B_{\lambda d}(x))$, $\lambda > 1$.

i) Let $u \in W_\infty^{r+1}(B_{\lambda d}(x))$, $r \geq 3$ and in (5.4) set

$$\alpha = \frac{\lambda^r}{\lambda^r - 1}, \beta = -\frac{1}{\lambda^r - 1}. \tag{5.6}$$

Then

$$|(u - v_h)(x)| \leq C \left(\ln \frac{d}{h} \right)^{\bar{r}} \left(h^{r+1} \|u\|_{W_\infty^{r+1}(B_{\lambda d}(x))} \right) + C d^{2-r-\frac{N}{2}} h^{2r-2}. \tag{5.7}$$

Here $\bar{r} = 1$ if $r = 3$ and $\bar{r} = 0$ otherwise.

ii) If $u \in W_\infty^{r+1}(B_{\lambda d}(x))$, $r \geq 2$ and in (5.4) is

$$\alpha = \frac{\lambda^{r-1}}{\lambda^{r-1} - 1}, \quad \beta = -\frac{1}{\lambda(\lambda^{r-1} - 1)}.$$

Then for any $1 \leq i \leq N$

$$\left| \frac{\partial(u - \tilde{v}_h)}{\partial x_i}(x) \right| \leq C \left(h^r \|u\|_{W_\infty^{r+1}(B_{\lambda d})} + d^{1-r-\frac{N}{2}} h^{2r-2} \right), \tag{5.8}$$

where

$$\frac{\partial \tilde{v}_h}{\partial x_i}(x) = \lim_{y \to x} \frac{\partial v_h(y)}{\partial x_i}$$

along any line through x on which $\dfrac{\partial v_h}{\partial x_i}$ exists.

REMARK 5.3   Thus, in general, if $d = 1$, one achieves a convergence rate of one higher order (modulo logarithms). Notice that it is not always necessary to take $d = 1$ to obtain this, e.g., if $N = 2$ and $r = 3$, so that

$$d^{1-r-\frac{N}{2}} h^{2r-2} = d^{-3} h^4.$$

Then $d^{-3} h^4 = h^3$ when $d = h^{1/3}$. Thus the estimate (5.8) yields h^3 for derivatives with similarity of the subspaces required only in an asymptotically small neighborhood of x. In general a lower gain in convergence is obtained for $h = 0(h^s)$ for $0 < s < 1$.

REMARK 5.4   The results may be extended to extrapolation on many many levels till an order of accuracy of h^{2r-2} is achieved both for $u - u_h$ or its first derivatives.

REMARK 5.5   The works cited in Remark 3.7 by Lin and other are important to extrapolation. The techniques used there are different than those used here and the results are different. Again we refer the reader to Lin and Xie [2] where references to other papers can be found and to Blum, Lin and Rannacher [1].

6   SOME SUPERCONVERGENCE RESULTS

Here Theorems 3.2 and 3.4 will again be applied, this time to obtain some new superconvergence results for meshes which are symmetric with respect to a point (introduced in Schatz, Sloan and Wahlbin [16]) and meshes which are translation invariant (as in Nitsche and Schatz [7]).

DEFINITION 6.1 Let x and $d > 0$ be given so that $B_d(x) \subset \Omega$. x is said to be a point of symmetry of the mesh if whenever $\varphi(y) \in S_r^h(B_d(x))$ then $\varphi(y) = \varphi(x + (y - x)) \in S_r^h(B_d(x))$. For a discussion of meshes which are symmetric with respect to a point we again refer the reader to [16]. We now give some special cases of results given in [13] which improve on the results in [16].

THEOREM 6.2 Suppose that x is a point of symmetry of the mesh, i.e., that Definition 6.1 is satisfied with $d = 1$.

i) If the conditions of Theorem 3.2 are satisfied, $n \geq 3$ is odd, and if for some $t > 0$ and $1 \leq p \leq \infty$

$$\|u - u_h\|_{W_p^{-t}(B_1(x))} \leq Ch^{r+1}, \qquad (6.1)$$

then

$$|(u - u_h)(x)| \leq Ch^{r+1}. \qquad (6.2)$$

ii) If instead, the conditions of Theorem 3.4 are satisfied, $r \geq 2$ is even, and for some $t \geq 0$ and $1 \leq p \leq \infty$

$$\|u - u_h\|_{W_p^{-t}(B_1(x))} \leq Ch^r, \qquad (6.3)$$

then

$$\left| \left(\frac{\partial u}{\partial x_i} - \frac{\partial \widetilde{u}_h}{\partial x_i} \right)(x) \right| \leq Ch^r. \qquad (6.4)$$

Here

$$\frac{\partial \widetilde{u}_h}{\partial x_i}(x) = \frac{\partial u_h}{\partial x_i}(x)$$

if the latter exists or in general is defined as the average of a left and right limits of $\dfrac{\partial u_h}{\partial x}$ along any line through x where $\dfrac{\partial u_h}{\partial x}$ exists (see [16]).

REMARK 6.3 (6.2) for r odd and (6.4) for r even predict superconvergence at symmetry points with a gain of one order of h for both $(u-u_h)(x)$ and $\left(\dfrac{\partial u}{\partial x_i} - \dfrac{\partial u_h}{\partial x_i} \right)(x)$, respectively. As simple examples take quadratics (where $r = 3$) and we have $|(u - u_h)(x)| \leq Ch^4$. For piecewise linears ($r = 2$) $\left| \left(\dfrac{\partial u}{\partial x_i} - \dfrac{\partial \widetilde{u}_h}{\partial x_i} \right)(x) \right| \leq Ch^2$.

REMARK 6.4 Notice that (6.2) does not give any information for derivatives when t is odd. If x is a symmetry point of a translation invariant mesh then using difference quotients for $r \geq 3$ and odd to approximate derivatives, we may achieve superconvergence of order h^{r+1}. This translates into h^2 better for first derivatives, h^3 better for second derivatives, etc. We will state the result without giving detailed definitions which can be found in [7].

THEOREM 6.5 Suppose that the conditions of part i) of Theorem 6.2 hold and in addition that the mesh is translation invariant of order h on $B_1(x)$. For γ a multi-index let $Q_h^\gamma u(x)$ be a difference operator which approximates $D^\gamma u(x)$ to order h^{r+1} with evaluations only at symmetry points then

$$|(D^\gamma u - Q_h^\gamma u)(x)| \leq Ch^{r+1}. \qquad (6.5)$$

REMARK 6.6 Thus e.g. we can obtain an h^4 approximation to any derivative at nodal points of quadratics or bi-quadratics on a regular grid if a sufficiently accurate difference approximation is used with evaluations only at nodal points.

REMARK 6.7 For an extensive list of publications on superconvergence we again refer the reader to [18].

REFERENCES

1. H. Blum, Q. Lin and R. Rannacher, *Asymptotic error expansions and Richardson extrapolation for linear finite elements*, Numer. Math. **49** (1986), 11–37.
2. Q. Lin and R.-F. Xie, *Error expansions for FEM and superconvergence under natural assumption*, J. Comput. Math. **7** (1989), 402–411.
3. J. Frehse and R. Rannacher, *Eine L^1-Fehlerabschätzung diskreter Grundlösungen in der Methode der Finiten Elemente*, Tagungsband, "Finite Elemente", Bonn. Math. Schriften **89** (1975), 92–114.
4. W. Hoffmann, A. H. Schatz and G. Wittum, *A local finite element error estimator for the finite element method for some problems*, In preparation.
5. F. Natterer, *Über die punktweise Konvergenz finiter element*, Numer. Math. **25** (1975), 67–77.
6. A. Nitsche, *L_∞ convergence of finite element approximations*, Proceedings Second Conference in Finite Elements, Rennes, France (1975).
7. J. A. Nitsche, *L_∞ convergence of finite element approximations*, in "Mathematical Aspects of Finite Element Methods, Lecture Notes in Math.," Springer, 1977, pp. 261–274.
8. J. A. Nitsche and A. H. Schatz, *Interior estimates for Ritz–Galerkin methods*, Math. Comp. **28** (1974), 937–958.
9. R. Rannacher and R. Scott, *Some optimal error estimates for piecewise linear finite element approximations*, Math. Comp. **38** (1982), 437–445.
10. A. H. Schatz, *Point error estimates and asymptotic error expansion inequalities for the finite element method on irregular grids: Part I. Global estimates*, preprint.
11. A. H. Schatz, *Point error estimates and asymptotic error expansion inequalities for the finite element method on irregular grids: Part II. Local estimates*.
12. A. H. Schatz, *A new approach to Richardson extrapolation in the finite element method*, preprint.
13. A. H. Schatz, *A mixture of superconvergence results in the finite element method*.
14. A. H. Schatz and L. B. Wahlbin, *Interior maximum norm estimates for finite element methods*, Math. Comp. **31** (1977), p. 414–442.
15. A. H. Schatz and L. B. Wahlbin, *Interior maximum norm estimates for finite element methods, Part II*, Math. Comp. **211** (1995), 907–928.
16. A. H. Schatz, I. Sloan and L. B. Wahlbin, *Superconvergence in the finite element method and meshes which are locally symmetric with respect to a point*, SIAM J. Numer. Anal. (to appear).
17. R. Scott, *Optimal L_∞ estimates for the finite element method on irregular grids*, Math. Comp. **30** (1970), 681–697.

18. L. B. Wahlbin, *Superconvergence in Galerkin Finite Element Methods*, in "Lecture Notes in Mathematics," Springer, 1995.

Approximate solution of problem on viscous flow with evaporating non-compact free boundary

ANTON A. SMOLIANSKI   University of Jyväskylä, Laboratory of Scientific Computing, P.O. Box 35, FIN–40351 Jyväskylä, Finland

Abstract   The stationary problem on planar flow of viscous incompressible fluid with evaporating non-compact free boundary is considered. The solvability of this problem and smoothness of its solution are studied. The technique of artificial restriction of the domain is used for obtaining the approximate solution. It is shown that if the restricted domain is sufficiently wide, then the solution on it differs little from that on the original domain. A finite-element approximation of the problem on the restricted domain is constructed. For this the curvilinear C^1-elements are used along the free boundary to approximate velocity and temperature and C^0-elements – for pressure. Special algorithm for mesh refinement in the vicinity of the corner point of the boundary is presented to obtain optimal estimates of approximation. Also estimates for the rate of convergence of the finite-element solution to the exact one are given.

We consider the problem on viscous flow with free evaporating surface, which has numerous applications in coating and drying processes encountered during the production of paper and polymers. The surface of the fluid is non-compact, and the gravitational field and heat sources are taken into account as external factors.

1   INTRODUCTION

The problem on steady flow of heavy viscous incompressible fluid leaking out of a narrow channel and spreading along an infinite bottom is studied. The bottom makes an angle α with the horizon. We assume that the motion of the fluid is plane-parallel and consider the problem in the plane $\mathbb{R}^2$ with fixed Cartesian coordinates. The Poiseuille flow is prescribed in the channel, and the bottom of the channel is moving with a given constant velocity. The flow takes place in a steady temperature field which is governed by a given distribution of heat sources and given temperature of the bottom, the upper wall of the channel and the environment in proximity to the free surface of the fluid.

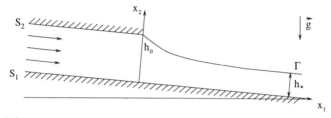

Figure 1

There are two types of singularities: the corner point and non-compactness of the free boundary.

Our problem is, in essence, close to the one whose classical solvability was considered by K. Pileckas ([4], [5]), however, it is complicated by the presence of (steady) temperature field and evaporation of the free surface of the fluid.

2   MATHEMATICAL MODEL AND SOLVABILITY OF THE PROBLEM

The setting of our problem is as follows. Let Ω be the domain bounded by the bottom of the channel S_1, its upper wall S_2 and the free surface of the fluid Γ (see Fig. 1). Note that Ω has two "exits" to infinity: at $x_1 \to -\infty$ and at $x_1 \to +\infty$.

It is required to find the velocity vector $\vec{v}(x)$, the pressure $p(x)$ and the temperature $\theta(x)$ satisfying in Ω the system of Navier–Stokes equations

$$-\nu\Delta\vec{v} + (\vec{v}\cdot\nabla)\vec{v} + \frac{1}{\rho}\nabla p = -\nabla G, \qquad \nabla\cdot\vec{v} = 0, \tag{1}$$

the heat conductivity equation

$$-\lambda\Delta\theta + \vec{v}\cdot\nabla\theta = f \tag{2}$$

and the boundary conditions

$$\vec{v}|_{S_1} = (R,0); \qquad\qquad \vec{v}|_{S_2} = (0,0);$$

$$\vec{v}\cdot\vec{n}|_\Gamma = \lambda_1\frac{\partial\theta}{\partial n}\bigg|_\Gamma; \qquad \vec{\tau}\cdot\mathbf{S}(\vec{v})\cdot\vec{n}|_\Gamma = 0; \tag{3}$$

$$\theta|_{S_1} = \theta_1 + \theta_\infty; \qquad\qquad \theta|_{S_2\cup\Gamma} = \theta_2 + \theta_\infty. \tag{4}$$

Here $G = g(-x_1\sin\alpha + x_2\cos\alpha)$ is the potential of gravitational field, $\mathbf{S}(\vec{v})$ is the strain tensor; R is a given constant velocity of the bottom of the channel; f, θ_1 and θ_2 are given functions; θ_∞ is a given constant temperature at infinity.

Note that the condition for the normal component of velocity on the free boundary (the so-called kinematic condition) simulates evaporation of the fluid.

It is also required to find the function $x_2 = \varphi(x_1)$, $x_1 > 0$ specifying the free boundary Γ and satisfying the equation (the so-called dynamic condition)

$$\sigma(\theta)\frac{d}{dx_1}\frac{\varphi'(x_1)}{\sqrt{1 + (\varphi'(x_1))^2}} = (-p(x) + \rho\nu\vec{n}\cdot\mathbf{S}(v(\vec{x}))\cdot\vec{n})|_{x_2=\varphi(x_1)}, \tag{5}$$

where $\sigma(\theta)$ is the coefficient of surface tension, and the left-hand side contains the curvature of Γ.

At that the boundary conditions for the function $\varphi(x_1)$ read

$$\varphi(0) = h_0, \quad \text{where } h_0 \text{ is the height of the channel,}$$

$$\lim_{x_1 \to +\infty} \varphi(x_1) = h_*, \quad \text{where } h_* \text{ is the elevation of fluid at infinity.} \tag{6}$$

h_* is a priori unknown and is derived from a physically justified requirement that the pressure is bounded at $x_1 \to +\infty$.

To prove the unique solvability of problem (1)–(6) in weighted Sobolev spaces, we use the method of splitting the complete problem into two auxiliary problems: (1)–(4) with a fixed boundary and (5)–(6) for finding the unknown boundary Γ (see, for example, [6], [7], [8], [9], [10], [14], [15], [13], [2], [1]). The analysis of solvability is carried out by a method proposed in [6], [12], [16] and is based on the coercive estimates for solutions of Stokes' problem and linearized problems for determining the temperature and the unknown boundary. This approach enables us to obtain the theorem below. Here we use the weighted Sobolev spaces $H_\mu^{(i)}(\Omega; b)$ $(i = 1, 2, \ldots)$ with power weight in vicinity of the corner point and exponential weight at infinity:

$$\|u\|_{H_\mu^{(i)}(\Omega;b)} = \|u\|_{H_\mu^{(i)}(\Omega_0)} + \|u \cdot \exp(-bx_1)\|_{H^i(\Omega_-)} + \|u \cdot \exp(bx_1)\|_{H^i(\Omega_+)}, \tag{7}$$

where $\Omega_0 = \{x \in \Omega \mid |x_1| < 2\}, \Omega_+ = \{x \in \Omega \mid x_1 > 1\}, \Omega_- = \{x \in \Omega \mid x_1 < -1\}$ and

$$\|u\|_{H_\mu^{(i)}(\Omega_0)} = \|u\|_{H^{i-1}(\Omega_0)} + \left(\sum_{|\gamma|=i} \int_{\Omega_0} d^\mu(x) |D^\gamma u(x)|^2 \, dx \right)^{1/2}.$$

$d(x)$ is a distance of point $x = (x_1; x_2)$ from the corner point $A = (0; h_0)$. Also we use the Slobodetski–Sobolev spaces $W_2^{l/2}(S_1; b), W_{2,\mu}^{l/2}(S_2 \cup \Gamma; b), W_{2,\mu}^{l/2}(\mathbb{R}_+^1; b)$ $(l = 1, 3, 5)$ the norms in which are defined analogously.

THEOREM 1   Let the numbers $h_0, h_* > 0$ satisfy the inequality $|h_0 - h_*| < \sqrt{\dfrac{2}{\beta_0}}$, where $\beta_0 = \dfrac{\rho g}{\sigma(\theta_\infty)}$. Let $f \in H_\mu^{(1)}(\Omega; b), \theta_1 \in W_2^{5/2}(S_1; b), \theta_2 \in W_{2,\mu}^{5/2}(S_2 \cup \Gamma; b)$.

Then there exist such numbers $b_* > 0, \delta_* \in (0; 1), R_*, Q_*, \alpha_*, f_*, \theta_{1*}, \theta_{2*}$ – all positive, that at

$$|R| < R_*, \quad |Q| < Q_*, \quad \alpha \in (0; \alpha_*), \quad \|f\|_{H_\mu^{(1)}(\Omega;b)} < f_*,$$

$$\|\theta_1\|_{W_2^{5/2}(S_1;b)} < \theta_{1*}, \quad \|\theta_2\|_{W_{2,\mu}^{5/2}(S_2 \cup \Gamma;b)} < \theta_{2*},$$

where $b \in (0; b_*), \mu > 2(1 - \delta), \delta \in (0; \delta_*)$, the problem (1)–(6) has a unique solution

$$\{\vec{v}(x), p(x), \theta(x), \varphi(x)\} \in H_\mu^{(2)}(\Omega; b) \times H_\mu^{(1)}(\Omega; b) \times H_\mu^{(2)}(\Omega; b) \times W_{2,\mu}^{5/2}(\mathbb{R}_+^1; b).$$

Note that the exact value of parameter δ determining the behaviour of solution in vicinity of the corner point is a priori unknown and depends on the magnitude of the angle.

3 RESTRICTION OF THE PHYSICAL DOMAIN AND CONNECTION BETWEEN SOLUTIONS ON THE RESTRICTED AND ORIGINAL DOMAINS

For solving our problem approximately, we restrict the original domain, so as to eliminate the exits to infinity. It can be shown that if we choose the bounded domain to be sufficiently wide, the solution will be close to the solution of the problem in the original unbounded domain.

So, let us consider in the domain $\Omega_L = \{x \in \Omega \mid |x_1| < L\}$ the problem: find the velocity vector $\vec{v}_L(x)$, the pressure $p_L(x)$, the temperature $\theta_L(x)$ and the function $\varphi_L(x)$ specifying the free boundary which satisfy the system of Navier–Stokes equations

$$-\nu\Delta\vec{v}_L + (\vec{v}_L \cdot \nabla)\vec{v}_L + \frac{1}{\rho}\nabla p_L = -\nabla G; \qquad \nabla \cdot \vec{v}_L = 0, \tag{8}$$

the heat conductivity equation

$$-\lambda\Delta\theta_L + \vec{v}_L \cdot \nabla\theta_L = f, \tag{9}$$

and the boundary conditions

$$\vec{v}_L|_{\Sigma_1} = (R,0); \quad \vec{v}_L|_{\Sigma_2} = (0,0); \quad \vec{v}_L \cdot \vec{n}|_{\Gamma_L} = \lambda_1 \left.\frac{\partial\theta_L}{\partial n}\right|_{\Gamma_L};$$

$$\vec{v}_L|_{\Sigma_3} = \vec{v}^-; \quad \vec{v}_L|_{\Sigma_4} = \vec{v}^+; \quad \vec{\tau} \cdot \mathbf{S}(\vec{v}_L) \cdot \vec{n}|_{\Gamma_L} = 0; \tag{10}$$

$$\theta_L|_{\Sigma_1} = \tilde{\theta}_1 + \theta_\infty; \quad \theta_L|_{\Sigma_2 \cup \Gamma_L} = \tilde{\theta}_2 + \theta_\infty;$$

$$\theta_L|_{\Sigma_3} = \theta_\infty; \quad \theta_L|_{\Sigma_4} = \theta_\infty; \tag{11}$$

$$\frac{d}{dx_1} \frac{\varphi'_L(x_1)}{\sqrt{1 + (\varphi'_L(x_1))^2}} = \frac{1}{\sigma(\theta_L)}(-p_L(x) + \rho\nu\vec{n} \cdot \mathbf{S}(\vec{v_L}) \cdot \vec{n})|_{\Gamma_L}; \tag{12}$$

$$\varphi_L(0) = h_0; \quad \varphi'_L(L) = 0. \tag{13}$$

Here $\Sigma_1 = S_1 \cap \partial\Omega_L$ is a part of the bottom of the cannel, $\Sigma_2 = S_2 \cap \partial\Omega_L$ is a part of its upper wall, Σ_3 and Σ_4 are "left" and "right" newly formed boundaries of our domain. Also $\Gamma_L = \Gamma \cap \partial\Omega_L$ is a part of free boundary. The functions $\vec{v}^-$ and $\vec{v}^+$ specify the asymptotics of velocity at $x_1 \to -\infty$ and $x_1 \to +\infty$, respectively. These functions as well as the asymptotics for pressure are found from the problem (1)–(4) on original domain with fixed boundary. The functions $\tilde{\theta}_1$ and $\tilde{\theta}_2$ are smooth truncations of θ_1 and θ_2, respectively, so that $\tilde{\theta}_1 = 0$ and $\tilde{\theta}_2 = 0$ when $|x_1| \geq L$.

The unique solvability of this problem under the assumption of smallness of initial data is proved as for the unbounded domain.

Let us now study the connection of solution of the problem in restricted domain with that in the original one.

The following theorem holds:

THEOREM 2   Let the conditions of Theorem 1 on unique solvability of the problem on unbounded domain hold. Let data of the problem be small enough:

$$|R| < R^0, \quad |Q| < Q^0, \quad \alpha \in (0; \alpha^0), \quad \|f\|_{H_\mu^{(1)}(\Omega;b)} < f^0,$$

$$\|\theta_1\|_{W_2^{5/2}(S_1;b)} < \theta_1^0, \quad \|\theta_2\|_{W_{2,\mu}^{5/2}(S_2\cup\Gamma;b)} < \theta_2^0,$$

where f^0, θ_1^0, θ_2^0, R^0, Q^0, α^0 are some positive numbers.

Then for any $\varepsilon > 0$ there can be found such sufficiently large $L = L(\varepsilon)$ that we have

$$\|\vec{v}_L - \vec{v}\|_{H_\mu^{(2)}(\Omega_L)} + \|p_L - p\|_{H_\mu^{(1)}(\Omega_L)} + \|\theta_L - \theta\|_{H_\mu^{(2)}(\Omega_L)} < \varepsilon$$

and

$$\|\varphi_L - \varphi\|_{W_{2,\mu}^{5/2}((0;L))} < \varepsilon,$$

where $\{\vec{v}_L, p_L, \theta_L, \varphi_L\}$ and $\{\vec{v}, p, \theta, \varphi\}$ are the solutions of problems in the bounded and original unbounded domain, respectively.

4   FINITE-ELEMENT SOLUTION OF PROBLEM ON THE RE-STRICTED DOMAIN

The last theorem enables us to limit ourselves to constructing an approximate solution in the restricted domain. First, we construct an approximation of the space $H_\mu^{(2)}(\Omega_L) \times H_\mu^{(1)}(\Omega_L) \times H_\mu^{(2)}(\Omega_L) \times W_{2,\mu}^{5/2}((0;L))$ containing the exact solution $\{\vec{v}_L, p_L, \theta_L, \varphi_L\}$ with appropriate finite element spaces. At that the space $H_\mu^{(2)}(\Omega_L)$ (i.e., the velocity and temperature) is approximated by the space $(H_\mu^{(2)})_h$ of C^1-elements which are Bell's triangular elements inside of the domain and curvilinear Ženíšek's triangular elements along the free boundary; the space $H_\mu^{(1)}(\Omega_L)$, i.e., the pressure, – by the space $(H_\mu^{(1)})_h$ of C^0-elements which are piecewise-linear inside of the domain (see [18], [19]). The free boundary is approximated by Hermite polynomials of 5-th degree.

For obtaining the optimal estimates of approximation, the grid must be constructed in a special way with mesh refinement in the vicinity of the corner point of the boundary.

The following theorem was proved in correspondence with A. Schatz's analysis ([17]):

THEOREM 3   Let

$$h \in (0;1) \quad \text{– the grid discretization parameter;}$$
$$T \quad \text{– any triangle of the triangulation;}$$
$$h_T \quad \text{– the length of the longest side of the triangle } T;$$
$$d_T \quad \text{– the distance of the triangle } T \text{ from the corner point } (0; h_0);$$
$$\delta \in (0;1) \quad \text{– the parameter determining the behaviour of solution}$$
$$\text{in vicinity of the corner point (see Th. 1).}$$

Let $v \in H^5(\Omega_L)$.

1. (Optimal grid in $L_2(\Omega_L)$). If

$$h_T \leq \begin{cases} Ch^{\frac{2}{1+\delta/2}} & \text{when } d_T = 0, \\ Chd_T^{\frac{1-\delta/2}{2}} & \text{when } d_T > 0, \end{cases}$$

then

$$\|v - v_I\|_{L_2(\Omega_L)} \le Ch^2 \|v\|_{H_\mu^{(2)}(\Omega_L)}.$$

2. (Optimal grid in $H^1(\Omega_L)$). If

$$h_T \le \begin{cases} Ch^{\frac{2}{\delta}} & \text{when } d_T = 0, \\ Chd_T^{1-\frac{\delta}{2}} & \text{when } d_T > 0, \end{cases}$$

then

$$\|v - v_I\|_{H^1(\Omega_L)} \le Ch \|v\|_{H_\mu^{(2)}(\Omega_L)}.$$

Here v_I is a $(H_\mu^{(2)})_h$-interpolant of function v and constants in the estimates are independent of both the longitudinal dimension L of the domain Ω_L and of v.

Note that the grid optimal in H^1 will also be optimal for L_2 but not vice versa.

Analogously one can obtain the interpolation estimates of first order with respect to h for $(H_\mu^{(1)})_h$-interpolant and $(W_{2,\mu}^{5/2})_h$-interpolant in L_2-norm and $W_2^{3/2}$-norm, respectively, considering the grid 2 of Theorem 3.

Using the result of Theorem 3 (see clause 2) we can construct the "optimal" grid on Ω_L.

At first, we describe the splitting of the interval $(0; L)$ with intervals Δ_j ($j = 1, 2, \dots$) to define an approximation of free boundary :

$$\Delta_0 = (0; h^{\frac{2}{\delta}}), \quad \Delta_j = (x_1^{(j)}; x_1^{(j+1)}), \ j = 1, 2, \dots,$$
$$x_1^{(0)} = 0, \quad x_1^{(1)} = h^{\frac{2}{\delta}}, \quad x_1^{(j+1)} = x_1^{(j)} + h(x_1^{(j)})^{1-\frac{\delta}{2}}. \tag{14}$$

We go on constructing Δ_j in this way until $x_1^{(j+1)}$ becomes greater than h_0, after which we cover the remaining part $(0; L)$ with intervals of length h. Then every function from $(W_{2,\mu}^{5/2})_h$ will be a Hermite polynomial of 5-th degree on each Δ_j.

Given the function $\varphi_h \in (W_{2,\mu}^{5/2})_h$ determining the free boundary of Ω_L, we construct the desired grid by the following scheme:

1. $d_0 = h^{\frac{2}{\delta}}$. We draw a circle of radius d_0 with center at the point $A = (0; h_0)$ and triangulate the domain $G_0 = \{x \in \Omega_L \mid \text{dist}(x, A) < d_0\}$ with triangles of size $h^{\frac{2}{\delta}}$.

2. For $j = 1, 2, 3 \dots$ we assume $d_j = d_{j-1} + h(d_{j-1})^{1-\frac{\delta}{2}}$ and triangulate the domain $G_j = \{x \in \Omega_L \mid d_{j-1} < \text{dist}(x, A) < d_j\}$ with triangles of size $h(d_{j-1})^{1-\frac{\delta}{2}}$. At that we substitute chords for all the circular arcs. We continue this process until d_j becomes greater than h_0, after which we cover the remaining part of Ω_L with rectilinear triangles of size h.

3. In the proximity to φ_h we alter the triangulation so that each point $(x_1; \varphi_h(x_1^{(j)}))$, where $x_1^{(j)}$ is an endpoint of Δ_j, becomes a node of the triangulation.

It is easy to see that all the curvilinear sides of the triangles located along the boundary $x_2 = \varphi_h(x_1)$ will be exact polynomials of degree no greater than 5-th

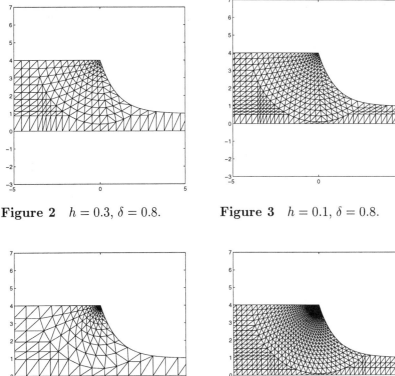

Figure 2 $h = 0.3$, $\delta = 0.8$.   **Figure 3** $h = 0.1$, $\delta = 0.8$.

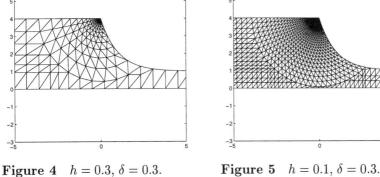

Figure 4 $h = 0.3$, $\delta = 0.3$.   **Figure 5** $h = 0.1$, $\delta = 0.3$.

that are known. This allows us to use an exact mapping of each curvilinear triangle onto the reference one.

It is also not difficult to ascertain that the total number of triangles will be of order Ch^{-2}, i.e., proportional to the number of triangles of a uniform triangulation.

Figures 2–5 corresponding to different values of discretization parameter h and refinement parameter δ illustrate the mesh refinement near the corner point.

Finally, for any function $v \in H_\mu^{(k)}(\Omega_L)$, $k = 1, 2$, we can obtain the estimate of approximation ([**3**], [**9**], [**10**])

$$\|v - v_h\|_{H^{k-1}(\Omega_L)} \leq Ch\|v\|_{H_\mu^{(k)}(\Omega_L)} \tag{15}$$

and also the estimate

$$\|\varphi - \varphi_h\|_{W_2^{3/2}((0;L))} \leq Ch\|\varphi\|_{W_{2,\mu}^{5/2}((0;L))} \tag{16}$$

for any $\varphi \in W_{2,\mu}^{5/2}((0;L))$, where v_h and φ_h are projections on the corresponding finite element spaces.

Approximate solution of our problem splits into two stages: approximate solution of auxiliary problem given a fixed boundary and approximate solution of the boundary-value problem for the free boundary itself.

The investigation of existence of the approximate solution and the rate of its convergence to the exact one was conducted by a theory worked out by V. Rivkind ([**9**], [**10**], [**11**]).

So, the following theorem holds

THEOREM 4   Let

$$\{\vec{v}, p, \theta, \varphi\} \in H_\mu^{(2)}(\Omega_L) \times H_\mu^{(1)}(\Omega_L) \times H_\mu^{(2)}(\Omega_L) \times W_{2,\mu}^{5/2}((0;L))$$

be the exact solution of the problem in the restricted domain Ω_L.

Then for h sufficiently small, there exists the approximate solution

$$\{\vec{v}_h, p_h, \theta_h, \varphi_h\} \in (H_\mu^{(2)})_h \times (H_\mu^{(1)})_h \times (H_\mu^{(2)})_h \times (W_{2,\mu}^{5/2})_h$$

such that

$$\|\vec{v} - \vec{v}_h\|_{H_\mu^{(2)}(\Omega_L)} + \|p - p_h\|_{H_\mu^{(1)}(\Omega_L)} + \|\theta - \theta_h\|_{H_\mu^{(2)}(\Omega_L)}$$
$$+ \|\varphi - \varphi_h\|_{W_{2,\mu}^{5/2}((0;L))} \le C_h, \qquad C_h \to 0 \quad (h \to 0)$$

and

$$\|\vec{v} - \vec{v}_h\|_{H_\mu^1(\Omega_L)} + \|p - p_h\|_{L_2(\Omega_L)} + \|\theta - \theta_h\|_{H_\mu^1(\Omega_L)}$$
$$+ \|\varphi - \varphi_h\|_{W_2^{3/2}((0;L))} \le Ch,$$

where the constants are independent of the longitudinal dimension L of the domain Ω_L.

The proof is based on the fact that the operator of the exact problem is contractive in some ball of the Banach space and approximate operator is also contractive in the finite-dimensional space. This fact allows us to make use of the method of successive approximations to solve our problem.

REFERENCES

1. Erunova I.B., *Solvability of a free boundary problem for two liquids in a container*, Vestnik Leningrad Univ. **19**, No. 2 (1986).

2. Neittaanmäki P., Rivkind V.Ya., *Mathematical modelling of liquid drops evaporation*, in "Jyväskylä–St.Petersburg Seminar on PDE and Numerical Methods," University of Jyväskylä, Dept. of Mathematics, Report 56, 1993, pp. 101–110.

3. Oganesian L.A. Rukhovec L.A., "Variational Difference Methods for the Solution of Elliptic Problems," Izd. Akad. Nauk ArSSR, Jerevan, 1979. (in Russian)

4. Pileckas K.I., *Solvability of a problem on the plane motion of a viscous incompressible liquid with a non-compact free boundary*, Differential equations and their applications, Vilnus **30** (1981), 57–96. (in Russian)

5. Pileckas K.I., *On the problem of motion of heavy viscous incompressible fluid with non-compact free boundary*, Lit. Mat. Sbornik **28, 2** (1988), 315–333. (in Russian)

6. Puhnačev V.V., *The plane stationary problem with a free boundary for the Navier–Stokes equations*, J. Appl. Mech. Thech. Phys. **13** (1972).

7. Puhnačev V.V., *On smoothness of stationary solutions of the Navier–Stokes equations near the free boundary*, Din. Spl. Sredy, Novosibirsk **15** (1973), 133–144. (in Russian)

8. Rivkind V.Ya., *A study of the problem of the stationary motion of a drop in the flow of a viscous incompressible fluid*, Dokl. Akad. Nauk SSSR **227**, no. 5 (1976), 1071–1073. (in Russian)

9. Rivkind V.Ya., *Computational Methods for fluid flows of viscous incompressible fluids with free boundaries*, Chislenn. Meth. Mech. Sploshnoi Sredy, Novosibirsk **12**, No. 4 (1981), 106–115.

10. Rivkind V.Ya., *Approximate Methods for solving problems of viscous fluid with a free boundary*, in "Proc. Numerical Methods and Applications," Sofia, 1985, pp. 91–98. (in Russian)

11. Rivkind V.Ya., *Numerical solution of coupled Navier–Stokes and Stefan equations*, in "Proc. of Conference on Numerical Methods for Free Boundary Problems," P. Neittaanmäki (ed), ISMM99, Birkhäuser, Basel, 1991, pp. 57–68.

12. Rivkind V.Ya., Fridman N.B., *The Navier–Stokes equations with discontinuous coefficients*, Zapiski Nauchn. Seminarov LOMI, Leningrad **38** (1973), 137–152.

13. Socolowsky J., *Solvability of a two-dimensional problem for the motion of two viscous non-compressible liquids with non-compact free boundaries*, Z. Angew. Math. Mech. **72, 7** (1992).

14. Solonnikov V.A., *Solvability of a problem on the plane motion of a heavy viscous incompressible capillary liquid partially filling a container*, Math. USSR Izvestija 4, No. 1 (1980).

15. Solonnikov V.A., *Solvability of the problem on flow of viscous incompressible fluid into an infinite open pool*, Trud. Matem. Inst. im. Steklov **13** (1988), 174–202. (in Russian)

16. Solonnikov V.A., Shchadilov V.E., *On the boundary value problem for the stationary Navier–Stokes equations system*, Trud. Matem. Inst. im Steklov, Leningrad (1973), 125–137. (in Russian)

17. Schatz A.H., Thomée V., Wendland W., "Mathematical Theory of Finite and Boundary Element Methods," Birkhäuser, Basel, 1990.

18. Ženíšek A., *Curved triangular C^m-elements*, Apl. Mat. **23** (1978), 346–377.

19. Ženíšek A., "Nonlinear Elliptic and Evolution Problems and their Finite Element Approximations," Academic Press, 1990.

Shape calculus and FEM in smooth domains

T. TIIHONEN   Department of Mathematics, University of Jyväskylä, P.O. Box 35, FIN–40351 Jyväskylä, Finland, e-mail: tiihonen@math.jyu.fi

Abstract   We consider the problem of deriving finite element error estimates in smooth domains. We propose to decompose the approximation error into two parts, a geometric part and a finite element part, by introducing an auxiliary problem defined in a polygonal domain approximating the original smooth domain. Some techniques to estimate the geometric part are then described making use of the techniques developed for shape optimization.

1   INTRODUCTION

In this paper we consider the dilemma of 'smooth polygonal domains' related to error analysis of the Finite Element Method. The dilemma is that the finite element methods are naturally formulated in polygonal (or more generally in piecewise polynomial) geometries. On the other hand, the abstract error estimates rely on interpolation error estimates that require smoothness of the solution that is typically achieved only in regular geometries.

In the literature this question has been treated in several ways. Perhaps the most popular approach is that of the above mentioned smooth polygonal domains. That is, the analysis is carried out assuming that the domain is polygonal and hence the grid fits exactly to the domain, and at the same time the solution is regular enough for the optimal interpolation estimates. At the other extreme are the works where the curved boundary is captured more or less exactly by introducing corresponding curved elements, [1], [2], [8], [9]. In this case the analysis can be made rigorous but the price to pay is more complicated local analysis and implementation. Finally, there exists quite a number of papers where the smooth domain is approximated by a polygonal one which is then triangulated. In many of the papers the error analysis is based on some specific feature, like the possibility to extend the FE-solution outside the polygonal domain. Also the analysis of the approximation of geometry is generally interwoven with the analysis of the FE approximation properties.

In this paper we introduce an approach where the approximation of geometry is detached from the FE-analysis. This means that we shall analyze the error made due to replacing the original problem in smooth domain by an auxiliary problem in (polygonal) approximate domain. Then, the error for the finite element approximation of the auxiliary problem is analyzed, bearing in mind that the auxiliary solution is close to the original, smooth one. The first error is estimated using the techniques familiar from shape optimization [5] where the question of continuous dependence with respect to variations in geometry is a key issue. The question of continuous dependence of the solution on the problem geometry has been discussed already in early 70's, [6] [7] with the motivation arising from finite element error estimates. Then, it seems, the issue was forgotten.

The contents of the paper can be briefly summarized as follows. In Chapter 2 we introduce the strategy in abstract framework. Then in Chapter 3 we concretize the abstract setting for H^1-estimates for second order elliptic problems. In the last chapter we consider some more refined methods for obtaining H^1 and L^∞ estimates.

2   ABSTRACT FORMULATION

The aim is to study the dependence between the solution u of the variational problem defined in a smooth domain Ω and its finite element approximation u_h defined in an approximate domain Ω_h. To be able to compare u and u_h we have to be able to prolongate one of the two to the domain of definition of the other. As prolongation of FE-functions is difficult (as FE-functions) in the general case we choose to prolongate the original solution from Ω to Ω_h by a prolongation operator P. The idea behind error estimation is to introduce an auxiliary problem defined in Ω_h to be able to separate the approximation of geometry from approximation by FE spaces.

Thus, let

$$a(u, w) = \langle f, w \rangle \qquad \forall w \in V \tag{1}$$

for $f \in V'$, $u \in V = V(\Omega)$ be the original problem. We introduce an auxiliary problem

$$\hat{a}(\hat{u}, w) = \langle \hat{f}, w \rangle \qquad \forall w \in \hat{V} \tag{2}$$

with $\hat{V} = \hat{V}(\hat{\Omega})$. The bilinear forms a and $\hat{a}$ are assumed to be continuous and V (resp. $\hat{V}$) elliptic. The auxiliary problem is defined so that its FE discretization gives the discrete problem

$$\hat{a}(u_h, w_h) = \langle \hat{f}, w_h \rangle \qquad \forall w_h \in V_h \subset \hat{V}. \tag{3}$$

We want now to estimate the error between u and u_h. Because of different domains of definition we have to consider the extension of u, Pu. Let now $\| \cdot \|$ be some norm for functions defined in Ω_h. Then by the triangle inequality

$$\|Pu - u_h\| \leq \|Pu - \hat{u}\| + \|\hat{u} - u_h\|.$$

Here the first part corresponds to the error due to geometry, whereas the second part is standard FE-approximation error (but now in polygonal domain).

Assume now that we have a quasi-optimality result with respect to $\|\cdot\|$ norm. That is,

$$\|\hat{u} - u_h\| \le C \inf_{w_h \in V_h} \|\hat{u} - w_h\|. \tag{4}$$

Then using the triangle inequality again we can estimate

$$\|\hat{u} - w_h\| \le \|\hat{u} - Pu\| + \|Pu - w_h\|.$$

Hence,

$$\|Pu - u_h\| \le C_1 \|Pu - \hat{u}\| + C_2 \inf_{w_h \in V_h} \|Pu - w_h\|. \tag{5}$$

This estimate is valid with coefficients C_1 and C_2 independents of h if the quasi-optimality estimate (4) is valid uniformly for all Ω_h. This is normally true at least for the energy norm related to $a(\cdot, \cdot)$.

In order to derive useful concrete error estimates from (5) we have to make sure that the following conditions are satisfied.

(1) $\hat{f}$, $\hat{a}$ and P are chosen so that we get an appropriate estimate for $\|Pu - \hat{u}\|$.

(2) P must be chosen so that Pu has the regularity needed for interpolation estimates. that is, Pu should be smooth when restricted to any element of the triangulation.

(3) $\hat{f}$ and $\hat{a}$ should be natural extensions of f and a as they will be used to construct the discrete problem.

Finally, let us remark that the above abstract estimate can be extended for example to cases where we compare the solution to the postprocessed discrete solution Ru, where Ru is some recovery operator, provided that we have an estimate of the type

$$\|\hat{u} - Ru_h\| \le C \inf_{w_h \in V_h} \|\hat{u} - Rw_h\|$$

that is valid uniformly in Ω_h:s.

3   ESTIMATE IN ENERGY NORM

To give an example of application of the idea presented above, we shall consider the error in energy norm for an elliptic problem of second order. To keep the notation as simple as possible we shall present the analysis only for the Laplace operator, extensions to more general cases being obvious.

Let us consider the Poisson problem with homogeneous Dirichlet condition. That is, assume that Ω is a C^2-domain with boundary Γ. We consider the problem

$$\begin{aligned} -\Delta u &= f &&\text{in } \Omega, \\ u &= 0 &&\text{on } \Gamma, \end{aligned}$$

with $f \in L^2(\Omega)$. Then it is well-known that there exists a unique strong solution $u \in H^2(\Omega) \cap H_0^1(\Omega)$.

We shall now construct a polygonal approximation Ω_h of Ω and in Ω_h a finite element approximation u_h of u. Denote by Γ_h a polygonal approximation of Γ, where h denotes the characteristic length of an edge of Γ_h. We assume that the distance between Γ and Γ_h is smaller than the radius of curvature of Γ. Hence, for any point x' of Γ_h there exists a unique closest point x on Γ with distance $v(x)$. This means that Γ_h can be represented as $\Gamma_h = \{x + n(x) \cdot v(x) \mid x \in \Gamma\}$. The domain Ω_h that is bounded by Γ_h is triangulated with simplicial elements of size h and the space V_h of continuous piecewise linear functions which vanish on Γ_h is associated with the triangulation.

Next we construct the extension Pu of the solution u. A natural choice would be, for example, the Calderon extension in H^2. However, we shall postpone the analysis of that extension to the next chapter as it requires some deeper results on shape calculus. Here we study an other type of 'extension' whose properties can be studied with quite elementary tools.

Let $V : \Omega \to R^n$ be a vector field that is Lipschitz continuous and satisfies $V \cdot n|_\Gamma = v|_\Gamma$. V induces a family of transformations $T_t : x \to x + tV(x)$ such that T_t is one to one for t small and $T_1(\Omega) = \Omega_h$. In fact it is possible to construct V such that $\|V\|_{L^\infty} \leq Ch^2$ and $\|V\|_{W^{1,\infty}} \leq Ch$. Moreover, V can be made piecewise C^2. Now we define $Pu = u \circ T_1^{-1}$. That is Pu is the transported solution under the deformation T_1.

As described in the previous chapter, the discrete problem is introduced as a discretization of an auxiliary problem defined in Ω_h. Let us formulate the auxiliary problem as follows

$$-\Delta \hat{u} = \hat{f} \qquad \text{in } \Omega_h,$$
$$\hat{u} = 0 \qquad \text{on } \Gamma_h,$$

where $\hat{f}$ is an extension of f to Ω_h. Here we can consider two different possibilities. Either we choose $\hat{f}$ as L^2 extension of f or we put $\hat{f} = f \circ T_1^{-1}$, i.e., we transport the data to Ω_h.

To be able to use the estimate (5) we have to check first that the quasi-optimality is valid uniformly with respect to h. This is indeed the case as from Cea's lemma we get

$$\|\nabla(\hat{u} - u_h)\|_{L^2} \leq \inf_{w_h \in V_h} \|\nabla(\hat{u} - w_h)\|_{L^2}.$$

Due to Poincare's inequality $\|\nabla \cdot \|_{L^2}$ is a norm in $H_0^1(\Omega_h)$ for all h. Hence the estimate (5) holds and we now only have to derive the estimates for $\|Pu - \hat{u}\|_{H^1}$ and $\inf_{w_h \in V_h} \|Pu - w_h\|_{H^1}$.

Let us estimate $\|Pu - \hat{u}\|_{H^1}$. For that we first construct an equation for $Pu - \hat{u}$. We notice that

$$\int_{\Omega_h} \nabla \hat{u} \nabla w = \int_{\Omega_h} \hat{f} w$$

and

$$\int_\Omega \nabla u \nabla w \circ T_1 = \int_\Omega f w \circ T_1 = \int_{\Omega_h} f \circ T_1^{-1} w |DT_1^{-1}|.$$

On the other hand

$$\int_\Omega \nabla u \nabla w \circ T_1 = \int_{\Omega_h} DT_1 \nabla Pu DT_1 \nabla w |DT_1^{-1}|.$$

Hence,

$$
\int_{\Omega_h} \nabla(Pu - \hat{u})\nabla w = \int_{\Omega_h} \nabla Pu \nabla w - DT_1 \nabla Pu DT_1 \nabla w |DT_1^{-1}|
$$
$$
+ \int_{\Omega_h} f \circ T_1^{-1} w |DT_1^{-1}| - \hat{f} w = I + II.
$$

Now, as $DT = I + DV$ and by construction $\|DV\|_{L^\infty} \le Ch$, it follows that when h is small enough DT is close to identity and hence invertible. Moreover, $|DT_1^{-1}| = 1 + O(h)$.

Hence, writing

$$
I = \int_{\Omega_h} \nabla Pu \nabla w (1 - |DT_1^{-1}|) + \int_{\Omega_h} DV \nabla Pu DT \nabla w |DT_1^{-1}|
$$
$$
+ \int_{\Omega_h} DT \nabla Pu DV \nabla w |DT_1^{-1}| + \int_{\Omega_h} DV \nabla Pu DV \nabla w |DT_1^{-1}|
$$

we easily conclude that

$$
I \le Ch \|Pu\|_{H^1} \|w\|_{H^1} \le Ch \|f\|_{H^{-1}} \|w\|_{H^1}. \tag{6}
$$

Similarly, for II we have

$$
II = \int_{\Omega_h} (f \circ T_1^{-1} - \hat{f}) w |DT_1^{-1}| + \int_{\Omega_h} \hat{f} w (|DT_1^{-1}| - 1).
$$

Now we can distinguish between the two different extensions to f. If $\hat{f} = f \circ T_1^{-1}$, we immediately obtain that

$$
II = \int_{\Omega_h} \hat{f} w (|DT_1^{-1}| - 1) \le Ch \|\hat{f}\|_{L^2} \|w\|_{L^2}. \tag{7}
$$

In the case where $\hat{f}$ is an ordinary L^2 extension of f a bit more delicate argument is needed: We can assume that there exists a domain D containing Ω and Ω_h such that T_1 maps D onto itself. Extending now w to $H^1(D)$ we can write

$$
II \le \left| \int_D \hat{f} \circ T_1^{-1} \hat{w} |DT_1^{-1}| - \hat{f} \hat{w} \right|.
$$

But now as T_1 maps D onto itself, we have that

$$
\int_D \hat{f} \circ T_1^{-1} \hat{w} |DT_1^{-1}| = \int_D \hat{f} \hat{w} \circ T_1.
$$

Hence

$$
II \le \left| \int_D \hat{f} (\hat{w} \circ T_1 - \hat{w}) \right| \le C \|\hat{f}\|_{L^2} \|\hat{w}\|_{H^1} h^2. \tag{8}
$$

Combining the above estimates, we can conclude that

$$
\|Pu - \hat{u}\|_{H^1} \le Ch \|\hat{f}\|_{L^2}. \tag{9}
$$

Next let us consider the interpolation error. As Ω is a C^2 domain and Ω_h is only polygonal, the transformation T can not be smooth globally. However, T (and T_1^{-1}) can be made piecewise smooth. Consequently, Pu is also piecewise smooth and, especially, Pu can be made smooth in any element. Hence,

$$\inf_{w_h \in V_h} \|Pu - w_h\|_{H^1} \leq Ch\|u\|_{H^2} \leq Ch\|f\|_{L^2}. \tag{10}$$

THEOREM 1   Let Ω_h be a polygonal approximation of Ω. Define the discrete problem as in (3). Then for the error between the discrete solution u_h and the prolongated solution Pu it holds

$$\|Pu - u_h\|_{H^1(\Omega_h)} \leq Ch\|f\|_{L^2(\Omega)}. \tag{11}$$

The technique of auxiliary problems is compatible also with the well-known duality technique that is used in proving L^2-estimates, for example. Namely, we have the following

LEMMA 1   If Ω_h is a polygonal approximation of Ω, then for the error between the auxiliary solution $\hat{u}$ and the discrete solution u_h we have

$$\|\hat{u} - u_h\|_{L^2} \leq Ch^2.$$

Proof: Let us denote by $\hat{w}$ the solution of the adjoint problem

$$-\Delta \hat{w} = \hat{u} - u_h \qquad \text{in } \Omega_h,$$
$$\hat{w} = 0 \qquad \text{on } \Gamma_h.$$

Then it holds by Galerkin orthogonality that

$$\|\hat{u} - u_h\|_{L^2}^2 = (\hat{u} - u_h, \hat{u} - u_h) = \int_{\Omega_h} \nabla \hat{w} \nabla (\hat{u} - u_h)$$
$$= \int_{\Omega_h} \nabla (\hat{w} - w_h) \nabla (\hat{u} - u_h),$$

where w_h is the finite element approximate of $\hat{w}$.
   Now,

$$\int_{\Omega_h} \nabla (\hat{w} - w_h) \nabla (\hat{u} - u_h) \leq \|\hat{u} - u_h\|_{H^1} \|\hat{w} - w_h\|_{H^1}$$
$$\leq \|\hat{u} - u_h\|_{H^1} (\|\hat{w} - Pw\|_{H^1} + \|Pw - w_h\|_{H^1}),$$

where w is the solution of an adjoint problem defined in Ω with data $\hat{u} - u_h$ extended to $L^2(\Omega)$. Now, as Ω is smooth, $w \in H^2(\Omega)$ and Pw is regular enough so that the interpolation estimates hold. Thus applying (9) and (10) we see that

$$\|\hat{w} - Pw\|_{H^1} + \|Pw - w_h\|_{H^1} \leq Ch\|\hat{u} - u_h\|_{L^2}.$$

This, together with the previous theorem concludes the proof. $\square$

Let us comment the above estimates and the related schema from the practical point of view. First we note that $\hat{a}$ is a natural extension of a as it is based on extending the elliptic coefficients outside Ω. In practice it can often be constructed by using FE as if the polygonal domain were the original one. The same holds true for $\hat{f}$ when L^2 extension is used. The version with transported data on the right hand side requires that the mapping T is constructed and applied in the finite element procedure.

The L^2 estimate is not treated satisfactorily as the analysis of the geometric part of the error is missing. This may be a quite difficult issue unless one has enough regularity to deduce the L^2 estimate from the corresponding geometric L^∞ estimate.

The estimate was for the difference between the finite element solution and the transported solution. Hence, in principle the FE solution should be transported to the original domain before the estimate can be used. If this is not done another error of order $O(h)$ in H^1-norm is introduced. This makes one to think weather the leading part in the 'geometrical' error is in fact due to the translation of the solution in which case other extension mechanisms might lead to improved estimates. This will be studied in the next chapter. Before that we want to mention that similar approach of transported solution and data was used by Lenoir [2]. However, he analyzed the geometric error simultaneously with the FE-error on the element level.

As we saw from the treatment of the right hand side, it is possible to construct the auxiliary problem in many ways. Obviously one could, for example, transport also the operator with mapping T_1 which would be more or less equivalent to introducing arbitrary curved elements.

4   SHAPE DERIVATIVE APPROACH

Let us recast the estimate (9) as a differentiability result. Recalling the definition of T_t, we can introduce a family of domains $\Omega_t = T_t(\Omega)$ and, correspondingly, a family of auxiliary problems with solutions u_t defined in Ω_t. Now using the techniques introduced above, it is possible to show that $(u - u_t \circ T_t)/t$ has a limit $\dot{u}$ in $H^1(\Omega)$. Thus

$$u - \hat{u} \circ T_1 = \dot{u} + \text{ higher order terms}$$

and $\|Pu - \hat{u}\|_{H^1} \approx \|u - \hat{u} \circ T_1\|_{H^1} \approx \|\dot{u}\|_{H^1}$. The quantity $\dot{u}$ is called the material derivative in the literature of shape optimization (as well as in continuum mechanics).

Let us now consider the case where the extension operator P is an ordinary extension, say a Lions type extension in H^2. We want to estimate the error $Pu - \hat{u} = Pu - u_1$. The related limit process is this time $\lim_{t \to 0+} (Pu - u_t)/t$, or more rigorously, $\lim_{t \to 0+} (u - \hat{P}u_t)/t$ where $\hat{P}$ is some extension operator from Ω_t (the value of the limit will not depend on the extension, however). The limit, when it exists is called the shape derivative of u and is denoted by u'. Moreover, it is known that $u' = \dot{u} - \nabla u \cdot V$ when all the limits and derivatives exist. As now $\|Pu - u_t\| \approx \|u'\| +$ higher order terms, we are led to study the properties of u'.

Thanks to regularity of the data, the existence of u' is well-known in our case, [5]. Also we know that u' depends only on the boundary values of the normal velocity

$V \cdot n|_\Gamma$. Hence, only on the shape of Ω_t:s. u' is the unique solution of the problem

$$-\Delta u' = 0 \qquad \text{in } \Omega,$$

$$u' = -\frac{\partial u}{\partial n} V \cdot n \qquad \text{on } \Gamma.$$

To achieve an error estimate for $\|Pu - u_1\|$ we have to be able to bound the norm of u'. For that we have the following

LEMMA 2   Let $u \in H^{5/2}(\Omega)$. Then for the shape derivative u' it holds, $\|u'\|_{H^1} \leq Ch^{3/2}$.

Proof: We observe easily that $\|u'\|_{H^1} = \left\| \frac{\partial u}{\partial n} V \cdot n \right\|_{H^{1/2}(\Gamma)}$. Now as $u \in H^{5/2}(\Omega)$,

we have that $\frac{\partial u}{\partial n}\Big|_\Gamma \in H^1(\Gamma)$. Thus

$$\left\| \frac{\partial u}{\partial n} V \cdot n \right\|_{H^1(\Gamma)} \leq C\|V \cdot n\|_{W^{1,\infty}}$$

and

$$\left\| \frac{\partial u}{\partial n} V \cdot n \right\|_{L^2(\Gamma)} \leq C\|V \cdot n\|_{L^\infty}.$$

Hence, as $H^{1/2}$ is an interpolation space of L^2 and H^1,

$$\|u'\|_{H^1(\Omega)} \leq C\|V \cdot n\|_{1,\infty}^{1/2}\|V \cdot n\|_\infty^{1/2} \leq Ch^{3/2}.$$

$\square$

Note that from the previous proof we can not conclude that $\|Pu - \hat{u}\|_{H^1} \leq Ch^{3/2}$ as we can not control the rate of convergence of $(Pu - u_t)/t$ to u' as t tends to zero. This would require a proof of Lipschitz continuous shape differentiability of u for $W^{1,\infty}$ vector fields which is difficult as the regularity of Ω is lost in the deformation.

One possible way to obtaining an error estimate using the concept of shape derivative is to regularize the perturbation in the geometry. Roughly speaking, one could decompose the perturbation to a regularized part and a small Lipschitz part. For the Lipschitz part one could then use different scaling to obtain a better estimate and for the regularized part there should be some hope to obtain continuous differentiability.

Let us note that the $O(h^{3/2})$ error estimate has be proved for this particular problem in the case of Ω being convex. The proof by Strang and Berger [6] is based on proving directly the finite element error estimate and deducing then the estimate for the continuous case a consequence. The proof in [7], on the other hand relies heavily on both convexity and maximum principle. Hence its generalization to other problems may be difficult.

Let us conclude by some remarks on L^∞-estimates. An appealing starting point is the quasi-optimality result by Rannacher and Scott, [3] which states that

$$\|\hat{u} - u_h\|_{W^{1,p}} \leq C \inf_{w_h \in V_h} \|\hat{u} - w_h\|_{W^{1,p}} \quad 2 \leq p \leq \infty.$$

The crucial question is whether the constant C can be chosen independently of Ω_h. This has not been addressed in [3]. Assuming that such a uniform estimate is true, the rest follows quite easily. The equation for u' immediately suggests an $O(h)$ estimate in $W^{1,\infty}$ for the geometric error. In fact now the required geometric error estimate can be deduced directly from the proof of shape differentiability in $W^{1,\infty}$. Other methods for obtaining the estimates can be found already in [7]. In L^∞ one obtains also easily an $O(h^2)$ estimate for the geometric part of the error. Hence, using the duality argument, optimal L^∞ estimates can be obtained as well (assuming always the uniform quasi-optimality result).

There exist some uniform quasi-optimality results like that of Schatz and Wahlbin [4] which states that under several technical conditions (including the condition that the discrete domains should be contained in the original one) one has the estimate

$$\|u - u_h\|_{L^\infty} \leq CX(h) \inf_{w_h \in V_h} \|w_h - u\|_{L^\infty}$$

where $X(h) = 1$ except in the two dimensional case when it equals to $-\ln h$. This proves that the general conclusion to which we are aiming is true. However, to prove the quasi-optimality result for $\hat{u}$ (that depends on h) may be even harder than to obtain the above estimate.

REFERENCES

1. Ciarlet P.G., Raviart P.A., *Interpolation theory over curved elements with applications to finite element methods*, Comp. Meth. Appl. Mech. Engrg. **1** (1972), 217–249.
2. Lenoir M., *Optimal isoparametric finite elements and error estimates for domains involving curved boundaries*, SIAM J. Numer. Anal. (1986), 562–580.
3. Rannacher R., Scott R., *Some optimal error estimates for piecewise linear finite element approximations*, Math. Comp. **38** (1982), 437–445.
4. Schatz A.H., Wahlbin L.B., *On the quasi-optimality in L^∞ of the H_0^1-projection into finite element spaces*, Math. Comp. **38** (1982), 1–22.
5. Sokolowski J. Zolésio J.P., "Introduction to shape optimization, shape sensitivity analysis," Springer, Berlin, 1992.
6. Strang G., Berger A. E., *The change in solution due to change in domain*, in "Proceedings of symposia in pure mathematics, XXIII," D.C. Spencer (ed.), AMS, 1973, pp. 199–205.
7. Thomée V., *Polygonal domain approximation in Dirichlet's problem*, Journal of IMA **11** (1973), 33–44.
8. Zlámal M., *Curved elements in the finite element method, I*, SIAM J. Numer. Anal. **10** (1973), 229–240.
9. Zlámal M., *Curved elements in the finite element method, II*, SIAM J. Numer. Anal. **11** (1974), 347–362.

General principles of superconvergence in Galerkin finite element methods

LARS B. WAHLBIN*   Department of Mathematics, Cornell University, Ithaca, NY 14853, USA

Abstract   In this talk I shall highlight four general principles (or, if you like, methods), giving them mnemonic monikers for reference:

COMPARE.u_I: Compare the finite element approximation to an "interpolant" u_I.
LOC\$YMM.MESH: A point about which the mesh is locally symmetric is a superconvergent point.
⊗PROD.ELEMTS: One-dimensional results translate to tensor product elements.
∂_H TRANSL.INV.MESH: Differencing is better than differentiating, in particular on translation invariant meshes.

(In response to a remark made at the conference, I shall also give, in the concluding remarks, the gist of the principle that "anything about superconvergence in linear problems translates to nonlinear problems".)

1   INTRODUCTION

We shall consider the standard h-method for approximating the solution of a scalar second order elliptic problem, and we shall formulate our principles in local terms. The subject of superconvergence is at present confused about what happens at boundaries of domains, and we shall therefore only treat interior domains. So, with $\Omega \subset\subset \mathbb{R}^N$ a basic domain, $S_h \subseteq H^1(\Omega)$ a parametrized family of finite element spaces on Ω, $\mathcal{D} \subset\subset \Omega$ and $\overset{\circ}{S}_h(\mathcal{D}) = \{\chi \in S_h : \operatorname{supp} \chi \subset \mathcal{D}\}$, let u and $u_h \in S_h$ be

*This work was supported by the National Science Foundation, USA.

two functions such that

$$\mathcal{A}(u - u_h, \chi) \equiv \int \left[\sum_{i,j=1}^{N} a_{ij}(x) \frac{\partial(u - u_h)}{\partial x_i} \frac{\partial \chi}{\partial x_j} + \sum_{i=1}^{N} b_i(x) \frac{\partial(u - u_h)}{\partial x_i} \chi \right.$$

$$\left. + c(x)(u - u_h)\chi \right] dx = 0, \qquad \text{for all } \chi \in \overset{\circ}{S}_h(\mathcal{D}). \qquad (1.1)$$

For our purposes, it is expedient to consider at once the more general situation of

$$\mathcal{A}(v - v_h, \chi) = F(\chi), \quad \text{for } \chi \in \overset{\circ}{S}_h(\mathcal{D}), \ v_h \in S_h. \qquad (1.2)$$

The following result is central and will be used over and over again in this talk. It is proven for general space dimension N in Schatz and Wahlbin [22]; see that paper for precise assumptions.

THEOREM 1.1 Given the bilinear form $\mathcal{A}(\cdot, \cdot)$, for any $s \geq 0$ and $1 \leq q \leq \infty$ there exist positive constants c_0 and C such that for $\mathcal{D}_0 \subset\subset \mathcal{D}$ with $\mathcal{D}_0 + d \subseteq \mathcal{D}$, $d \geq c_0 h$, and v and v_h satisfying (1.2), we have

$$\|\nabla(v - v_h)\|_{L_\infty(\mathcal{D}_0)} \leq C \min_{\chi \in S_h} \left(\|\nabla(v - \chi)\|_{L_\infty(\mathcal{D})} + d^{-1}\|v - \chi\|_{L_\infty(\mathcal{D})} \right)$$

$$+ Cd^{-1-s-N/q}\|v - v_h\|_{W_q^{-s}(\mathcal{D})} + C \ln(d/h)\|F\|_{-1,\infty,\mathcal{D}},$$

where

$$\|F\|_{-1,\infty,\mathcal{D}} = \sup_{\substack{\varphi \in \overset{\circ}{S}_h(\mathcal{D}) \\ |\varphi|_{W_1^1(\mathcal{D})}=1}} F(\varphi).$$

When the finite element spaces S_h are such that on each element τ which is in, or close to, $\mathcal{D}$, we have

$$S_h(\tau) = \{\chi|_\tau : \chi \in S_h\} \supseteq \Pi_{r-1}(\tau), \qquad (1.3)$$

the polynomials of total degree $r - 1$ on τ, and when, on $\mathcal{D}$, h is comparable to the diameters of elements, then under various assumptions on what happens outside of $\mathcal{D}$, one may conclude from (1.1) and Theorem 1.1 (with $v = u$, $v_h = u_h$ and $F \equiv 0$) that

$$\|\nabla(u - u_h)\|_{L_\infty(\mathcal{D}_0)} \leq Ch^{r-1}, \qquad (1.4)$$

where now C further depends on the local smoothness of u, in particular.

If $S_h(\tau)$ does not contain $\Pi_p(\tau)$ for any $p > r - 1$, the estimate (1.4) is in general best possible with respect to order of convergence. However, there may be points $\eta = \eta(h)$, i.e., they may vary with h, such that

$$|\nabla(u - u_h)(\eta)| \leq Ch^{r-1+\sigma}, \quad \text{with } \sigma > 0. \qquad (1.5)$$

Such points will be referred to as *superconvergent* points for gradients. For brevity, we shall only treat superconvergence in gradients.

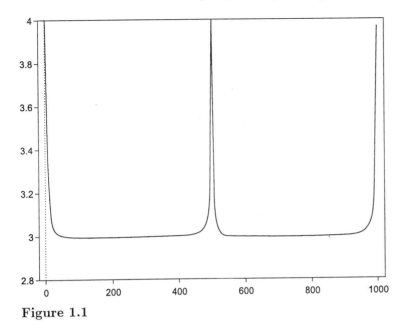

Figure 1.1

Superconvergent rates are sharply localized phenomena. Figure 1.1 depicts the rates of convergence for a two-point boundary value problem approximated by the Galerkin method with Hermite cubics, $r = 4$, on a uniform mesh. What is shown is a conglomerate picture of how the rate of convergence varies on a typical mesh interval away from the endpoints of the basic interval. Cf. Wahlbin [**29**, Section 1.13] for details on how this was computed.

Our main problem is, of course, to identify superconvergent points a priori. In principle, we could talk about superconvergent points for a particular problem or even for a particular solution u. Happily, most theories turn out to be more general in scope. In practice, second order problems are often approximated using C^0 elements, and then $\nabla u_h(\eta)$ does not in general have a unique value if η lies on the boundary between two or more elements. In this case we may define $\nabla_h u_h(\eta)$ as the simple equal weighted average limit coming in towards η from two (or more) elements. Having admitted this tiny step away from what a purist may call "natural" superconvergence, we may as well admit other postprocessors $\widetilde{\nabla}_h u_h(\eta)$, at least as long as they are felt to be simple to implement. In my opinion, it is unnatural to make a big distinction between "natural" superconvergence and postprocessing; indeed, many postprocessing methods are easier to implement on a computer than is evaluating $\nabla u_h(\eta)$.

The plan of this talk is as follows. In Section 2 we shall describe the method COMPARE.u_I for showing superconvergence; this method comes first due to its historical precedence. As we shall see, it involves a painstaking case-by-case analysis for each particular type of finite element and is at present restricted to low order elements. In Section 3 we consider the recent discovery LOC$YMM.MESH, a general principle which is extremely simple to apply. It is independent of polynomial degree of elements and independent of interelement continuity degree. In particular, it explains, for simplicial elements, most, if not all, results known by COMPARE.u_I

and much, much more. In Section 4 we delve into yet another easy-to-apply general principle, ⊗PROD.ELEMTS: If you know superconvergent points in a two-point boundary value problem for $-u'' = f$, then you know them for tensor-product elements in any space dimension N.

In Section 5 we then take a reality check of how the principles LOC\$YMM.MESH and ⊗PROD.ELEMTS fare against a careful numerical study.

The principles LOC\$YMM.MESH and ⊗PROD.ELEMTS are mainly devoted to investigating natural superconvergence or superconvergence in a simple limit average $\nabla_h u_h$. Section 6 describes the powerful general principle ∂_H TRANSL.INV.MESH of taking difference quotients. In concluding, in Section 7 we also atone for some sins of omission by listing.

The present talk is very much based on my recent lecture notes Wahlbin [29], which may be consulted for further detail and a comprehensive bibliography. To my knowledge there is no other monograph in the English language; brief surveys are in Křížek and Neittaanmäki [13] and Wahlbin [28, Chapter VII]. The Chinese language is better endowed with the treatises Chen [5], Zhu and Lin [31] and Chen and Huang [6]

2   COMPARE.U_I

The subject of superconvergence in finite element methods emerged in the late sixties, see e.g. Stricklin [25], Zienkiewicz and Cheung [32], Filho [11], Stricklin [26] and Tong [27]. (A notable early example of an enduring method for gaining efficiency by postprocessing in a numerical scheme is in Nyström [17][1].) The germ for actually proving something about superconvergence in our context, and in particular the germ for the method COMPARE.u_I, is in Oganesyan and Rukhovetz [18, eqn. (3.18)]. We shall describe it in a local setting and with maximum-norm gradient errors (eqn. (3.18) is in the global H^1-norm).

Let u_I be a "standard" locally defined "interpolant" of u. Then from (1.1) we have

$$\mathcal{A}(u_h - u_I, \chi) = \mathcal{A}(u - u_I, \chi), \quad \text{for } \chi \in \overset{\circ}{S}_h(\mathcal{D}). \qquad (2.1)$$

Assume now that we could somehow show (this is the hard part!) that

$$|\mathcal{A}(u - u_I, \chi)| \le Ch^{r-1+\sigma} \|\chi\|_{W_1^1}, \quad \text{for } \chi \in \overset{\circ}{S}_h(\mathcal{D}). \qquad (2.2)$$

Via (2.1) it would then follow from Theorem 1.1 (with $v = 0$, $v_h = u_I - u_h$) that

$$\|\nabla(u_I - u_h)\|_{L_\infty(\mathcal{D}_0)} \le C\|u_I - u_h\|_{W_q^{-s}(\mathcal{D})} + C(\ln 1/h)h^{r-1+\sigma}. \qquad (2.3)$$

(In this section we shall not trace the dependence on d.) Since

$$\|u_I - u_h\|_{W_q^{-s}(\mathcal{D})} \le \|u - u_I\|_{W_q^{-s}(\mathcal{D})} + \|u - u_h\|_{W_q^{-s}(\Omega)}, \qquad (2.4)$$

it is easy to furnish conditions which ensure that

$$\|\nabla(u_I - u_h)\|_{L_\infty(\mathcal{D}_0)} \le C(\ln 1/h)h^{r-1+\sigma}. \qquad (2.5)$$

---

[1]Nyström graduated from High School in Jyväskylä Stenij [24].

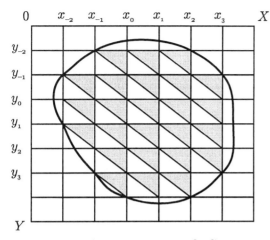

Figure 2.1 (From Schellbach [23])

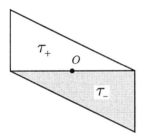

Figure 2.2

Superconvergence in $\nabla(u - u_h)$ is now reduced to considering $\nabla(u - u_I)$, a locally defined quantity which is frequently trivial to analyze.

In applications, the hard part is to verify (2.2), if it indeed holds! As an example of what goes into proving it, let us consider the case of the Laplacian in two dimensions so that

$$A(u - u_I, \chi) = \int \frac{\partial}{\partial x}(u - u_I)\frac{\partial \chi}{\partial x} + \frac{\partial}{\partial y}(u - u_I)\frac{\partial \chi}{\partial y} \equiv I_x + I_y, \qquad (2.6)$$

and let us take piecewise linear triangles on a three-directional mesh with two directions axis-parallel (see Figure 2.1).

Considering here I_x, we may pair off black-white triangles (see Figure 2.2).

Letting O be the origin, if $u \in C^{2+\sigma}(\mathcal{D})$, it is easy to see by Taylor expansion that it suffices to consider the case of

$$\tilde{u}(x, y) = Ax^2 + 2Bxy + Cy^2 \quad \text{on } \tau_+ \cup \tau_-. \qquad (2.7)$$

Now, with the pairing as above, $\dfrac{\partial \chi}{\partial x}$ is a constant and the same in τ_+ and τ_-. So,

$$\int_{\tau_+ \cup \tau_-} \frac{\partial}{\partial x}(\tilde{u} - \tilde{u}_I)\frac{\partial \chi}{\partial x} = \left[\int_{\tau_+ \cup \tau_-} \frac{\partial}{\partial x}(\tilde{u} - \tilde{u}_I)\right]\frac{\partial \chi}{\partial x}. \qquad (2.8)$$

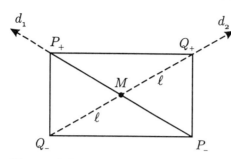

Figure 2.3

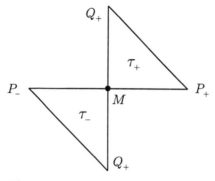

Figure 2.4

With $\tilde{u}_I$ the standard piecewise linear interpolant of $\tilde{u}$, $\tilde{u}_I\big|_{y=0}$ is a constant so $\dfrac{\partial \tilde{u}_I}{\partial x} = 0$ on $\tau_+ \cup \tau_-$. Further, $\dfrac{\partial \tilde{u}}{\partial x} = 2Ax + 2By$ and, by elementary symmetry considerations, $\displaystyle\int_{\tau_+\cup\tau_-} x = 0 = \int_{\tau_+\cup\tau_-} y$. Hence, the quantity in (2.8) vanishes and, after a similar argument for I_y, (2.2) obtains with $r = 2$.

To conclude the argument, one scrutinizes $u - u_I$ locally (the easy part!). For the midpoint M of a common side, one considers the two directional derivatives in the directions d_1 and d_2, as indicated (see Figure 2.3). Using the centered difference quotients at the given points $P_\pm$, $Q_\pm$, respectively, and writing

$$\frac{\partial u_I}{\partial d_2} = \frac{u_I(Q_+) - u_I(Q_-)}{2\ell} = \frac{1}{2}\left[\frac{u_I(Q_+) - u_I(M)}{\ell} + \frac{u_I(M) - u_I(Q_-)}{\ell}\right], \quad (2.9)$$

one immediately realizes that the average $\nabla_h u_h$ is superconvergent to ∇u at M. Similarly, for M a meshpoint, the average $\nabla_h u_h$ from τ_+ and τ_- is superconvergent at M (see Figure 2.4).

We note that with averaging each triangle has six superconvergence points, while without averaging, only the derivative in the side direction at the midpoints will be superconvergent (see Figure 2.5).

The method COMPARE.u_I, as briefly described above, can be pushed through also for uniform quadratic Lagrange elements, showing that Gauss-points on each side of a triangle are superconvergent for the derivative in that side-direction.

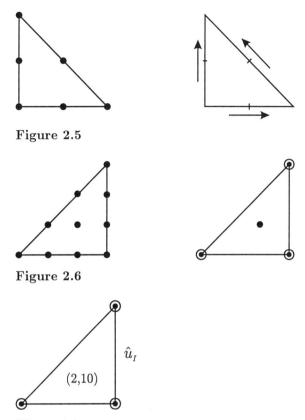

Figure 2.5

Figure 2.6

$\hat{u}_I$

(2,10)

Figure 2.7

With Li [**14**], the method appeared to have come to a grinding halt. He gave two counterexamples, cubic Lagrange elements and Hermite cubics, respectively, showing that the standard interpolants, as indicated, on uniform meshes are not superclose to u_h even in the H^1-norm. However, on a tedious case-by-case basis, the method can perhaps be resurrected. E.g., for Hermite cubics, if one considers, instead of the standard interpolant as depicted in Figure 2.6, the following $\hat{u}_I$, where the interior node is replaced by the mean value condition (see Figure 2.7)

$$\int_\tau \nabla(u - \hat{u}_I) \cdot \nabla(\lambda_1 \lambda_2 \lambda_3) = 0 \quad (\lambda_i \text{ area coordinates}), \tag{2.10}$$

then, indeed, $\nabla(u_h - \hat{u}_I)$ is "supersmall" on uniform meshes, Lin, Yan and Zhou [**15**].

As will be apparent in Section 5, for higher order elements superconvergence in $\nabla(u - u_h)$ is *not* ruled by what happens in $\nabla(u - u_I)$ for u_I a "reasonably standard" interpolant, let alone the interpolant that traditionally defines the finite element in say the engineering literature, cf. Ciarlet [**7**, Section 10]. It appears that the method COMPARE.u_I is a genuinely limited one, also calling for ingenuity on the part of the user in verifying (2.2) and in tweaking u_I to the special case at hand.

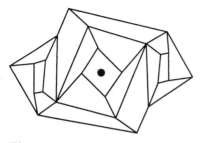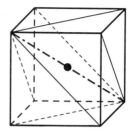

Figure 3.1

3   LOC\$YMM.MESH

This powerful, easy to apply and general principle was given in Schatz, Sloan and Wahlbin [20] (and, based on that paper, in Wahlbin [29]). We consider symmetry about a point S under the antipodal mapping $x \to \overline{x} = 2S - x$; e.g., Figure 3.1 enjoys such symmetry ($S = \bullet$).

Assume now that in a ball B_{2d} of radius $2d$ around S we have the following, with $\overline{\chi}(x) = \chi(\overline{x})$:

$$\text{If } \chi \in S_h(B_{2d}), \text{ then } \overline{\chi} \in S_h(B_{2d}). \tag{3.1}$$

If the finite element spaces are "the same" on each element, this is simply a local symmetry condition on the meshes.

To give the fundamental reason why such a symmetry point S is superconvergent (for gradients if r is even), consider first a symmetric form with constant coefficients,

$$\mathcal{A}(v, w) = \int \left(\sum_{i,j=1}^{N} a_{ij} \frac{\partial v}{\partial x_i} \frac{\partial w}{\partial x_j} + cvw \right). \tag{3.2}$$

Then if $\mathcal{A}(u - u_h, \chi) = 0$, we have also by a change of variables, using (3.1),

$$\mathcal{A}(\overline{u} - \overline{u}_h, \chi) = (-1)^N \mathcal{A}(u - u_h, \overline{\chi}) = 0, \quad \text{for } \chi \in \overset{\circ}{S}_h(B_{2d}). \tag{3.3}$$

Hence, with $u_{\text{odd}} = \dfrac{u - \overline{u}}{2}$, $u_{h,\text{odd}} = \dfrac{u_h - \overline{u}_h}{2}$ (which belongs to $S_h(B_{2d})$ by (3.1)),

$$\mathcal{A}(u_{\text{odd}} - u_{h,\text{odd}}, \chi) = 0, \quad \text{for } \chi \in \overset{\circ}{S}_h(B_{2d}). \tag{3.4}$$

From Theorem 1.1 (with $v = u_{\text{odd}}$, $v_h = u_{h,\text{odd}}$ and $F \equiv 0$) it then follows that

$$\|\nabla(u_{\text{odd}} - u_{h,\text{odd}})\|_{L_\infty(B_d)}$$

$$\leq C \min_{\chi \in S_h} \left(\|\nabla(u_{\text{odd}} - \chi)\|_{L_\infty(B_{2d})} + \frac{1}{d}\|u_{\text{odd}} - \chi\|_{L_\infty(B_{2d})} \right)$$

$$+ Cd^{-1-s-N/q}\|u - u_h\|_{W_q^{-s}(B_{2d})}. \tag{3.5}$$

Assuming a standard approximation theory estimate, if $u \in \mathcal{C}^{r+1}$ locally and r *is even*, it follows that

$$\min_{\chi \in S_h} \left(\|\nabla(u_{\text{odd}} - \chi)\|_{L_\infty(B_{2d})} + \frac{1}{d}\|u_{\text{odd}} - \chi\|_{L_\infty(B_{2d})} \right)$$

$$\leq Ch^{r-1}|u_{\text{odd}}|_{W_\infty^r(B_{2d}+ch)} \leq Cdh^{r-1}. \tag{3.6}$$

For the last term in (3.5), various global conditions now have to be imposed. A most favorable case is a problem with natural boundary conditions, in which case it is not totally unrealistic to assume that, with $q = \infty$, $s = r - 2$, we have

$$\|u - u_h\|_{W_\infty^{-(r-2)}(\Omega)} \leq C(\ln 1/h)h^{2(r-1)}. \tag{3.7}$$

Then (3.5) gives, with $d = h^{1-1/r}$,

$$\|\nabla(u_{\text{odd}} - u_{h,\text{odd}})\|_{L_\infty(B_d)} \leq (C\ln 1/h)(dh^{r-1} + d^{1-r}h^{2(r-1)})$$
$$\leq C(\ln 1/h)h^{r-1+(1-1/r)}. \tag{3.8}$$

If now ∇u_h is continuous at S, we have from this

$$|\nabla(u - u_h)(S)| \leq C(\ln 1/h)h^{r-1+(1-1/r)}, \tag{3.9}$$

i.e., superconvergence with $\sigma = 1 - 1/r$ (almost), *provided only the mesh is symmetric in a small neighborhood of radius $h^{1-1/r}$ around S.* This fact makes the results applicable to meshes in which there is only local symmetry, see, e.g., Glowinski and Lions [**12**, p. 315].

If ∇u_h is not continuous at S, i.e., S is on a boundary between two C^0 elements, form an equally weighted average limit $\nabla_h u_h$ and deduce (3.9) for $(\nabla u - \nabla_h u_h)(S)$.

If we don't enjoy the full power of (3.7), but we have a better rate of convergence than $r - 1$ in *some* norm, similar superconvergent results obtain.

If the mesh is symmetric around S in a full $0(1)$-size neighborhood, then superconvergence of order (almost) $\sigma = 1$ can be proven, Schatz [**19**].

For nonsymmetric and variable coefficient forms $\mathcal{A}$, we no longer have (3.4). Instead,

$$\mathcal{A}(u_{\text{odd}} - u_{h,\text{odd}}, \chi) = F(\chi), \tag{3.10}$$

where, under various conditions,

$$|F(\chi)| \leq Ch^{r-1+\sigma}\|\chi\|_{W_1^1}. \tag{3.11}$$

This implies superconvergence essentially as before now by use of the full power of Theorem 1.1.

For r odd, a similar argument using $u_{\text{even}} = \dfrac{u + \overline{u}}{2}$ and so on establishes superconvergence in the function values of $(u - u_h)(S)$, if global influences permit.

4 ⊗PROD.ELEMTS

This principle is from Douglas, Dupont and Wheeler [**9**]. In the slightly more general form given in Wahlbin [**29**, Chapter 6] it says the following. Consider the N one-dimensional problems

$$-u''(x_i) = f(x_i), \quad i = 1, \ldots, N, \tag{4.1}$$

(with, say, Dirichlet boundary conditions). If η_i is an interior superconvergent point for u' in (4.1) when approximated with a finite element space S_h^i, then if S_h is such that locally on an N-dimensional box $\mathcal{D} = \prod\limits_{i=1}^{N} \mathcal{J}_i$,

$$S_h(\mathcal{D}) = \mathop{\otimes}\limits_{i=1}^{N} S_h^i(\mathcal{J}_i), \tag{4.2}$$

we have that $(\eta_1, \eta_2, \ldots, \eta_N)$ is a superconvergent point for $\nabla(u - u_h)$ of (1.1), global influences permitting. There is a further condition: The form $\mathcal{A}$ must have its principal part in scalar divergence form, i.e., with $a(x)$ a scalar function,

$$\mathcal{A}(v, w) = \int a(x) \nabla v \cdot \nabla w + \cdots. \tag{4.3}$$

We note that the S_h^i, $i = 1, \ldots, N$, do not have to be the same type element spaces. Nor do the one-dimensional meshes have to be uniform or symmetric around η_i.

The basic reason for this result is the following. Consider first the Dirichlet form $\mathcal{A}(v, w) = \int \nabla v \cdot \nabla w$. With R_j denoting the one-dimensional Ritz projection in the jth variable x_j and I_j the corresponding identity operator on the interval $\mathcal{J}_j$, let $W_h = \left(\mathop{\otimes}\limits_{j=1}^{N} R_j \right) u$. Then, for $\chi \in \overset{\circ}{S}_h(\mathcal{D})$,

$$\mathcal{A}(u_h - W_h, \chi) = \mathcal{A}(u - W_h, \chi)$$

$$= \sum_{i=1}^{N} \int \cdots \int \frac{\partial}{\partial x_i} \left(I - \mathop{\otimes}\limits_{j=1}^{N} R_j \right) u \frac{\partial \chi}{\partial x_i} dx_1 \cdots dx_N$$

$$= \sum_{i=1}^{N} \int \cdots \int dx_1 \cdots dx_{i-1} dx_{i+1} \cdots dx_N$$

$$\cdot \left[\int \frac{\partial}{\partial x_i} (I - R_1 \otimes \cdots R_{i-1} \otimes I_i \otimes R_{i+1} \otimes \cdots R_N) u \frac{\partial \chi}{\partial x_i} dx_i \right]$$

$$= \sum_{i=1}^{N} \int \cdots \int dx_1 \cdots dx_{i-1} dx_{i+1} \cdots dx_N$$

$$\cdot \left[\int (I - R_1 \otimes \cdots R_{i-1} \otimes I_i \otimes R_{i+1} \otimes \cdots R_N) \frac{\partial u}{\partial x_i} \frac{\partial \chi}{\partial x_i} dx_i \right]. \tag{4.4}$$

Assuming h^r rates of convergence in function values for the one-dimensional projections, it follows for u smooth enough that

$$|\mathcal{A}(u_h - W_h, \chi)| \leq Ch^r \|\chi\|_{W_1^1}, \quad \text{for } \chi \in \overset{\circ}{S}_h(\mathcal{D}). \tag{4.5}$$

Hence, by Theorem 1.1 (with $v = 0$, $v_h = u_h - W_h$),

$$\|\nabla(u_h - W_h)\|_{L_\infty(\mathcal{D}_0)} \leq C \|u_h - W_h\|_{W_q^{-s}(\mathcal{D})} + C(\ln 1/h)h^r. \tag{4.6}$$

(We shall not trace the d-dependence.) Under suitable assumptions on global behavior we then have for, say, $\dfrac{\partial}{\partial x_1}$,

$$\left| \frac{\partial}{\partial x_1}(u - u_h)(x) \right| \leq \left| \frac{\partial}{\partial x_1}(u - W_h)(x) \right| + C(\ln 1/h)h^r. \tag{4.7}$$

Writing $u - W_h = \left(I - \overset{N}{\underset{j=1}{\otimes}} R_j \right)u$ and

$$\frac{\partial}{\partial x_1}\left(I - \overset{N}{\underset{j=1}{\otimes}} R_j \right) = \left(\frac{\partial}{\partial x_1}(I_1 - R_1) \right) \otimes (I_2 \otimes \cdots \otimes I_N)$$

$$+ \left(I - I_1 \otimes \overset{N}{\underset{j=2}{\otimes}} R_j \right) \frac{\partial}{\partial x_1}$$

$$+ \left(\frac{\partial}{\partial x_1}(R_1 - I_1) \right) \otimes \left(\overset{N}{\underset{j=2}{\otimes}} I_j - \overset{N}{\underset{j=2}{\otimes}} R_j \right), \tag{4.8}$$

one finds that any point $x = (\eta_1, x_2, \ldots, x_N)$ such that η_1 is a superconvergent point for derivatives in the first dimension will be superconvergent for $\dfrac{\partial}{\partial x_1}(u - u_h)$.

The result with lower order terms is easy, and the result for variable coefficients operators of the type $-div(a(x)\nabla) + \cdots$ with $a(x)$ a scalar function follows by an argument using superapproximation and Theorem 1.1 (see Wahlbin [29, Section 6.4]). We point out that the latter proof fails if $a(x)$ is a general matrix.

For one-dimensional C^0 elements, all superconvergence points in (4.1) are well known, Chen [4]. For C^1-spaces, they are given for locally uniform meshes in Wahlbin [29, Sections 1.8 and 1.10]. A complete determination of all superconvergent points for locally uniform meshes and all C^μ-spaces, $\mu \geq 2$, will be given in Dunlap [10].

5 A COMPUTATIONAL INVESTIGATION

A thorough computational investigation of superconvergent points for periodically repeated C^0 element configurations in the case of Poisson's equation in two dimensions is given in Babuška, Strouboulis, Upadhyay and Gangaraj [1]. The authors there reduce the determination of superconvergence points in the cases at hand to that of finding the roots of a finite number of polynomials, i.e., no asymptotics with respect to h is involved in the actual determination. The key in their proof is the observation that the error is basically periodic and then use of Theorem 1.1 (and there are further clever technicalities).

5.1 Triangular elements

The authors consider four different mesh configurations, locally periodically repeated (see Figure 5.1).

Running through Lagrange elements with $r = 2, \ldots, 8$, but considering only strictly natural superconvergence for $\dfrac{\partial u}{\partial x}$, they find:

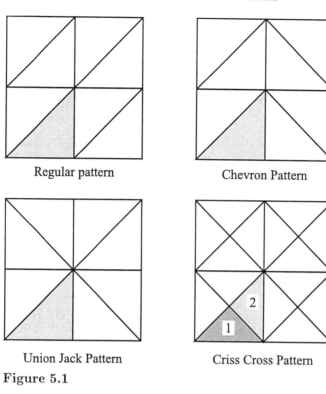

Regular pattern   Chevron Pattern

Union Jack Pattern   Criss Cross Pattern

Figure 5.1

For the regular pattern:

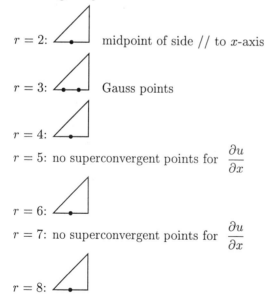

$r = 2$:   midpoint of side $/\!/$ to x-axis

$r = 3$:   Gauss points

$r = 4$:

$r = 5$: no superconvergent points for $\dfrac{\partial u}{\partial x}$

$r = 6$:

$r = 7$: no superconvergent points for $\dfrac{\partial u}{\partial x}$

$r = 8$:

We immediately note that all results for $r = 2$, 4–8 are completely explained by our symmetry theory. In particular, for $r = 5$ and 7 there are no superconvergent points.

(For $r = 3$, we know from symmetry that mesh and midpoints are superconvergent for function values; it easily follows that the Gauss points are superconvergent for derivatives along edges.)

Also note that, for r even, by not averaging one looses five of the six superconvergent points.

Similarly, for the Chevron pattern, the Union Jack mesh and the Criss–Cross grid, the symmetry theory completely explains all points found (or not there) in the numerical determination (see Wahlbin [29, Section 12.3] for details).

It is, indeed, astonishing that such a simple principle as symmetry explains so much.

5.2   Quadrilateral elements

Babuška, Strouboulis, Upadhyay and Gangaraj [1] also consider superconvergence in $\dfrac{\partial u}{\partial x}$ for some uniform quadrilateral C^0 elements. For tensor-product elements, the numerical results are exactly as predicted by the $\otimes$PROD.ELEMTS principle, no more, or less superconvergent points (actually, for $\dfrac{\partial u}{\partial x}$ alone, they are lines, cf. (4.8) et seq.). For the intermediate (incomplete) family, the numerically found superconvergent points are as for tensor-products, a result our theory is largely silent about. (For $r = 2$, intermediate is the same as tensor product. For $r = 3$, our symmetry theory gives nine points for function values which is enough to determine the derivative points found.) A complete theoretical justification has recently been given in Zhang [30].

Finally, for the Serendipity family, $r = 2$ and 3 are the same as the intermediate case. For $r = 5, 6, 7$ and 8, the LOC$YMM.MESH theory again completely explains what was found (and not found) in the numerical investigation. For $r = 4$, our symmetry theory only predicts three points (if averaging is allowed, there would be nine), Babuška, Strouboulis, Upadhyay and Gangaraj [1] found four points and a line (see Figures 5.2 and 5.3). This is again explained in Zhang [30].

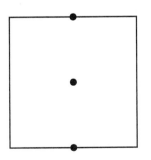

Figure 5.2   Figure 5.3

6 ∂_H TRANSL.INV.MESH

Let us start with a trivial observation which is valid also for irregular meshes. If

$$\|u - u_h\|_{L_\infty(\mathcal{D})} \leq Ch^r, \tag{6.1}$$

then one may construct a superconvergent approximation to $\dfrac{\partial u}{\partial x}$ (or, any derivative) simply by taking "long" different quotients. For, with H a parameter, let ∂_H with basic step H be a difference quotient which is accurate to order s, i.e.,

$$\left| \left(\frac{\partial u}{\partial x} - \partial_H u \right)(x) \right| \leq CH^s \tag{6.2}$$

for u smooth. Such a difference quotient would also typically satisfy

$$|\partial_H v(x)| \leq CH^{-1}\|v\|_{L_\infty}, \tag{6.3}$$

and so, using (6.1),

$$\left| \frac{\partial u}{\partial x} - \partial_H u_h \right| = \left| \left(\frac{\partial u}{\partial x} - \partial_H u \right) + \partial_H (u - u_h) \right| \leq C(H^s + H^{-1}h^r). \tag{6.4}$$

Hence, if (e.g.) $H = h^{\frac{r}{s+1}}$, we have an approximation of order $h^{r\left(\frac{s}{s+1}\right)}$ for $\dfrac{\partial u}{\partial x}$. This is superconvergent if $s > r - 1$.

On locally translation invariant meshes, i.e.,

$$T_{\pm H}\chi(x) = \chi(x \pm He_i) \in S_h, \quad \text{for } \chi \in \overset{\circ}{S}_h, \tag{6.5}$$

Nitsche and Schatz [16, Section 6] noted the following: For $\mathcal{A}$ a constant coefficient symmetric form, from (6.5) and (1.1) we have

$$\mathcal{A}(T_H(u - u_h), \chi) = (-1)^N \mathcal{A}(u - u_h, T_{-H}\chi) = 0, \quad \text{for } \chi \in \overset{\circ}{S}_h, \tag{6.6}$$

and hence for ∂_H a difference operator built up from such translations, locally on a region $\mathcal{D}$,

$$\mathcal{A}(\partial_H(u - u_h), \chi) = 0, \quad \text{for } \chi \in \overset{\circ}{S}_h(\mathcal{D}). \tag{6.7}$$

Then Schatz and Wahlbin [21] a counterpart of Theorem 1.1 with $F \equiv 0$ for function values, gives, again without tracing the dependence on d,

$$\|\partial_H(u - u_h)\|_{L_\infty(\mathcal{D}_0)} \leq C(\ln 1/h)^{\bar{r}} \min_{\chi \in S_h} \|\partial_H u - \chi\|_{L_\infty(\mathcal{D})}$$
$$+ C\|\partial_H(u - u_h)\|_{W_q^{-s}(\mathcal{D})}. \tag{6.8}$$

Here $\bar{r} = 0$ if $r \geq 3$, $\bar{r} = 1$ if $r = 2$. Hence, under suitable assumptions,

$$\left\| \frac{\partial u}{\partial x} - \partial_H u_h \right\|_{L_\infty(\mathcal{D}_0)} \leq C\left(H^s + \left(\ln \frac{1}{h}\right)^{\bar{r}} h^r \right). \tag{6.9}$$

The argument extends to nonsymmetric variable coefficient operators by use of Schatz and Wahlbin [**22**, Theorem 1.1]. In many practical translation invariant situations, H can be taken as a small multiple of h (cf. Figure 5.1).

We also note that the arguments above extend to any order derivative. By use of a local averaging operator, the K-operator Bramble and Schatz [**2**], one may even construct approximations to any derivative which are of the same order of accuracy as that of $\|u - u_h\|_{W_q^{-s}(\Omega)}$, for any s or q.

In conclusion of this section, difference quotients can always be used to construct superconvergent approximations to derivatives, and they are particularly powerful on translation invariant meshes.

7   CONCLUDING REMARKS

We have briefly reported on two general techniques for predicting natural (and almost natural) superconvergent points, LOC\$YMM.MESH and $\otimes$PROD.ELEMTS. The theory for computationally finding superconvergent points for $\mathcal{C}^0$-elements given in Babuška, Strouboulis, Upadhyay and Gangaraj [**1**], already applied there also to harmonic functions and solutions of elasticity problems, will certainly lead to more theoretical challenges, cf. Zhang [**30**]. (The theory tells you that you can nail down the points by rootfinding, but it does not explain why they are there or how many there are.) Construction of superconvergent approximations to derivatives by postprocessing is always possible by use of difference quotients.

The situation is less fortunate for systems, fourth order equations and mixed methods. What is lacking is a counterpart of Theorem 1.1, and most investigations have therefore been limited to either L_2-based investigations or investigations on translation invariant meshes (where one can use a form of Sobolev's inequality involving difference quotients). As already mentioned, the theory of superconvergence up to boundaries is not well developed. Superconvergence frequently fails close to boundaries, but no general principle appears to be available for guidance in this respect; the scant literature is mainly devoted to case-by-case investigations.

Let us finally mention some topics which are fairly well developed but not considered above. Extensions to nonlinear problems with smooth solutions basically follow Douglas and Dupont [**8**], cf. Wahlbin [**29**, Chapter 9]. The main idea is as follows. Let u denote the solution to the nonlinear problem and u_h the finite element solution to the discrete nonlinear problem. With $\widetilde{u}_h$ the projection of u onto S_h with respect to an associated elliptic bilinear form $\widetilde{\mathcal{A}}[u](v,w)$, which hence corresponds to a linear problem with coefficients depending on u, we have that u_h and $\widetilde{u}_h$ are superclose,

$$\|u_h - \widetilde{u}_h\|_{W_\infty^1(\mathcal{D})} \le C(\ln 1/h)h^{2r-2}. \tag{7.1}$$

The investigation of superconvergence in derivatives is thus reduced to a linear case. For superconvergence in function values the same argument applies for $r > 2$.

Extension of the translation invariant theory for difference quotients to piecewise polynomial mappings of such meshes is done in Cayco, Schatz and Wahlbin [**3**], see Wahlbin [**29**, Chapter 10] for a description. It is possible to treat meshes which are systematically refined towards a singularity. Several results on superconvergence have been extended to time-dependent problems, generally on an ad hoc basis from the elliptic case (and often only in space dimension $N = 2$ or 1).

REFERENCES

1. I. Babuška, T. Strouboulis, C. S. Upadhyay and S. K. Gangaraj, *Computer-based proof of the existence of superconvergence points in the finite element method; superconvergence of derivatives in finite element solutions of Laplace's, Poisson's, and the elasticity equations*, Numer. Methods Partial Differential Equations **12** (1996), 347–392.

2. J. H. Bramble and A. H. Schatz, *Higher order local accuracy by averaging in the finite element method*, Math. Comp. **31** (1977), 94–111.

3. M.-E. Cayco, A. H. Schatz and L. B. Wahlbin, to appear.

4. C.-M. Chen, *Superconvergent points of Galerkin's method for two-point boundary value problems*, Numer. Math. J. Chinese Univ. **1** (1979), 73–79.

5. C.-M. Chen, "The Finite Element Method and Its Analysis in Improving Accuracy," Hunan Science Press, 1982.

6. C.-M. Chen and Y. Huang, "High Accuracy Theory of Finite Element Methods," Hunan Science and Technology Press, Changsha, 1995.

7. P. G. Ciarlet, *"Basic Error Estimates for Elliptic Problems"*, in "Handbook of Numerical Analysis," P. G. Ciarlet and J. L. Lions (eds.), vol. II (Finite Element Methods, Part 1), Elsevier, 1991, pp. 17–351.

8. J. Douglas Jr. and T. Dupont, *A Galerkin method for a nonlinear Dirichlet problem*, Math. Comp. **29** (1975), 689–696.

9. J. Douglas Jr., T. Dupont and M. F. Wheeler, *An L^∞ estimate and a superconvergence result for a Galerkin method for elliptic equations based on tensor products of piecewise polynomials*, RAIRO Modél. Math. Anal. Numér. **8** (1974), 61–66.

10. R. Dunlap, "Superconvergence points in locally uniform finite element meshes for second order two-point boundary value problems," Thesis, Department of Mathematics, Cornell University, 1996.

11. F. V. Filho, *Comment on "Computation of stress resultants from the element stiffness matrices"*, AIAA Journal **6** (1968), 571–572.

12. R. Glowinski and J. L. Lions, *Exact and approximate controllability for distributed parameter systems*, Acta Numerica, Cambridge (1995), 159–333.

13. M. Křížek and P. Neittaanmäki, *On superconvergence techniques*, Acta Appl. Math. **9** (1987), 175–198.

14. B. Li, *Superconvergence for higher-order triangular finite elements*, Chinese J. Numer. Math. Appl. **12** (1990), 75–79.

15. Q. Lin, N. Yan and J. Zhou, *Integral-identity argument and natural superconvergence for cubic elements of Hermite type*, Beijing Mathematics **1** (1995), 19–26.

16. J. A. Nitsche and A. H. Schatz, *Interior estimates for Ritz–Galerkin methods*, Math. Comp. **28** (1974), 937–958.

17. E. J. Nyström, *Über die praktische Auflösung von Integralgleichungen mit Anwendungen auf Randwertaufgaben*, Acta Math. **54** (1930), 185–204.

18. L. A. Oganesyan and L. A. Rukhovetz, *Study of the rate of convergence of varia-*

tional difference schemes for second order elliptic equations in a two-dimensional field with a smooth boundary, USSR Comput. Math. and Math. Phys. **9** (1969), 158–183.

19. A. H. Schatz, *Pointwise error estimates and asymptotic error expansion inequalities for the finite element method on irregular grids: Part II. Interior estimates*, to appear.

20. A. H. Schatz, I. H. Sloan and L. B. Wahlbin, *Superconvergence in finite element methods and meshes that are symmetric with respect to a point*, SIAM J. Numer. Anal. **33** (1996), 505–521.

21. A. H. Schatz and L. B. Wahlbin, *Interior maximum norm estimates for finite element methods*, Math. Comp. **31** (1977), 414–442.

22. A. H. Schatz and L. B. Wahlbin, *Interior maximum-norm estimates for finite element methods, Part II*, Math. Comp. **64** (1995), 907–928.

23. K. Schellbach, *Probleme der Variationsrechnung*, J. Reine Angew. Math. **41** (1851), 293–363.

24. S. E. Stenij, *Evert Johannes Nyström in Memoriam*, Normat **8** (1960), 105–109.

25. J. A. Stricklin, *Computation of stress resultants from the element stiffness matrices*, AIAA Journal **4** (1966), 1095–1096.

26. J. A. Stricklin, *Reply to F. V. Filho*, AIAA Journal **6** (1968), p. 572.

27. P. Tong, *Exact solutions of certain problems by finite-element method*, AIAA Journal **7** (1969), 178–180.

28. L. B. Wahlbin, *"Local Behavior in Finite Element Methods"*, in "Handbook of Numerical Analysis," P. G. Ciarlet and J. L. Lions (eds.), vol. II (Finite Element Methods, Part 1), Elsevier, 1991, pp. 353–522.

29. L. B. Wahlbin, "Superconvergence in Galerkin Finite Element Methods," Springer Lecture Notes in Mathematics 1605, 1995.

30. Z. Zhang, *Derivative superconvergence points in finite element solutions of Poisson's equation for the serendipity and intermediate families – a theoretical justification*, to appear in Math. Comp.

31. Q.-D. Zhu and Q. Lin, "The Hyperconvergence Theory of Finite Elements," Hunan Science and Technology Publishing House, Changsha, 1989.

32. O. C. Zienkiewicz and Y. K. Cheung, "The Finite Element Method in Structural and Continuum Mechanics," McGraw-Hill, 1967.

A survey of superconvergence techniques in finite element methods

QIDING ZHU*   Department of Mathematics and Mechanics, Southeast University, Nanjing 210018, People's Republic of China

Abstract   The purpose of this paper is to give a survey of superconvergence techniques in finite element methods, such as Green function method, tensor product method, Nitsche trick and convolution mean method, energy orthogonal method, the synthetical method of the Green function and energy orthogonality, local superconvergence estimate technique, etc.

1   INTRODUCTION

The theory of superconvergence has been investigated since the beginning of the 1970s. Superconvergence phenomena were first discovered by engineers in practical calculations (cf. [**41**]). When trying to solve differential equations with the finite element method, they found to their surprise that the finite element solutions had an inconceivable superconvergent rate at some special points. This attracted the attention of many mathematicians. Douglas, Dupont, de Boor and Swartz (cf. [**9**], [**10**]) first discussed one-dimensional problem ($n = 1$). For multidimensional problems ($n \geq 2$), until the end of the 1970s no satisfactory results were obtained.

It is obvious that the research in superconvergence phenomena is of practical and theoretical importance. At the Vancouver International Congress of Mathematicians, Strang (cf. [**26**], [**27**]) regarded it as a new phase in the research of the finite element theory. Now, the theory of superconvergence begins to draw to its completion. Křížek and Neittaanmäki (cf. [**16**]) made a report on superconvergence titled "On superconvergence techniques" in which they introduced not only the superconvergent results for elliptic boundary value problems but also for hyperbolic and parabolic problems, and they also introduced results for integral equations and least square method. However, they did not mention how to prove these superconvergent results.

---

*The work was partly supported by National Natural Science Foundation of China.

This paper aims to introduce techniques for proving superconvergence results comprehensively in order to help those people who want to do some further research work in superconvergence.

Inductively, the usual research techniques for superconvergence are as follows:

1. Green function method;
2. Tensor product method;
3. Nitsche trick and convolution mean method;
4. Energy orthogonal method;
5. The synthetical method of the Green function and energy orthogonality;
6. Local superconvergence estimate technique.

In this paper, we shall in turn introduce the above techniques. For the sake of simplicity, we only consider the following variational problem: Find $u \in H_0^1(\Omega)$ such that

$$a(u, v) = (f, v) \quad \forall v \in H_0^1(\Omega),$$

where $\Omega \subset R^n (n \geq 1)$ denotes a bounded open domain, $(\cdot, \cdot)$ is the inner product in $L^2(\Omega)$ and

$$a(u, v) = \int_\Omega \left(\sum_{i,j=1}^n a_{ij} \partial_i u \partial_j v + Quv \right) dx_1 \, dx_2 \cdots dx_n. \tag{1}$$

We shall assume throughout this paper that $a(\cdot, \cdot)$ is $H_0^1(\Omega)$-elliptic, the coefficients a_{ij}, Q (≥ 0) and the right hand side f are as smooth as necessary on $\bar{\Omega}$ for our analysis to carry though. Furthermore, we also demand that (1) has a unique generalized solution $u \in H_0^1(\Omega) \cap C(\bar{\Omega})$.

Let S^h be the finite element space of Lagrange pattern (or Hermite pattern) based on subdivision T^h and satisfy

$$\inf_{v \in S^h} \|u - v\|_m \leq ch^{k+1-s-m} \|u\|_{k+1-s}, \quad m = 0, 1, \ 0 \leq s \leq k. \tag{2}$$

Let u^h be the finite element approximation which satisfies

$$a(u^h, v) = a(u, v) \quad \forall v \in S^h. \tag{3}$$

The symbol u^I ($\in S^h$) denotes the standard interpolant. As far as the Lagrange element is concerned, u^I satisfies

$$u^I(P) = u(P) \quad \forall P \in T,$$

where T denotes the set of nodes.

2   GREEN FUNCTION METHOD

The Green function method is generally used to make L^∞-error estimates for finite elements. Douglas and Dupont (cf. [10]) first used this method to get a superconvergence estimate. They discussed the following two-point boundary value problem:

$$-u'' + a(x)u' + b(x)u = f(x), \quad x \in (0, 1), \tag{4}$$
$$u(0) = u(1) = 0.$$

The corresponding energy is

$$a(u, v) = \int_0^1 (u'v' + a(x)u'v + b(x)uv)\, dx.$$

Let $T^h = \{I_i\}$ be a sequence of uniform subdivisions of $I = (0, 1)$. Now we set

$$S_k^h(I) = \{v \in C_0(\overline{I}) : v|_{I_i} \in P_k(I_i),\ I_i \in T^h\},$$

where $P_k(E)$ denotes the polynomials of degree $\leq k$ on E.

To any $z \in (0, 1)$, let G_z be the Green function such that

$$a(v, G_z) = v(z) \quad \forall v \in H_0^1(I).$$

The so-called Green function method defines the error in the form of energy:

$$u(z) - u^h(z) = a(u - u^h, G_z).$$

Note that when $n = 1$, as long as the coefficients $a(x)$ and $b(x)$ in (4) are sufficiently smooth, we have

$$G_z \in H_0^1(I) \cap H^{k+1}([0, z]) \cap H^{k+1}(I).$$

Hence, when z is a node, there exists G_z^I ($\in S_k^h(I)$) such that

$$\|G_z - G_z^I\|_1 \leq Ch^k(\|G_z\|_{k+1,[0,z]} + \|G_z\|_{k+1,[z,1]}). \tag{5}$$

Thus, in total, from (2), (3) and (5),

$$
\begin{aligned}
|u(z) - u^h(z)| &= |a(u - u^h, G_z - G_z^I)| \\
&\leq C\|u - u^h\|_1 \|G_z - G_z^I\|_1 \\
&\leq Ch^{k+s}\|u\|_{s+1}, \qquad 0 \leq s \leq k.
\end{aligned}
\tag{6}
$$

The proof of the superconvergent result (6) is very simple and the result, too, turns out to be very elegant. However, with this method, we cannot obtain the superconvergence results of the inner part of every element I_i. Moreover, this method cannot be generalized to n-dimensional problem ($n \geq 2$), for z is a singular point of the Green function G_z. Using

$$\|G_z - G_z^h\|_{1,1} \leq Ch|\ln h|, \tag{7}$$

Scott (cf. [25]) obtained

$$
\begin{aligned}
|u(z) - u^h(z)| &= |a(u - u^h, G_z - G_z^h)| \\
&\leq Ch^{k+1}|\ln h|\, \|u\|_{k+1,\infty}.
\end{aligned}
$$

Then, the problem of L^∞-estimate was solved.

3 TENSOR PRODUCT METHOD

In order to extend the results to n-dimensional problem ($n \geq 2$), Douglas, Dupont and Wheeler (cf. [12]) discussed the following two-dimensional Poisson equation

$$-\Delta u = f \quad \text{in } \Omega, \tag{8}$$
$$u = 0 \quad \text{on } \partial\Omega,$$

where $\Omega = (0,1) \times (0,1) = I \times I$. Now the finite element space is the following tensor product space

$$S^h = S_k^h(I) \otimes S_k^h(I), \quad k \geq 3, \tag{9}$$

where $S_k^h(I)$ (cf. Section 2) is a one-dimensional finite element space. Now setting Galerkin projection

$$P: \; H_0^1(I) \longrightarrow S_k^h(I), \tag{10}$$

we have

$$\int_I (u - Pu)' v' \, dx = 0 \quad \forall v \in S_k^h(I).$$

Form the following two-dimensional tensor product operator

$$P \otimes P: \; H_0^1(\Omega) \longrightarrow S^h, \tag{11}$$

namely, if $u \in H_0^1(\Omega)$, then $W = P \otimes Pu$ satisfies

$$W(\cdot, y) = Pu(\cdot, y) \quad \forall y \in I,$$
$$W(x, \cdot) = Pu(x, \cdot) \quad \forall x \in I. \tag{12}$$

In other words, $W = P \otimes Pu$ satisfies

$$\int_I (I \otimes Pu - W)'_x \cdot v'_x \, dx = 0 \quad \forall v(x) \in S_k^h(I),$$
$$\int_I (P \otimes Iu - W)'_y \cdot v'_y \, dy = 0 \quad \forall v(y) \in S_k^h(I). \tag{13}$$

Thus, when x_i, x_j are subdivisional points of interval I, we have (cf. [11], [12])

$$|u(x_i, x_j) - W(x_i, x_j)| = O(h^{k+2}). \tag{14}$$

In order to obtain the superconvergent estimate of $u - u^h$ at point (x_i, x_j), we only need to estimate

$$U = u^h - W. \tag{15}$$

Since

$$(\nabla U, \nabla v) = (\nabla(u - W), \nabla v) = -(\eta, v) \quad \forall v \in S^h. \tag{16}$$

Here

$$\eta = I \otimes (I - P)u''_{xx} + (I - P) \otimes Iu''_{yy}, \tag{17}$$

Then, this follows

$$\|U\|_1 \leq C\|\eta\|_{-1} \tag{18}$$

and

$$|(\nabla U, \nabla v)| \leq C\|\eta\|_{-2}\|v\|_2 \quad \forall v \in V_h, \tag{19}$$

where

$$V_h = S^h \cap H^2(\Omega).$$

Using the negative norm estimate (cf. (25))

$$\|(I - P)u\|_{-s} \leq Ch^{k+1+s}\|u\|_{k+1}, \quad 0 \leq s \leq k-1,$$

and (17) we thus find that

$$\|\eta\|_{-s} \leq Ch^{k+1+s}\|u\|_{k+3}, \quad s = 1, 2, \ k \geq 3. \tag{20}$$

This and (18) shows that

$$\|U\|_1 \leq Ch^{k+2}\|u\|_{k+3}, \quad k \geq 3. \tag{21}$$

Now, find φ such that

$$-\Delta\varphi = U \quad \text{in } \Omega, \qquad \varphi = 0 \quad \text{on } \partial\Omega.$$

Then, from (19) there exists $v \in V_h$ so that

$$\begin{aligned}
\|U\|_0^2 &= (\nabla U, \nabla\varphi) \\
&= (\nabla U, \nabla(\varphi - v)) + (\nabla U, \nabla v) \\
&\leq Ch\|U\|_1\|\varphi\|_2 + C\|\eta\|_{-2}\|\varphi\|_2.
\end{aligned}$$

Here we have used $\varphi \in H^2(\Omega)$ and the well-known estimate

$$\inf_{v \in V_h} \{h^{-1}\|\varphi - v\|_1 + \|\varphi - v\|_2\} \leq C, \quad k \geq 3. \tag{22}$$

So, from a priori estimate

$$\|\varphi\|_2 \leq C\|U\|_0$$

and (20), (21) we get

$$\|U\|_0 \leq Ch^{k+3}\|u\|_{k+3}, \quad k \geq 3.$$

Finally, noting the inverse estimate

$$\|U\|_{0,\infty} \leq Ch^{k+2}\|u\|_{k+3}, \tag{23}$$

and by (14), (15) we obtain the following superconvergence estimate

$$|u(x_i, x_j) - u^h(x_i, x_j)| \leq Ch^{k+2}\|u\|_{k+3}, \quad k \geq 3. \tag{24}$$

In the above argument, Douglas, Dupont and Wheeler made use of a subsidiary function $W = P \otimes Pu$, which made the whole argumentation possible in Hilbert space H^m and very convenient. However, estimate (22) hold only for $k \geq 3$, but might be improved. Actually, there exists superconvergence for the case of $k = 2$.

We must also note that Douglas et al have already obtained superconvergence estimate

$$\|U\|_0 = O(h^{k+3}),$$

which indicates that at subdivisional points superconvergence may not be $O(h^{k+2})$ but $O(h^{k+3})$, or even better. (At least, as may be true in the sense of average.)

4 NITSCHE TRICK AND CONVOLUTION MEAN METHOD

Assume that the boundary of $\Omega \subset \mathbb{R}^n$ $(n \geq 1)$ is sufficiently smooth. To any $\varphi \in H^s(\Omega)$ there exists $\tilde{\varphi} \in H^{s+2}(\Omega) \cap H_0^1(\Omega)$ such that

$$\begin{aligned}
|(u - u^h, \varphi)| &= |a(u - u^h, \tilde{\varphi})| \\
&= |a(u - u^h, \tilde{\varphi} - \tilde{\varphi}^I)| \\
&\leq C\|u - u^h\|_1\|\tilde{\varphi} - \tilde{\varphi}^I\|_1.
\end{aligned}$$

From (2) we have

$$\begin{aligned}
|(u - u^h, \varphi)| &\leq Ch^{k+s+1}\|u\|_{k+1}\|\tilde{\varphi}\|_{k+2} \\
&\leq Ch^{k+s+1}\|u\|_{k+1}\|\varphi\|_s, \qquad 0 \leq s \leq k-1.
\end{aligned}$$

We thus obtain

$$\|u - u^h\|_{-s} \leq Ch^{k+s+1}\|u\|_{k+1}, \quad 0 \leq s \leq k-1, \tag{25}$$

which is the well-known Nitsche trick. The negative norm estimate can reach the optimal convergence rate of $O(h^{2k})$.

Using spline function, Bramble and Schatz (cf. [2]) introduced an average kernel $K_h(x)$ which satisfies

K1) The support of K_h is a spherical domain centered at origin and of radius Ch;

K2) $K_h(x)$ is a piecewise polynomial function and the continuity of its $k-1$th-order derivative does not depend on the equation;

K3) The Fourier transform of $K_h(x)$ satisfies

$$\hat{K}_h(\xi) = 1 + O(h^{2k}|\xi|^{2k}),$$
$$|\hat{K}_h(\xi)| \leq C(1 + |\xi|^2)^{-\frac{p}{2}},$$

where p $(\geq k-1)$ is a positive integer and

$$\hat{f}(\xi) = \int_{\mathbb{R}^n} f(x)e^{-i\xi \cdot x}\, dx,$$

where $x = (x_1, x_2, \cdots, x_n)$, $\xi = (\xi_1, \xi_2, \cdots, \xi_n)$, $\xi \cdot x = \sum_{i=1}^{n} \xi_i x_i$.

Form the following convolution:

$$K_h * u^h(x) = \int_{\mathbb{R}^n} K_h(x - \eta)u^h(\eta)\, d\eta,$$

and let it be the approximation of u. Since the radius of the support of K_h is Ch, the convolution $K_h * u^h$ exists in any fixed inner subregion of Ω. Now we need to estimate the following error:

$$u - K_h * u^h = u - K_h * u + K_h * (u - u^h).$$

On the one hand, from condition K3 we have

$$\begin{aligned}
|u - \widehat{K_h * u}| &= |\hat{u} - \hat{K}_h \hat{u}| \\
&= |1 - \hat{K}_h||\hat{u}| \\
&\leq Ch^{2k}|\xi|^{2k}|\hat{u}|.
\end{aligned}$$

When u is smooth enough and fastly decreasing, from the Parseval equality (cf. [29]), we can get

$$\begin{aligned}
\|u - K_h * u\|_0 &= \|u - \widehat{K_h * u}\|_0 \\
&\leq Ch^{2k}\| |\xi|^{2k}|\hat{u}| \|_0 \\
&\leq Ch^{2k}\|u\|_{2k}.
\end{aligned}$$

If $u(\in H_0^1(\Omega))$ is sufficiently smooth only in inner subregion $\Omega_1 \subset\subset \Omega$ and if we set

$$\tilde{u} = \omega u,$$

where $\omega \in C_0^\infty(\Omega)$ and $\omega \equiv 1$ in Ω_1, (here $\Omega_0 \subset\subset \Omega_1 \subset\subset \Omega$) then we obtain

$$\begin{aligned}
\|u - K_h * u\|_{0,\Omega_0} &\leq \|\tilde{u} - K_h * \tilde{u}\|_0 \\
&\leq Ch^{2k}\|\tilde{u}\|_{2k} \\
&\leq Ch^{2k}\|u\|_{2k,\Omega_1}.
\end{aligned}$$

On the other hand,

$$\begin{aligned}
\|K_h * (u - u^h)\|_0 &= \|K_h * \widehat{(u - u^h)}\|_0 \\
&= \|\hat{K}_h \widehat{(u - u^h)}\|_0 \\
&\leq \|(1 + |\xi|^2)^{-\frac{p}{2}}\widehat{|u - u^h|}\|_0 \\
&\leq C\|u - u^h\|_{-p},
\end{aligned}$$

This and (25) show that

$$\|u - K_h * u^h\|_{0,\Omega_0} \leq Ch^{2k}\|u\|_{2k} + C\|u - u^h\|_{-p} \leq Ch^{2k}[\|u\|_{2k,\Omega_1} + \|u\|_{k+1}]. \quad (26)$$

As for derivative, Thomée (cf. [28]) obtained the similar results by using difference ratio.

In order to understand the nature of the above method easily, let us take the one-dimensional case for example to show how to form the Bramble kernel.

Set

$$g_0(t) = \begin{cases} 1, & \text{if } -\frac{1}{2} \le t \le \frac{1}{2}; \\ 0, & \text{else} \end{cases}$$

$$g_l(t) = g_0 * g_{l-1}(t), \quad l = 1, 2, 3, ..., k.$$

and

$$K_h(t) = \frac{1}{h} g_{k-1}\left(\frac{t}{k}\right).$$

First, note that

$$\hat{g}_0(\xi) = \int_{-\infty}^{+\infty} g_0(t) e^{-i\xi t} \, dt = \frac{\sin \frac{\xi}{2}}{\frac{\xi}{2}},$$

$$\hat{g}_{k-1}(\xi) = [\hat{g}_0(\xi)]^k = \left(\frac{\sin \frac{\xi}{2}}{\frac{\xi}{2}}\right)^k = 1 + O(|\xi|^{2k}).$$

Therefore,

$$\begin{aligned} \hat{K}_h(\xi) &= \int_{-\infty}^{+\infty} \frac{1}{h} g_{k-1}\left(\frac{t}{h}\right) e^{-i\xi t} \, dt \\ &= \int_{-\infty}^{+\infty} g_{k-1}(u) e^{-iuh\xi} \, du \\ &= \hat{g}_{k-1}(h\xi) \\ &= 1 + O(h^{2k}|\xi|^{2k}). \end{aligned}$$

Finally, note that

$$\begin{aligned} |\hat{K}_h(\xi)| &= |\hat{g}_{k-1}(h\xi)| = \left| \left(\frac{\sin \frac{h\xi}{2}}{\frac{h\xi}{2}}\right)^k \right| \\ &= \left| 2h^{-1} \sin \frac{h\xi}{2} \right|^k \left[\frac{(1+|\xi|^2)^{\frac{1}{2}}}{|\xi|} \right]^k \cdot (1+|\xi|^2)^{-\frac{k}{2}} \\ &\le C(1+|\xi|^2)^{-\frac{k}{2}}. \end{aligned}$$

The above argumentation shows that K_h satisfies conditions K1, K2 and K3, so $K_h(x)$ is the Bramble kernel.

5   ENERGY ORTHOGONAL METHOD

Let the subdivision T^h satisfy some condition, say, let it be piecewise uniform. Then we have the following weak estimate

$$|a(u - u^I, v)| \le Ch^{k+1} \|u\|_{k+2} \|v\|_1 \quad \forall v \in S^h. \tag{27}$$

Since

$$a(u^h - u, v) = 0 \quad \forall v \in S^h,$$

we have

$$\|u^h - u^I\|_1^2 \leq Ca(u^h - u^I, u^h - u^I)$$
$$= Ca(u - u^I, u^h - u^I)$$
$$\leq Ch^{k+1}\|u\|_{k+2}\|u^h - u^I\|_1.$$

Thus we obtain

$$\|u^h - u^I\|_1 \leq Ch^{k+1}\|u\|_{k+2},$$

which is one order higher than the usual error estimate

$$\|u - u^h\|_1 = O(h^k).$$

The above method is called energy orthogonal method which was used by Zlámal, Chen and Zhu (cf. [3], [4], [5], [6], [30], [31], [42], [43]). However, mathematicians were not satisfied with the above results, for we could only obtain, by using energy orthogonal method, average superconvergence estimate of gradient

$$\left(\frac{1}{\gamma}\sum_{x \in G_h}|\nabla(u - u^h)(x)|^2\right)^{\frac{1}{2}} = O(h^{k+1}), \quad k \geq 1. \tag{28}$$

where G_h denotes the so-called "optimal points of the stresses" set (cf. [40]) and γ denotes the total number of G_h.

In order to obtain the superconvergence estimate for the error itself, we (cf. [33]) introduce the negative norm estimate

$$\|u - u^h\|_{-1} \leq Ch^{k+2}\|u\|_{k+1}, \quad k \geq 2.$$

This leads to

$$\|u - u^I\|_{-1} \leq Ch^{k+2}\|u\|_{k+2}, \quad k \geq 2$$

and then we get

$$\|u^h - u^I\|_{-1} \leq Ch^{k+2}\|u\|_{k+2}.$$

Using (27), we thus find that

$$\|u^h - u^I\|_0^2 = (u^h - u^I, u^h - u^I)$$
$$\leq \|u^h - u^I\|_{-1}\|u^h - u^I\|_1$$
$$\leq Ch^{2k+3}\|u\|_{k+2}.$$

Finally, we obtain

$$\|u^h - u^I\|_0 = O(h^{k+1+\frac{1}{2}}), \quad k \geq 2. \tag{29}$$

Although (29) is not the best superconvergence estimate, it first shows that there exists superconvergence for the quadratic element, and at least there exists the superconvergence estimate

$$\left(\frac{1}{\gamma}\sum_{x \in T}|(u - u^h)(x)|^2\right)^{\frac{1}{2}} = O(h^{\frac{7}{2}}), \tag{30}$$

where γ denotes the total number of nodes of the set T.

Using the Nitsche trick, we can find $W \in H^2(\Omega) \cap H_0^1(\Omega)$ such that

$$a(W, v) = (u^h - u^I, v) \quad \forall v \in H_0^1(\Omega),$$
$$\|W\|_2 \leq C \|u^h - u^I\|_0.$$

Making use of the energy orthogonality relation, we have

$$\begin{aligned}
\|u^h - u^I\|_0^2 &= (u^h - u^I, u^h - u^I) = a(W, u^h - u^I) \\
&= a(W - W^I, u^h - u) + a(W - W^I, u - u^I) + a(W^I, u - u^I) \\
&\leq Ch^{k+1}\|u\|_{k+1} \cdot h\|W\|_2 + |a(u - u^I, W^I)|.
\end{aligned}$$

Note that for the piecewise C-uniform subdivision, there exists the second weak estimate (cf. [40])

$$|a(u - u^I, W^I)| \leq Ch^{k+2}\|u\|_{k+2}\|W\|_2, \quad k \geq 2. \tag{31}$$

Thus

$$\begin{aligned}
\|u^h - u^I\|_0^2 &\leq Ch^{k+2}\|u\|_{k+2}\|W\|_2 \\
&\leq Ch^{k+2}\|u\|_{k+2}\|u^h - u^I\|_0.
\end{aligned}$$

Finally we have

$$\|u^h - u^I\|_0 \leq Ch^{k+2}\|u\|_{k+2}, \quad k \geq 2. \tag{32}$$

The above method, which can be found in [4], [32], and [33], combines the merits of the Nitsche trick with those of the energy orthogonality method. However, the above results are still average superconvergence estimates, not point-by-point estimates. We recall (cf. [32], [34]) the second weak estimate

$$|a(u - u^I, W^I)| \leq Ch^{k+2}\|u\|_{k+2,q}\|W\|_{2,p}, \tag{33}$$

where $k \geq 2$, $\frac{1}{p} + \frac{1}{q} = 1$, $2 < q \leq \infty$, and the first weak estimate

$$|a(u - u^I, v)| \leq Ch^{k+1}\|u\|_{k+2,q}\|v\|_{1,p}, \tag{34}$$

where $k \geq 2$, $\frac{1}{p} + \frac{1}{q} = 1$, $2 < q \leq \infty$.

For $\varphi \in L^p(\Omega)$, use the Nitsche trick and find $W \in W^{2,p}(\Omega) \cap W_0^{1,p}(\Omega)$ such that

$$\begin{aligned}
|(u^h - u^I, \varphi)| &= |a(u^h - u^I, W)| \\
&\leq |a(u^h - u^I, W - W^I)| + |a(u^h - u^I, W^I)|.
\end{aligned}$$

Using the energy orthogonality relation once again, we find that

$$\begin{aligned}
|a(u^h - u^I, W - W^I)| &= |a(u^h - u^I, W^I - W^h)| \\
&= |a(u - u^I, W^I - W^h)| \\
&\leq Ch^{k+1}\|u\|_{k+2,q}\|W^I - W^h\|_{1,p} \\
&\leq Ch^{k+2}\|u\|_{k+2,q}\|W\|_{2,p}.
\end{aligned}$$

From the a priori estimate $\|W\|_{2,p} \leq C\|\varphi\|_{0,p}$ and (33), we find that

$$|(u^h - u^I, \varphi)| \leq Ch^{k+2}\|u\|_{k+2,q}\|\varphi\|_{0,p}.$$

Therefore,

$$\|u^h - u^I\|_{0,q} \leq Ch^{k+2}\|u\|_{k+2,q}, \quad 2 < q < \infty.$$

If q is large enough, by making use of the inverse estimate

$$\|u^I - u^h\|_{0,\infty} \leq Ch^{-\frac{2}{q}}\|u^I - u^h\|_{0,q} \leq Ch^{k+2-\frac{2}{q}}\|u\|_{k+2,q}, \tag{35}$$

we obtain (cf. [35])

$$\left(\frac{1}{\gamma}\sum_{x \in T}|(u - u^h)(x)|^q\right)^{\frac{1}{q}} \leq Ch^{k+2}\|u\|_{k+2,q}, \tag{36}$$

and

$$|(u - u^h)(x_0)| \leq Ch^{k+2-\frac{2}{q}}\|u\|_{k+2,\infty} \quad \forall x_0 \in T, \ 2 < q < \infty. \tag{37}$$

Here γ denotes the total number of nodes of the set T. Theoretically speaking, the above method is still not perfect.

6   THE SYNTHETICAL METHOD OF THE GREEN FUNCTION AND ENERGY ORTHOGONALITY

Since Douglas and Dupont (cf. [10]) solved one-dimensional superconvergence problem by using the Green function method, there has been a few mathematicians who tried to study superconvergence by using this method for the simple reason that Green function G_z has a singular point z when n is not less than 2. Generally speaking, we only have

$$G_z \in W^{1,p}(\Omega), \quad 1 \leq p < 2,$$

and there even exists $G_z \notin H_0^1(\Omega)$.

In [34], [35], we used a discrete Green function to make a point-by-point superconvergence estimate. Using the estimate (cf. [34])

$$\|G_{z+\Delta z}^h - G_z^h\|_{1,1} \leq Ch|\ln h|, \quad |\Delta z| \leq Ch,$$

and the first weak estimate, we find that

$$\left|\frac{(u^h - u^I)(z + \Delta z) - (u^h - u^I)(z)}{\Delta z}\right|$$
$$= \left|\frac{a(u^h - u^I, G_{z+\Delta z}^h - G_z^h)}{\Delta z}\right|$$
$$\leq Ch^{k+1}|\ln h|\,\|u\|_{k+2,\infty}, \qquad Ch \leq |\Delta z| \leq h, \ k \geq 1.$$

Therefore, for any $z \in \Omega$, we have

$$|\nabla(u^h - u^I)(z)| \leq Ch^{k+1}|\ln h|\,\|u\|_{k+2,\infty}, \quad k \geq 1. \tag{38}$$

Thus if x_0 is an optimal stress point, in other words, if x_0 satisfies

$$|\nabla(u - u^I)(x_0)| = O(h^{k+1}),$$

we have

$$|\nabla(u - u^h)(x_0)| \leq Ch^{k+1}|\ln h| \, \|u\|_{k+2,\infty}. \tag{39}$$

Let us introduce (cf. [37]) a discrete δ-function δ_z^h ($\in S^h$) and its derivative $\partial_z \delta_z^h$ ($\in S^h$) which satisfy

$$\begin{aligned}
(u^h - u^I, \delta_z^h) &= (u^h - u^I)(z), \\
(u^h - u^I, \partial_z \delta_z^h) &= \partial_z[(u^h - u^I)(z)].
\end{aligned} \tag{40}$$

Assuming the equation satisfies a regular condition, we thus can find a unique G_z^* and a unique $\partial_z G_z^* \in H_0^1(\Omega)$ such that

$$\begin{aligned}
a(G_z^*, \varphi) &= (\delta_z^h, \varphi) & \forall \varphi \in H_0^1(\Omega), \\
a(\partial_z G_z^*, \varphi) &= (\partial_z \delta_z^h, \varphi) & \forall \varphi \in H_0^1(\Omega).
\end{aligned}$$

Then let $g = G_z^*$. This leads to $g^h = G_z^h$ and $g^I = (G_z^*)^I$. Thus,

$$\begin{aligned}
|(u^h - u^I)(z)| &= |a(u^h - u^I, g^h)| \\
&\leq |a(u - u^I, g^h - g^I)| + |a(u - u^I, g^I)| \\
&\leq Ch^{k+1}\|u\|_{k+2,\infty}\|g^h - g^I\|_{1,1} + Ch^{k+2}\|u\|_{k+2,\infty}\|g\|_{2,1}.
\end{aligned} \tag{41}$$

Since (cf. [37])

$$\|g^h - g^I\|_{1,1} \leq Ch|\ln h|, \quad \|g\|_{2,1} \leq C|\ln h|,$$

we have

$$|(u^h - u^I)(z)| \leq Ch^{k+2}|\ln h| \, \|u\|_{k+2,\infty} \quad \forall z \in \Omega. \tag{42}$$

If $z \in T$, we have

$$|(u - u^h)(z)| \leq Ch^{k+2}|\ln h| \, \|u\|_{k+2,\infty}, \quad k \geq 2. \tag{43}$$

Substituting $\partial_z G_z^*$ for g in (41) and using the weak estimate, we obtain

$$|\partial_z(u^h - u^I)(z)| \leq Ch^{k+1}|\ln h|^\lambda \|u\|_{k+2,\infty}, \quad k \geq 1, \tag{44}$$

where

$$\lambda = \begin{cases} 1 & \text{if } k = 1, \\ 0 & \text{if } k > 1. \end{cases}$$

We thus obtain the point-by-point superconvergence estimate of gradient

$$|\nabla(u - u^h)(x_0)| \leq Ch^{k+1}|\ln h|^\lambda \|u\|_{k+2,\infty} \quad \forall x_0 \in G_h. \tag{45}$$

The reader can find the point-by-point superconvergence estimate of gradient for the quadratic triangular element in [1], [7], [32], [34], and [35]. The reader can also find the point-by-point superconvergence error estimate for the quadratic element in [35], [36], and [40].

For a superconvergence analysis of the derivative for triangular linear element, the reader is referred to [1], [3], [13], [14], [15], [16], [17], [18], [19], [20], [24], [30]. The superconvergence analysis of the derivative for the bilinear element has appeared in [6], [8], [21], [30], [42], [43]. The reader interested in the point-by-point superconvergence analysis of derivatives for the triangular linear element and the bilinear element can find them in [20], [21], [22], [23], [34], [35], [40].

7   LOCAL SUPERCONVERGENCE ESTIMATE TECHNIQUE

The above superconvergence techniques, except that of Bramble et al, are based on the assumption that solution u is sufficiently smooth. This assumption does not hold for a general boundary value problem, especially for domains with corners or mixed boundary conditions. Now we only assume that the solution u is smooth enough in a local part of the domain Ω and that the corresponding local subdivision T^h is C-uniform. Under this weak condition we obtain (cf. [35], [38], [40]) superconvergence results in the subdomain.

Fix a point z in Ω and assume that solution u is smooth enough in a fixed neighborhood of z:

$$U_\rho = \{x \in \Omega : |x - z| < \rho\}.$$

Choose a function $\omega \in C_0^\infty(\Omega)$ such that $\omega \equiv 1$ in $D_0 = U_{\frac{\rho}{2}}$, and $\operatorname{supp} \omega \subset U_\rho$. Hence, $u_1 = \omega u$ is smooth enough in Ω and thus we have

$$|(u_1^h - u_1^I)(z)| \le Ch^{k+2}|\ln h|\,\|u_1\|_{k+2,\infty} \le Ch^{k+2}|\ln h|\,\|u\|_{k+2,\infty,U_\rho}. \tag{46}$$

Let $u_2 = u - u_1$. Obviously, $u_2^I - u_2 \equiv 0$ in D_0, and therefore,

$$
\begin{aligned}
|(u_2^h - u_2^I)(z)| &= |(u_2^h - u_2)(z)| \\
&= a(u_2^h - u_2, G_z) \\
&= a(u_2^h - u_2, G_z - G_z^h) \\
&= a(u_2^I - u_2, G_z - G_z^h),
\end{aligned}
$$

and

$$|(u_2^h - u_2^I)(z)| \le \|u_2^I - u_2\|_{1,1,\Omega\backslash D_0}\|G_z - G_z^h\|_{1,\infty,\Omega\backslash D_0}.$$

Using the local estimate (cf. [40])

$$\|G_z - G_z^h\|_{1,\infty,\Omega\backslash D_0} \le Ch^k,$$

we get

$$
\begin{aligned}
|(u_2^h - u_2^I)(z)| &\le Ch^{k+s}\|u_2\|_{s+1,1,\Omega\backslash D_0}, \\
&\le Ch^{k+s}\|u\|_{s+1,1,\Omega\backslash D_0}, \qquad 1 \le s \le k.
\end{aligned}
\tag{47}
$$

From (46) and (47) we obtain

$$|(u^h - u^I)(z)| \le Ch^{k+2}|\ln h|\big[\|u\|_{k+2,\infty,U_\rho} + \|u\|_{3,1}\big], \tag{48}$$

where $k \geq 2$. Similarly, we have (cf. [**40**])

$$|\nabla(u^h - u^I)(z)| \leq Ch^{k+1}|\ln h|^\lambda \big[\|u\|_{k+2,\infty,U_\rho} + \|u\|_{3,1}\big], \qquad (49)$$

where $k \geq 1$, and

$$\lambda = \begin{cases} 1, & \text{if } k = 1, \\ 0, & \text{if } k \geq 2. \end{cases}$$

Relations (48), (49) hold only when Ω is a smooth domain. If Ω is a convex domain with corners, the corresponding results are (cf. [**40**])

$$|(u^h - u^I)(z)| \leq C\big[h^{k+2}|\ln h|\,\|u\|_{k+2,\infty,U_\rho} + h^{\min(k+2,\beta_M+2)-\varepsilon}\|u\|_{3,1}\big], \qquad (50)$$

where $k \geq 2$, and

$$|\nabla(u^h - u^I)(z)| \leq C\big[h^{k+1}|\ln h|^\lambda\|u\|_{k+2,\infty,U_\rho} + h^{\min(k+2,\beta_M+2)-\varepsilon}\|u\|_{3,1}\big], \qquad (51)$$

where $k \geq 1$.

Relations (48), (49), (50), and (51), which are the most perfect local superconvergence estimates, hold only under the assumption that the subdivision T^h is C-uniform (or piecewise C-uniform). The assumption is applicable in all cases.

REFERENCES

1. A. B. Andreev, *Error estimate of type superconvergence of the gradient for quadratic triangular elements*, C. R. Acad. Bulgare Sci. **37** (1984), 1179–1182.
2. J. H. Bramble and A. H. Schatz, *Higher order local accuracy by averaging in the finite element method*, Math. Comp. **31** (1977), 94–111.
3. C.-M. Chen, *Optimal points of the stresses for triangular linear element*, Numer. Math. J. Chinese Univ. **2** (1980), 12–20. (in Chinese)
4. C.-M. Chen, *Superconvergence of finite element solutions and their derivatives*, Numer. Math. J. Chinese Univ. **3** (1981), 118–125. (in Chinese)
5. C.-M. Chen, "The Finite Element Method and Its Analysis in Improving Accuracy," Hunan Science Press, 1982. (in Chinese)
6. C.-M. Chen and Q.-D. Zhu, *A new estimate in the finite element method and the theorem of optimal points of the stresses*, Natur. Sci. J. Xiangtan Univ. **1** (1978). (in Chinese)
7. H.-S. Chen, *The optimal points of stresses for quadratic triangular elements*, a paper for master degree. (in Chinese)
8. H.-S. Chen, *A rectangular finite element analysis*, Natur. Sci. J. Xiangtan Univ. **11** (1989), 1–11.
9. C. de Boor and B. Swartz, *Local piecewise polynomial projection methods for an ODE which give high-order convergence at knodes*, Math. Comp. **36** (1981), 21–33.
10. J. Douglas Jr. and T. Dupont, *Some superconvergence results for Galerkin methods for the approximate solution of two-point boundary value problems*, in "Topics in Numerical Analysis," Academic Press, 1973, pp. 89–92.

11. J. Douglas Jr., T. Dupont and M. F. Wheeler, *Some superconvergence results for an H^1 Galerkin procedure for the heat equation*, in "Lecture Notes in Comput. Sci.," vol. 10, Springer, 1974, pp. 288–311.

12. J. Douglas Jr., T. Dupont and M. F. Wheeler, *An L^∞ estimate and a superconvergence result for a Galerkin method for elliptic equations based on tensor products of piecewise polynomials*, RAIRO Anal. Numér. **8** (1974), 61–66.

13. M. El Hatri, *Superconvergence in the finite element method for a degenerated boundary value problem*, in "Constructive Theory of Functions," Bulgarian Academy of Sciences, 1984, pp. 328–333.

14. E. Hinton and J. S. Campbell, *Local and global smoothing of discontinuous finite element functions using a least squares method*, Internat. J. Numer. Methods Engrg. **8** (1974), 461–480.

15. X. F. Huang, *The optimal point of the gradient of finite element solution*, Appl. Math. Mech. **7** (1986), 785–794.

16. M. Křížek and P. Neittaanmäki, *On superconvergence techniques*, Acta Appl. Math. **9** (1987), 175–198.

17. M. Křížek and P. Neittaanmäki, *On a global superconvergence of the gradient of linear triangular elements*, J. Comput. Appl. Math. **18** (1987), 221–233.

18. N. Levine, *Superconvergent recovery of the gradient from piecewise linear finite element method*, Numerical analysis report 6/83, Univ. of Reading, 1983.

19. Q. Lin, T. Lu and S.-M. Shen, *Asymptotic expansions for finite element approximations*, Research Report IMS-11, Inst. Math. Sci. Chengdu Branch of Acad. Sinica, 1983, pp. 1–6..

20. Q. Lin and J. C. Xu, *Linear finite elements with high accuracy*, J. Comput. Math. **3** (1985), 115–133.

21. Q. Lin and Q.-D. Zhu, *Unidirectional extrapolations of finite difference and finite elements*, J. Engineering Math. **1** (1984), No. 2.

22. Q. Lin and Q.-D. Zhu, *Asymptotic expansions for the derivative of finite elements*, J. Comput. Math. **2** (1984), 361–364.

23. Q. Lin and Q.-D. Zhu, *Local asymptotic expansion and extrapolation for finite elements*, J. Comput. Math. **4** (1986), 263–265.

24. G. Marchuk and V. Shaidurov, "Difference Methods and Their Extrapolations," Springer, Berlin, 1983.

25. R. Scott, *Optimal L^∞ estimates for the finite element method on irregular meshes*, Math. Comp. **30** (1976), 681–697.

26. G. Strang, Comp. Appl. Math. **12** (1977), 52–53.

27. G. Strang, in "Proceedings of the International Congress of Mathematicians," Vancouver, 1973, V. 5, pp. 429–435.

28. V. Thomée, *High order local approximations to derivatives in the finite element method*, Math. Comp. **31** (1977), 652–660.

29. K. Yosida, "Functional Analysis," Springer, Berlin, 1978.

30. Q.-D. Zhu, *The theorem of optimal points of stresses under isoparametric transformation in the finite element method*, Natur. Sci. J. Xiangtan Univ. **1** (1978), 91–100.

31. Q.-D. Zhu, *Point by point estimation and maximum norm inner estimation in the finite element method*, J. Comp. Math. **1** (1981), 87–90.

32. Q.-D. Zhu, *The derivative good point for the finite element method with 2-degree triangular element*, Natur. Sci. J. Xiangtan Univ. **1** (1981), 36–45.

33. Q.-D. Zhu, *A superconvergence result for the finite element method*, Numer. Math. J. Chinese Univ. **1** (1981), 50–55.

34. Q.-D. Zhu, *Uniform superconvergence estimates of derivative for the finite element method*, Numer. Math. J. Chinese Univ. **4** (1983), 311–318.

35. Q.-D. Zhu, *Natural inner superconvergence for the finite element method*, in "Proc. China–France Sympos. on F. E. M.," Beijing, 1982, pp. 935–960.

36. Q.-D. Zhu, *Uniform superconvergence estimates for the finite element method*, Natur. Sci. J. Xiangtan Univ. **2** (1985), 10–26.

37. Q.-D. Zhu, *Local estimates for the finite element method*, Science exploration **7** (1987), 189–196.

38. Q.-D. Zhu, *Some L^∞ estimates and interior superconvergence estimates for piecewise linear finite element approximations*, Natur. Sci. J. Xiangtan Univ. **1** (1987).

39. Q.-D. Zhu, *Local superconvergence estimate for the finite element method*.

40. Q.-D. Zhu and Q. Lin, "The Superconvergence Theory of Finite Elements," Hunan Science and Technology Publishing House, Changsha, China, 1989. (in Chinese)

41. O. C. Zienkiewicz and Y. K. Cheung, "The Finite Element Method in Structural and Continuum Mechanics," McGraw-Hill, London, 1967.

42. M. Zlámal, *Some superconvergence results in the finite element method*, in "Lecture Notes in Mathematics 606," Springer, Berlin, 1977, pp. 353–362.

43. M. Zlámal, *Superconvergence and reduced integration in the finite element method*, Math. Comp. **32** (1978), 663–685.

Adaptive procedure with superconvergent patch recovery for linear parabolic problems

S. ZIUKAS and N.-E. WIBERG   Department of Structural Mechanics, Chalmers University of Technology, S-412 96 Göteborg, Sweden

Abstract   Performance of Zienkiewicz–Zhu error estimator is considered for parabolic equations. Estimate of the local elliptic discretization error in energy norm is obtained by constructing a recovered gradient field of a higher order of accuracy at each time step using superconvergent properties of the solution. The adaptive procedure presented controls the local time error as a fraction of elliptic projection errors that are estimated using ZZ error estimator and Superconvergent Patch Recovery.

1 INTRODUCTION

The method of lines (MOL) is a very popular method to solve initial-boundary value problems in engineering. In this approach a problem is first discretized in space using, for example, Galerkin method, and then the resulting system of ordinary differential equations (ODE) is solved in time. The finite difference method (FDM) provides the most simple discretization concept for transforming the differential problem into an algebraic one. Due to the decomposition to space and time discretizations the approach is obviously simple and practical.

A detailed treatment of Galerkin finite element methods for parabolic equations can be found in Thomée [1]. The smoothness of a solution to a parabolic problem may vary considerably in space and time. In order to resolve phenomena in an efficient way, discretizations with variable mesh sizes in both space and time are needed. Thus, an efficient and accurate error estimates are required. Error estimates based on the residual of the computed solution for semidiscrete parabolic problem were derived by Bieterman and Babuška, see [2]. Adjerid and Flaherty suggested estimates obtained as solutions of either local parabolic or local elliptic finite element problems using piecewise polynomial corrections, cf. [3]. These estimates have been proven to converge to the exact error under mesh refinement. A posteriori error estimates in $L_2(\Omega)$ norm based on variational formulation were suggested by Eriksson and Johnson [4].

Error estimates developed for elliptic equations can be used to estimate space discretization errors for the parabolic problems. One of such estimators based on the recovered solutions was developed by Zienkiewicz and Zhu, see [5]. The estimator has been remarkably successful for elliptic problems. The main advantages of using ZZ error estimator are the simplicity of its implementation and its cost effectiveness. In the post-processing stage some recovery procedure is employed in order to achieve more acceptable approximations to the gradients. Using recovered gradients, the ZZ error estimator can be calculated at a fraction of the total cost of computation. Babuška et al [6] have shown by numerical experiments that ZZ error estimator based on superconvergent patch recovery (SPR) is the most robust one for sufficiently smooth solutions.

2   MODEL PROBLEM AND OBJECTIVES

Let us consider a parabolic initial-boundary value problem for the heat equation (groundwater flow)

$$
\begin{aligned}
\frac{\partial u}{\partial t} + Lu &= f(\mathbf{x}, t) && \text{in } (\mathbf{x}, t) \in \Omega \times I,\ I = (0, T], \\
u(\mathbf{x}, t) &= 0 && \text{on } \partial\Omega \times I, \\
u(\mathbf{x}, 0) &= u_0(\mathbf{x}) && \text{in } \Omega,
\end{aligned}
\tag{2.1}
$$

where L is an elliptic operator defined as

$$
Lu = -\sum_{i,j=1}^{d} \frac{\partial}{\partial x_i}\left(a_{ij}(\mathbf{x})\frac{\partial u}{\partial x_j}\right) \equiv -\nabla \cdot [A(\mathbf{x})\nabla u]
\tag{2.2}
$$

and Ω is a bounded domain in $\mathbb{R}^d$ with smooth boundary $\partial\Omega$, $a_{ij}(\mathbf{x}) = a_{ji}(\mathbf{x}) \geq \gamma > 0$, $a_{ij}(\mathbf{x}) \in C^\infty(\Omega)$, and T is some positive number.

Formulating the standard Galerkin method in $H = L_2(\Omega)$ with $V = H_0^1(\Omega)$ leads to the weak formulation: find $u(t) \in L^2(0, T; V) \cap C^0(0, T; L^2(\Omega))$ such that

$$
\begin{aligned}
(u_t(t), v) + a\left(u(t), v\right) &= (f(t), v), && \forall v \in V,\ t \in (0, T], \\
u(0) &= u_0,
\end{aligned}
\tag{2.3}
$$

where u_t denotes the partial derivative $\partial u/\partial t$. The bilinear form $a(\cdot, \cdot)$ is defined as

$$
a(u, v) = \int_\Omega \sum_{i,j=1}^{n} a_{ij}(\mathbf{x})\frac{\partial u(\mathbf{x})}{\partial x_i}\frac{\partial v(\mathbf{x})}{\partial x_j}\, d\mathbf{x}.
\tag{2.4}
$$

For the finite element method we partition the domain Ω into triangular elements τ defined by the mesh T_h. Let

$$
S_h(\Omega) = \left\{ \chi \in H_0^1(\Omega) \mid \chi|_\tau \in \mathcal{P}_n\ \forall \tau \in T_h \right\}.
\tag{2.5}
$$

The approximate semidiscrete problem is to find $u_h(t) \in S_h(\Omega)$ such that

$$
\begin{aligned}
(u_{h,t}, \chi) + a(u_h, \chi) &= (f, \chi), && \forall \chi \in S_h(\Omega),\ t \in (0, T], \\
u(0) &= u_0.
\end{aligned}
\tag{2.6}
$$

The result is a system of N ordinary differential equations. We shall analyze the backward difference approximation for the time variable introducing a time step $k_n = t_n - t_{n-1}$, replacing the time derivative by a backward difference quotient $\bar{\partial}U^n = (U^n - U^{n-1})/k_n$ and seeking approximations U^n in S_h of $u(t_n)$

$$(U^n, \chi) + k_n a(U^n, \chi) = (U^{n-1} + k_n f(t_n), \chi), \quad \forall \chi \in S_h, \ n > 0. \tag{2.7}$$

Our goal is to discuss an estimate of the space discretization error in the 'energy' norm

$$\|v\|_{E,\Omega} = a(v,v)^{1/2} = |v|_{1,\Omega} \tag{2.8}$$

using Zienkiewicz–Zhu error estimator, i.e., using the local error indicator

$$\eta_\tau(t) := \left(\int_\tau (Gu_h - A\nabla u_h)^t A^{-1} (Gu_h - A\nabla u_h) \right)^{\frac{1}{2}} \tag{2.9}$$

as an approximation to $\|e\|_{E(\Omega)}$. The symbol G denotes some recovery operator which acts on the finite element solution u_h to give an approximation to the gradient $A\nabla u$. The a posteriori error estimator is defined by

$$E(t) := \left(\sum_{\tau \in T_h} \eta_\tau^2 \right)^{\frac{1}{2}}. \tag{2.10}$$

To assess performance of the estimator $E(t)$, we define the effectivity index

$$\theta(t) = \frac{E(t)}{\|e(\cdot, t)\|_{E(\Omega)}}. \tag{2.11}$$

Estimator is said to be asymptotically exact if

$$\|e\|_{E(\Omega)} \sim \{1 + O(h^\gamma)\} E \quad \text{as } h \to 0, \tag{2.12}$$

where $\gamma > 0$ is independent of h. This means that it tends to estimate the true error exactly as size of elements tends to zero.

3   ERROR ANALYSIS

Solving parabolic problems with the finite element method of lines it is reasonable to study the total discretization error in two steps – to analyze separately the finite element method error and the error incurred in solving the system of ODE's

$$u(\mathbf{x}, t) - U(\mathbf{x}, t) = u(\mathbf{x}, t) - u_h(\mathbf{x}, t) + u_h(\mathbf{x}, t) - U(\mathbf{x}, t) = e(h, \mathbf{x}, t) + e(k, \mathbf{x}, t). \tag{3.1}$$

Our intention here is to separate the error arising from the finite element discretization and consider methods of estimating it only. Thus, we shall assume that the system of ODE's can be solved exactly, i.e., the total discretization error is equal to the space discretization error

$$e(\mathbf{x}, t) = e(h, \mathbf{x}, t) = u(\mathbf{x}, t) - u_h(\mathbf{x}, t). \tag{3.2}$$

A behaviour of the error can be characterized by an initial-boundary value problem analogous to (2.3). Replacing u by $u_h + e$, the discretization error satisfies the Galerkin formulation: find $e \in H_0^1(\Omega)$ such that

$$(v, e_t) + a(v, e) = (v, f) - a(v, u_h) - (v, u_{h,t}), \quad \forall v \in H_0^1, \ \forall t > 0, \qquad (3.3)$$
$$a(v, e) = a(v, u_0 - u_h), \qquad \forall v \in H_0^1, \ t = 0. \qquad (3.4)$$

Error estimates can be established by solving or approximating equations (3.3) and (3.4) to find a magnitude of the error.

We shall cite several results reflecting basic error estimates for parabolic equations. We denote by $\| \cdot \|$ the norm in $L_2(\Omega)$, and by $\| \cdot \|_r$ the norm in the Sobolev space $H^r(\Omega)$. It is well-known that gradient of the finite element solution is $O(h^{r-1})$ as in the elliptic case

$$\|\nabla(u_h(t) - u(t))\| \leq C\|\nabla(u_h(0) - u(0))\|$$
$$+ Ch^{r-1}\left\{ \|u_0\|_r + \|u(t)\|_r + \int_0^T \|u_t\|_{r-1}\, ds \right\}. \qquad (3.5)$$

We introduce the so-called elliptic (Ritz) projection $R_h\colon H_0^1 \to S_h$ as the orthogonal projection with respect to the bilinear form $a(\cdot, \cdot)$, so that

$$\|u - R_h u\|_{E(\Omega)} = \inf_{\psi \in S_h} \|u - \psi\|_{E(\Omega)} \quad \forall \psi \in S_h. \qquad (3.6)$$

In fact, $R_h u$ is the finite element approximation of the solution of the corresponding elliptic problem. Comparison of a solution of the semidiscrete problem to the elliptic projection of the exact solution

$$u - u_h = (u - R_h u) + (R_h u - u_h) = \rho + \Theta \qquad (3.7)$$

reveals that if $u_h(0) = R_h u(0)$ then

$$\|\nabla\Theta(t)\| \leq Ch^r \left(\int_0^T \|u_t\|_r^2 \right)^{1/2}. \qquad (3.8)$$

Hence, the gradient of Θ is of order $O(h^r)$, whereas the gradient of the total error is only $O(h^{r-1})$. It can be concluded that in the energy norm $\Theta(\mathbf{x}, t)$ is negligible in comparison with $u - R_h u$ and the Galerkin error satisfies

$$a(u - u_h, u - u_h)^{1/2} \approx a(u - R_h u, u - R_h u)^{1/2} \leq Ch^{r-1}\|u\|_r. \qquad (3.9)$$

We are going to show that estimator for the local elliptic projection $\|\rho\|$ is also an asymptotically exact estimator for $\|e\|$, i.e., $\|e\| \approx \|\rho\|$ up to higher order terms. Using the triangle inequality, we have

$$\|u - u_h\|_{E(\Omega)} - \|R_h u - u_h\|_{E(\Omega)} \leq \|\rho\|_{E(\Omega)} \leq \|u - u_h\|_{E(\Omega)} + \|R_h u - u_h\|_{E(\Omega)}, \quad (3.10)$$

and comparing $\|\rho\|_{E(\Omega)}$ to $\|e\|_{E(\Omega)}$, we get

$$1 - \frac{\|\theta\|_{E(\Omega)}}{\|e\|_{E(\Omega)}} \leq \frac{\|\rho\|_{E(\Omega)}}{\|e\|_{E(\Omega)}} \leq 1 + \frac{\|\theta\|_{E(\Omega)}}{\|e\|_{E(\Omega)}}. \tag{3.11}$$

Let E_1 be asymptotically exact error estimator for $\|\rho\|$. Then the estimator is asymptotically exact for $\|e\|$ also:

$$\|e\|_{E(\Omega)} \sim \left\{ 1 + O(h^{\min(\gamma, 1)}) \right\} E_1 \quad \text{as } h \to 0. \tag{3.12}$$

Thus, by solution of local elliptic problems instead of (3.3) we greatly reduce complexity of the error estimator and have asymptotically exact error estimator. We neglect the time rate of change of the error, and thereby, seek an approximate solution to the local elliptic problem: find a magnitude of $e \in H_0^1(\Omega)$ in $\|\cdot\|_{E(\Omega)}$ such that

$$a(v, e) = (v, f) - a(v, u_h) - (v, u_{h,t}) \quad \forall v \in H_0^1, \ \forall t > 0 \tag{3.13}$$

interpreting $u_{h,t}$ as given data.

4 ZIENKIEWICZ–ZHU ERROR ESTIMATOR

Zienkiewicz–Zhu error estimator was originally devised for elliptic problems. The basic concept is to replace a gradient of the exact solution in an analytical expression of the discretization error in energy norm

$$\|u - u_h\|_{E(\Omega)}^2 = \int_\Omega (\nabla u - \nabla u_h)^t A (\nabla u - \nabla u_h) \, d\Omega \tag{4.1}$$

by in some way recovered gradient Gu_h

$$\|u - u_h\|_{E(\Omega)}^2 \approx \int_\Omega (Gu_h - A\nabla u_h)^t A^{-1} (Gu_h - A\nabla u_h) \, d\Omega. \tag{4.2}$$

The approximation defines Zienkiewicz–Zhu error estimator. The next step is to get the recovered gradient that is sufficiently good approximation to ∇u, i.e., the recovered gradient must be much better approximation to ∇u than ∇u_h

$$\|A\nabla u - Gu_h\|_{L^2(\Omega)} \ll \|u - u_h\|_{E(\Omega)}. \tag{4.3}$$

Obviously, reliability of the estimator is dependent on the properties of the recovered gradient Gu_h. Zienkiewicz and Zhu have shown in [13] that the effectivity index θ approaches unity as the ratio of errors $\|e^*\| = \|A\nabla u - Gu_h\|_{L^2(\Omega)}$ and $\|e\| = \|u - u_h\|_{E(\Omega)}$ tends to zero

$$\left(1 - \frac{\|e^*\|}{\|e\|} \right) \leq \theta \leq \left(1 + \frac{\|e^*\|}{\|e\|} \right). \tag{4.4}$$

The condition is achieved if $\|e^*\|$ converges at a higher rate than $\|e\|$. It follows that if $\|e^*\|$ is superconvergent then asymptotic exactness of the error estimator is assured.

Ainsworth and Craig (cf. [7]) have shown that under suitable assumptions the discretization error in energy norm can be bounded using the estimator $E(t)$ considered. For our case the following bound will hold

$$\|u - u_h\|_{E(\Omega)}^2 \leq (1 + C_1 h^\gamma) \left(E(t) + \frac{h^\alpha}{C_2} \|\nabla \cdot G u_h + f - u_{h,t}\|_{L^2(\Omega)}^2 \right), \qquad (4.5)$$

where $\alpha \in (0, 2)$, γ, C_1, C_2 are some constants.

5   SUPERCONVERGENT PATCH RECOVERY

Several schemes for a recovery of a superconvergent gradient field from the finite element solution $G \colon S_h \to [S_h]^d$ have been an object of an intensive research in the last decade. They share a common feature to use superconvergent properties of the finite element solution. We will mention briefly some results known for us.

Thomée et al [8] have considered maximum norm estimates of the gradient for domains with smooth boundary and proved the following estimate

$$\|\nabla(u_h - R_h u)(t)\|_{L_\infty} \leq C h^2 \left(\log \frac{1}{h} \right)^{1/2} \left\{ \|u_t(0)\|_{H^2} + \int_0^t \|u_{tt}\|_{H^2} \, ds \right\}. \qquad (5.1)$$

Superconvergent error estimates in $l_\infty(H^1)$ and $l_2(H^1)$ for recovered gradients of finite difference in time/piecewise linear Galerkin approximations in space for linear and nonlinear problems were derived by Wheeler and Whiteman, see [9]. They proved that the recovery operator for regular meshes (regular pattern with exactly six elements)

$$\nabla u_h^*(k) = \sum_{i=1}^6 w_i (\nabla u_h)_{\tau_k^i} \qquad (5.2)$$

at a vertex (node) $\mathbf{x} \in \tau$ surrounded by six triangles produces a superconvergent gradient field $\nabla u_h^* \in S_h$ provided the exact solution is sufficiently regular. Different weights w_i were suggested for different configurations of triangular patches, see [11]. Piecewise linear interpolants are then fitted to these pointwise recovered gradients to obtain $O(h^2)$ estimates for the recovered gradient in $L_2(\Omega)$ norm.

Zienkiewicz and Zhu proposed a superconvergent patch recovery procedure (cf. [12]) which became very popular in engineering community. The SPR is used to obtain the gradient of higher accuracy and ensure narrow bounds of the effectivity index [13]. The authors showed numerically that the recovered gradient is also superconvergent in $L_2(\Omega)$ norm. The procedure is a simplified version of the those mentioned above and represents a problem of finding $G u_h \in [S_h]^d$ such that

$$\|G u_h - \nabla u_h\|_{L^2(\Omega)} = \inf \|G u_h - \nabla u_h\|_{L^2(\Omega)}. \qquad (5.3)$$

It is based on a least squares fit of a local polynomial $\mathbf{P}$ to the gradient values at superconvergent points $\{\mathbf{x}_i\}_{i=1}^{nsp}$. Assuming that

$$\nabla u^* = \mathbf{P} \mathbf{a} \qquad (5.4)$$

minimization condition results in the solution of

$$\sum_{i=1}^{nsp} \mathbf{P}^t(x_i)\mathbf{P}(x_i)\mathbf{a} = \sum_{i=1}^{nsp} \mathbf{P}^t(x_i)\nabla u_h(x_i) \tag{5.5}$$

at some points of a patch ω of elements where the gradient of the finite element solution exhibits superior accuracy.

6   LOCAL TIME ERRORS IN THE METHOD OF LINES

Local integration error of the Euler method can be estimated by constructing new higher order approximation to the solution or by evaluation of the second derivative of the solution with respect to the time in the *a-priori* estimate.

For the construction of a higher order approximation Crank–Nicolson scheme can be used as well and if the computations are arranged in a suitable way, then no extra derivative calculations need actually be performed. The second order correction η^n is used as the estimate of the local truncation error

$$(\mathbf{B} + k_n\mathbf{A})\eta^n = \frac{k_n}{2}\left\{\mathbf{A}(\mathbf{U}^n - \mathbf{U}^{n-1}) - (\tilde{f}(t_n) - \tilde{f}(t_{n-1}))\right\}. \tag{6.1}$$

Besides those methods, there are ones which make use of the higher derivatives of the solution. Assumption that second derivatives U_{tt} vary linearly yields a simple posteriori local error estimate e_n

$$|e_n| = \left|\left(\mathbf{U}_{tt}^n - 4\mathbf{U}_{tt}^{n+\frac{1}{2}}\right)\frac{k_n^2}{6}\right|. \tag{6.2}$$

Unknown second derivatives U_{tt}^n may be obtained by derivation of the original initial-boundary value problem.

7   NUMERICAL EXAMPLES

The test example is a problem in which the source function $f(x,t)$ and Dirichlet boundary conditions are chosen such that the exact solution is given by

$$u(x,y,t) = \exp\left[-80((x - r(t))^2 + (y - 0.5)^2)\right], \tag{7.1}$$

where $r(t) = 0.5 - 0.25\exp(-1000t)$. This represents a cone which is centered at $(0.5, 0.25)$ at $t = 0$ and moves with a continuously decreasing speed towards the center of the domain $(0.5, 0.5)$ where steady state conditions are reached. The solution was computed from $t = 0$ to $t = 0.1$. Backward Euler method was used for an integration of ODE system in time with a uniform time step size $k = 0.0002$ (500 steps totally).

In order to investigate efficiency and applicability of the estimator, the problem was solved using a sequence of uniform triangular meshes using 10, 20, 30, and 40 linear finite elements per side. The convergence of the global error in energy

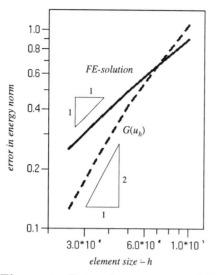

Figure 1 Rate of convergence of FE-solution and recovered gradient $G(u_h)$ in energy norm. Uniform triangulation.

norm of the finite element solution $\|u - u_h\|_{E(\Omega)}$ and the recovered gradient field $\|A\nabla u - G(u_h)\|_{L^2(\Omega)}$ by original SPR procedure at $t = 0.1$ is shown in Figure 1. Error $\|A\nabla u - G(u_h)\|_{L^2(\Omega)}$ is superconvergent with convergence rate of almost $O(h^2)$ for all time steps which assures asymptotic exactness of the estimator. The global effectivity index versus time for a 30×30 discretization is shown in Figure 2. Values of θ are close to unity for all time steps except some oscillation in the several starting steps. It is an effect of too large initial time step size. A growth of the global error occurs in that case and the global effectivity index decreases to 0.5. The accumulated error decays rather fast and after approximately 20 integration steps the global discretization error is almost equal to the local. Integrating with a uniform time step size $k = 0.00001$ yields the global effectivity index very close to unity. Distribution of elemental effectivity indices for the initial projection and the final state is presented in Figure 3. We notice some overestimation of the discretization error for the initial projection. Elemental effectivity indices become worse as time passes in regions near to the boundaries, i.e., where the exact solution and its gradient are small, however, in regions with a high solution gradient elemental effectivity indices are near unity for all time steps. That explains the fact that the global effectivity index remains almost unchanged with even small tendency for overestimation of the error.

The performance of Zienkiewicz–Zhu estimator was tested also in the context of adaptive analysis. The problem was solved using a variable time step aiming $TolS = 12\%$ accuracy relative error in energy norm in space and $TolT = 2\%$ accuracy of local temporal error. The error control strategy used aims to select the spatial grid accordingly to some tolerance and to integrate the ODE system in time with just sufficient accuracy so that the temporal error does not significantly corrupt the spatial accuracy. We use the standard higher order procedure to control

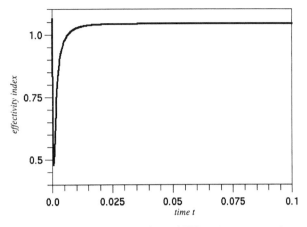

Figure 2   Effectivity index of ZZ estimator vs time. Uniform triangulation 30×30 with linear elements.

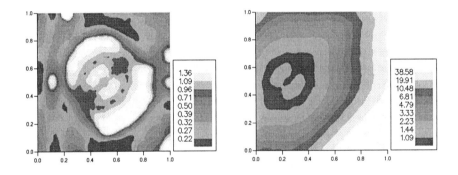

Figure 3   Elemental effectivity indices for ZZ error estimator in energy norm at $t = 0$ and $t = 0.1$. Uniform triangulation 20×20 with linear elements.

the local time error per step with respect to supplied accuracy tolerance. Starting from a triangular 20×20 grid, three additional grids were used, see Figure 6. Time step sizes used in the analysis are presented in Figure 4. Due to the adaptive time step sizes the global effectivity index has less oscillation in the starting steps (see Figure 5.). Elemental effectivity indices reveals better performance of the estimator comparing to uniform discretization with approximately the same number degrees of freedom. This fact confirms that the SPR performs better for 'almost' equilibrated meshes.

8   CONCLUSIONS

In this paper, performance of Zienkiewicz–Zhu error estimator with Superconvergent Patch Recovery procedure is tested for linear parabolic problems. Estimate of the local elliptic discretization error in energy norm is obtained by constructing

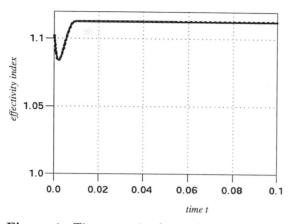

Figure 4   Time step size k_n used in the adaptive analysis of the problem.

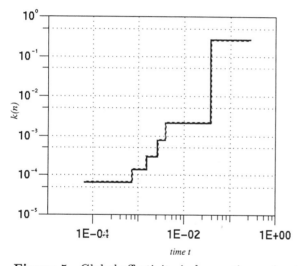

Figure 5   Global effectivity index vs time using the adaptive analysis.

a recovered gradient field of a higher order of accuracy at each time step using superconvergent properties of the solution.

It is shown that such error estimator is asymptotically exact for sufficiently regular solutions that is confirmed by a numerical example. The original Superconvergent Patch Recovery procedure gives a recovered gradient field which is more accurate and have a higher convergence rate $O(h^2)$ than the finite element solution $O(h^1)$ in energy norm. The global effectivity index does not differ from unity for uniform meshes as well as for those used in the adaptive analysis. The error estimator does not yield an upper bound of the error in general, however, it can be noticed that the estimator tends slightly to overestimate the discretization error for triangular meshes.

The error estimator requires only a small fraction of the computational cost used

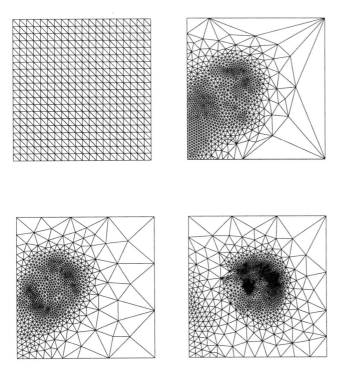

Figure 6 A sequence of meshes used in the adaptive analysis aiming 12% accuracy relative error.

for the finite element analysis.

Performance of the estimator can be improved by forcing the recovered gradient field to satisfy the local equilibrium equation as in the elliptic case, (see [**14**]).

The estimator presented does not take into account t-dependence of the error evolution. Thus, it might not be reliable for problems with significant accumulation of the global error. However, this estimator is more efficient and still can be applied to problems with sufficiently high error decay rates as in the numerical example presented. The shortcoming can be solved by summing the error estimators at each time step or in the same way as in [**2**], i.e., by introducing exponential decay rates of the error with lower bound for eigenvalue of the elliptic operator L.

The behaviour of the estimator for the numerical example presented reveals a rather good evaluation of the space discretization error. We believe that with some improvements mentioned it can be robust as well as for elliptic problems.

REFERENCES

1. V. Thomée,, "Galerkin Finite Element Methods for Parabolic Problems," Lecture Notes in Mathematics 1054, Springer, 1984.
2. M. Bieterman and I. Babuška, *The finite element method for parabolic equations. Part I. A posteriori error estimation*, Numer. Math. **40** (1982), 330–371.

3. S. Adjerid, J.E. Flaherty and Y.J. Wang, *A posteriori error estimation with finite element methods of lines for one-dimensional problems*, Numer. Math. **65** (1993), 1–21.

4. K. Eriksson and C. Johnson, *Adaptive finite element methods for parabolic problems I: A linear model problem*, SIAM J. Numer. Anal. **28** (1991), 43–77.

5. O.C. Zienkiewicz and J.Z. Zhu, *A simple error estimator and adaptive procedure for practical engineering analysis*, Internat. J. Numer. Methods Engrg. **24** (1987), 337–357.

6. I. Babuška, T. Strouboulis, C.S. Upadhyay, S.K. Gangaraj and K. Copps, *Validation of a posteriori error estimators by numerical approach*, Internat. J. Numer. Methods Engrg. **37** (1994), 1073–1123.

7. M. Ainsworth and A. Craig, *A posteriori error estimators in the finite element method*, Numer. Math. **60** (1992), 429–463.

8. V. Thomée, J.C. Xu and N.Y. Zhang, *Superconvergence of the gradient in piecewise linear finite element approximation to a parabolic problem*, SIAM J. Numer. Anal. **26** (1989), 553–573.

9. M.F. Wheeler and J.R. Whiteman, *Superconvergence of recovered gradients of discrete time/piecewise linear Galerkin approximations for linear and nonlinear parabolic problems*, Numer. Methods Partial Differential Equations **10** (1994), 271–294.

10. M.F. Wheeler and J.R. Whiteman, *Superconvergent recovery gradients on subdomains from piecewise linear finite element approximations*, Numer. Methods Partial Differential Equations **3** (1987), 357–374.

11. G. Goodsell and J.R. Whiteman, *A unified treatment of superconvergent recovered gradient functions for piecewise linear finite element approximations*, Internat. J. Numer. Methods Engrg. **27** (1989), 469–481.

12. O.C. Zienkiewicz and J.Z. Zhu, *The superconvergent patch recovery and a posteriori error estimates. Part 1: The recovery technique*, Internat. J. Numer. Methods Engrg. **33** (1992), 1331–1364.

13. O.C. Zienkiewicz and J.Z. Zhu, *The superconvergent patch recovery and a posteriori error estimates. Part 2: Error estimates and adaptivity*, Internat. J. Numer. Methods Engrg. **33** (1992), 1365–1382.

14. N.-E. Wiberg and F. Abdulwahab, *Patch recovery based on superconvergent derivatives and equilibrium*, Internat. J. Numer. Methods Engrg. **36** (1993), 2703–2724.

Bibliography on superconvergence

MICHAL KŘÍŽEK* Mathematical Institute, Academy of Sciences, Žitná 25,
CZ–11567 Prague 1, Czech Republic, e-mail: krizek@math.cas.cz

PEKKA NEITTAANMÄKI Department of Mathematics, University of Jyväskylä,
P.O. Box 35, FIN–40351 Jyväskylä, Finland, e-mail: pn@math.jyu.fi

Abstract   This paper contains an annotated bibliography to superconvergence
phenomena that arise in solving differential and integral equations and other problems of applied mathematics and mathematical physics by numerical methods.

1 INTRODUCTION

Superconvergence of the finite element method is a quickly and dynamically developing field of research. In 1987, we published a survey paper [**324**] on superconvergence techniques with 200 references. Since that time at least 400 new papers have
been written on this subject. Great progress has been made especially in design-
ing superconvergence meshes, in introducing many post-processing methods leading
to superconvergence and in applying various superconvergence phenomena to an a
posteriori error estimation.

This paper contains 600 references to superconvergence phenomena which arise
when solving differential and integral equations, variational inequalities and other
problems of mathematical physics by numerical methods; in particular, by Galerkin
and finite element methods, finite difference methods, collocation methods, boundary element methods, box methods, etc. Some other related references can be found
in the below-mentioned monographs and survey papers.

The bibliography is equipped with the *Mathematical Reviews* reference numbers
available in August 1997. It comprises, however, also many papers which are not
included in MR.

There are already six monographs written by Chen [**126**], Chen and Huang [**139**],

---

*The work was supported by grant No. A1019601 of the Grant Agency of the Academy of Sciences
of the Czech Republic. This support is gratefully acknowledged. The authors also wish to thank
to Jari Toivanen for his help.

Lin and Yan [371], Lin and Zhu [381], Wahlbin [524], and Zhu and Lin [588] dealing with superconvergence of finite element approximations of partial differential equations. Let us further mention the Lecture Notes [523], the Proceedings [328] devoted to superconvergence phenomena and a large contribution [521] contained in Handbook of Numerical Analysis. Survey papers on the existing literature on superconvergence include Chen [131], Goldberg [242], Křížek [318], [319], Křížek and Neittaanmäki [324], [325], Wahlbin [521], and Whiteman and Goodsell [544]. Main ideas of proofs of various superconvergence techniques are introduced in Zhu [585].

2 SUPERCONVERGENCE OF THE FINITE ELEMENT METHOD FOR SOLVING DIFFERENTIAL EQUATIONS

Most papers in the reference list examine superconvergence phenomena that arise in solving differential equations by the finite element method. Throughout the paper we assume that the true solution is sufficiently smooth.

2.1 Two-point problem

In the first place we deal with a one-dimensional second order elliptic equation

$$-u'' + a(x)u' + b(x)u = f(x), \quad x \in (0,1),$$

with the homogeneous Dirichlet boundary conditions $u(0) = u(1) = 0$. Consider its standard Galerkin approximation based on continuous piecewise polynomial functions, which are of degree k in each element K (a one-dimensional subinterval of [0,1]). A thorough numerical analysis of a superconvergence phenomenon for this case has been firstly done by Douglas and Dupont in [187], [188]. They proved that the Galerkin approximation yields the $\mathcal{O}(h^{2k})$-superconvergence at *nodes* whereas $\mathcal{O}(h^{k+1})$ is the optimal global convergence rate in the maximum norm, which cannot be improved.

Chen [121] and later independently Bakker [55] have proved that the error is $\mathcal{O}(h^{k+2})$ at $k+1$ Lobatto points of each element K.

In 1979, Chen [121] and also Lesaint and Zlámal [339] have shown the $O(h^{k+1})$-superconvergence of the derivatives at the k *Gauss–Legendre points* of each element K. Note that $\mathcal{O}(h^k)$ is the optimal global convergence rate for the first derivatives.

Superconvergence at the *Jacobi points* is investigated in Nakao [415] for a special finite element scheme, see also Carey, Humphrey and Wheeler [107].

According to Strang and Fix [497], a piecewise linear Galerkin approximation is even equal to the classical solution at all nodal points provided $a = b = 0$. Note that a similar result is proved in Křížek and Neittaanmäki [326] for a fourth order one-dimensional problem approximated by the cubic Hermite elements.

Early papers established superconvergence at isolated points. Later, however, there has been a growing literature on superconvergence obtained by simple post-processing techniques based on averaging, interpolating, iteration, convolution and integral smoothing operators, see, for instance, Babuška and Miller [40], and Bramble and Schatz [76]. Further references on superconvergence of the two-point problem include [8], [12], [15], [21], [36], [56], [59], [102], [108], [130], [155], [161], [166], [178], [180], [189], [201], [230], [283], [286], [288], [346], [386], [396], [403], [415], [420], [451], [455], [457]–[459], [476], [497], [500], [516], [522], [534], [535], [566], [575], [576], [583].

2.2   Superconvergence of function values

Now we consider the standard Galerkin approximation of the two-dimensional Dirichlet problem

$$-\Delta u = f \quad \text{in } \Omega,$$
$$u = 0 \quad \text{on } \partial\Omega,$$

where $\Omega \subset \mathbb{R}^2$ is for simplicity a convex polygonal domain.

For *linear elements* we get the $\mathcal{O}(h^4)$-superconvergence at nodal points provided all triangular elements are equilateral (see Blum, Lin and Rannacher [71], and Makarenko [399]). Let us point out that the optimal global rate of convergence cannot be better than $\mathcal{O}(h^2)$ in the maximum norm. The high accuracy at nodes is based on an asymptotic error expansion theorem by Lin and Wang [364], which characterizes all the behaviour of the finite element solution (cf. also [361], [366], [369], [380], [587]). To extend the above $\mathcal{O}(h^4)$-superconvergence result to three-dimensional domains seems to be impossible, since the regular tetrahedron is not a space-filler.

According to Křížek et al [321], [325], a special averaged finite element scheme produces the $\mathcal{O}(h^4)$-superconvergence of piecewise bilinear finite element approximations at nodes.

Zhu [577], [578], [581] and later also Makarenko [399] proved the $\mathcal{O}(h^4)$-superconvergence at vertices for *quadratic elements* over *uniform triangulations*, i.e., when any two adjacent triangles form a parallelogram. Remember that the optimal global rate of convergence is $\mathcal{O}(h^3)$ in the maximum norm. We have also the $\mathcal{O}(h^4)$-superconvergence at midpoints of each side (compare [326]).

Superconvergence of rectangular *biquadratic elements* at vertices has been shown by Chen [147]. For the tensor product of one-dimensional spaces of piecewise polynomial functions of degree k, Douglas, Dupont and Wheeler [191] proved the $\mathcal{O}(h^{k+2})$-superconvergence at nodal points.

Schatz, Sloan and Wahlbin [470] developed an interesting superconvergence theory for triangulations which are locally symmetric with respect to one point. For other superconvergence results of function values, see also [75], [103], [347].

2.3   Superconvergence of derivatives

Now we introduce some results concerning superconvergent approximations of the derivatives for the Dirichlet problem from the foregoing Section 2.2. In 1969, Oganesjan and Ruhovec [440] proved that the difference between the Galerkin approximation over uniform triangulations and the interpolation of the weak solution has order $\mathcal{O}(h^2)$ in the H^1-norm, whereas $\mathcal{O}(h)$ is the optimal interpolation order. However, they did not know that this result is very close to get the superconvergence (cf., e.g., [322], [585]). Later it has been shown by Chen [122], [128], [142] that the tangential component of the gradient of *linear elements* has very good approximation properties at midpoints of sides on uniform triangulations. The average of gradients of any two adjacent triangles then leads to the $O(h^2)$-superconvergence at midpoints. Various generalizations of this property are presented, e.g., in Andreev [16], Goodsell and Whiteman [252], Levine [343], [344], Lin et al [361], [362], [369], Neittaanmäki and Křížek [425]. For a three-dimensional case, see Chen [123], [126].

A superconvergence phenomenon arising from averaging of gradients of linear elements at nodes has been proved by Křížek, Neittaanmäki [322], [323] and later generalized to many directions by Goodsell and Whiteman [251], [252], Hlaváček et al [280], [281]; for a three-dimensional case, see Goodsell [250], and Kantchev and Lazarov [309].

An integral smoothing operator for continuous piecewise linear finite element functions has been suggested by Oganesjan and Ruhovec in [441]. It yields the $\mathcal{O}(h^{3/2})$-superconvergence of the gradient in the L^2-norm and this order can be probably still improved. Higher order accuracy by integral averaging was obtained also in the key paper by Bramble and Schatz [76].

A superconvergent recovery of the gradient of linear elements at centroids is proposed by El Hatri [214] and Levine [343]. For rectangular *bilinear elements*, the sampling at centroids leads directly to the $\mathcal{O}(h^2)$-superconvergence of the first derivatives, see Lesaint and Zlámal [339], (cf. also [14], [397]). Superconvergence of the second mixed derivatives at centroids of bilinear elements is proved in Chen [136].

Zhu in [577] and independently Andreev in [17] proved that the tangential component of the gradient at the two Gaussian points of each side of the triangular *quadratic elements* is superconvergent. Superconvergence of the gradient of quadratic elements is also investigated in Andreev and Lazarov [20], Goodsell and Whiteman [254], [255]. For the three-dimensional case, see Pehlivanov [447].

Superconvergence of the gradient at the Gaussian points of rectangular Serendipity elements with incomplete polynomials of the third degree has been obtained in Zlámal [595], see also Chen [124], [138], Huang [292], and Leyk [345]. For a generalization to the three-dimensional case, see Zlámal [597].

For a superconvergence of derivatives coming from the use of rectangular *biquadratic elements* we refer to Jin and Huang [303], and Wu and Huang [551].

A general theory for superconvergence of function values and derivatives over triangulations which are *locally symmetric* with respect to one point is introduced in Schatz, Sloan and Wahlbin [470], and Wahlbin [523]–[525]. This theory includes, for example, all uniform or piecewise uniform triangulations. For *quasiuniform triangulations* (i.e, when any two adjacent triangles form only an approximate parallelogram), see, e.g., Hlaváček et al [280], [281], Levine [343], for *nonuniform quadrilateral meshes*, see Lin et al [352], [365], for *locally periodic meshes*, see Babuška, Strouboulis et al [46]. In Babuška, Strouboulis et al [43], [45], a new definition of superconvergence – the $\eta\%$-superconvergence – is introduced, which generalizes the classical idea of superconvergence to *general meshes*.

Superconvergent approximations of derivatives can also be found in [14], [44],

[**47**], [**57**], [**58**], [**104**], [**145**], [**154**], [**228**], [**249**]–[**255**], [**292**], [**298**], [**321**]–[**328**], [**331**], [**351**]–[**383**], [**390**], [**396**], [**421**], [**422**], [**427**], [**430**], [**433**], [**477**], [**501**]–[**503**], [**536**]–[**548**], [**569**], [**570**], [**579**], [**580**], [**586**], [**593**]–[**599**].

Douglas, Dupont and Wheeler [**192**] propose and analyse a method to compute a superconvergence approximation to the *boundary flux*. Other efficient superconvergence methods for boundary flux calculations can be found in [**106**], [**107**], [**163**], [**323**], [**338**], [**450**].

2.4 Various extensions

The above results for the Poisson equation over a convex polygonal domain can be extended to many directions. For instance, to a scalar elliptic equation with variable coefficients, see, e.g., [**16**], [**292**], [**343**], [**344**], [**349**], [**437**], [**468**], [**469**], nonconvex domains with reentrant corners [**72**], [**135**], [**151**], [**293**], irregular meshes [**13**], [**281**], [**317**], [**320**], [**341**], [**352**], [**365**], [**390**], boundary stresses [**276**], [**434**], eigenvalue problems [**19**], [**199**], [**357**], [**456**], [**490**], [**558**], [**559**], axisymmetric problems [**129**], [**213**], singular and degenerated problems [**214**]–[**216**], [**222**], [**566**], singularly perturbed problems [**28**], [**31**], [**32**], [**443**], [**463**], [**498**]. We can also include *numerical integration* and still observe superconvergence, see, e.g., [**58**], [**166**], [**174**], [**215**], [**280**], [**325**], [**397**], [**597**].

For superconvergence phenomena which arise when solving a system of second order elliptic equations, which describes, e.g., the *linear elasticity problem*, see Babuška et al [**42**], [**44**], [**47**], Chen [**126**], Goodsell [**250**], Goodsell and Whiteman [**253**], Hlaváček and Křížek [**275**]–[**279**], Lazarov [**337**], Milner [**407**], Morley [**410**], and Whiteman and Goodsell [**540**], [**541**]. Superconvergent approximations of *fourth order elliptic problems* are examined, e.g., in [**152**], [**171**], [**174**], [**290**], [**326**].

Some superconvergence results hold also for many nonstandard finite elements, for example, isoparametric elements [**339**], [**565**], [**597**], hybrid elements [**261**], the Raviart–Thomas elements [**224**], the Wilson element [**150**], [**160**], [**371**], [**453**], and other nonconforming elements [**22**], [**236**], [**313**], [**385**], [**393**], [**394**], [**452**].

Superconvergence effects analogous to that for elliptic problems can be expected also for *parabolic problems*, see [**1**]–[**3**], [**16**], [**18**], [**23**], [**37**], [**54**], [**81**], [**143**], [**149**], [**160**], [**190**], [**192**], [**212**], [**223**], [**291**], [**295**], [**301**], [**302**], [**348**], [**409**], [**413**], [**419**], [**424**], [**431**], [**444**], [**445**], [**493**], [**494**], [**504**]–[**507**], [**509**], [**538**]. For *hyperbolic problems* we refer to [**156**], [**157**], [**168**], [**182**], [**183**], [**193**], [**231**], [**239**], [**256**], [**291**], [**336**], [**363**], [**381**], [**419**], [**526**], [**527**], [**573**].

Moreover, the superconvergence has been observed in solving various special equations, e.g., the Stokes equations [**335**], [**528**], [**529**], [**571**], the Navier–Stokes equations [**153**], [**159**], the Boussinesq equation [**549**], the convection-diffusion equation [**167**], [**384**], [**466**], [**496**], the Maxwell equations [**81**], [**371**], [**408**], the delay differential equation [**63**], the Euler–Poisson–Darboux equation [**239**], the Sobolev equation [**24**], [**375**], [**418**], [**464**], the viscoelasticity type equation [**375**], and equations describing porous media problem [**156**], [**157**], [**185**], [**186**], [**226**], [**530**]. We can also get superconvergence in solving the Stefan problem [**304**], [**444**]. Further results with superconvergence of the finite element method for nonlinear differential equations were obtained by Arnold and Douglas [**23**], Chen [**125**], [**127**], [**130**],

[133], [138], [140], Chou and Li [160], Chow et al [161], [162], Kantchev [307], [308], Li [348], Nakao [418], Sanz-Serna and Abia [467], Wahlbin [524], Wang [526], Wheeler and Whiteman [538], and Zhang [568].

Besides the standard Galerkin finite element method, also other finite element methods are investigated in connection with superconvergence. See, e.g., [107], [109], [417] for the *Galerkin-collocation methods*, [200] for the *discontinuous finite element method*, [195] for the *primal hybrid method*, and [22], [78]–[85], [146], [148], [152], [193], [194], [196]–[198], [202], [223]–[227], [236]–[238], [400], [410], [423], [424], [449], [495], [531], [557], [571] for *mixed methods*, which are applied especially to fourth order problems.

Superconvergence is mostly established in L^2, L^∞ or H^1 norms. However, it can also be measured in various *special norms*, in a discrete L^2 norm [597], negative norms [304], [414], [506], difference norms [315], mesh norms [172], [174], [314], etc. A great amount of papers deals with (local) interior estimates of superconvergence type, see [77], [132], [194]–[196], [278]–[280], [414], [432], [448], [471], [508], [521], [536], [537], [582], [589]. A global superconvergence has been achieved in [323], [348], [356], [359], [363], [374], [375].

Babuška et al [42]–[47] propose a computer-based approach to find all superconvergence points for any kind of two-dimensional finite elements. A survey of a computational investigation of superconvergence is contained in Wahlbin [524].

Superconvergence effects also may arise by higher order interpolation on several adjacent lower order elements, see Lin [353], and Lin, Yan and Zhou [372], [373]. Acceleration of convergence by one additional iteration is proposed by Dzhishkariani et al [206], [207]. For other postprocessing techniques yielding superconvergence, we quote, e.g., [22], [39], [40], [78], [134], [162], [169], [207], [326], [381], [390], [426], [495], [545], [550], [584], [594]. They may give practically "good" results even though the true solution does not satisfy high regularity assumptions guaranteeing superconvergence theoretically (cf. [323]).

Superconvergence is thus of considerable interest not only from a theoretical but also from a practical viewpoint. Indeed, there are several important applications of superconvergence results, e.g., we may increase the accuracy in computation of the magnetic flux, see Křížek, Neittaanmäki [326], or in computation of mechanical stresses in a dam, see Chen and Huang [139]. We can also use superconvergence in creating adaptive mesh refinements, see Adjerid and Flaherty [2], [3], Babuška et al [38], [49], and Zienkiewicz and Zhu [592]. Various superconvergence techniques are also developed to get reliable *a posteriori error estimates* in Ainsworth and Craig [5], Babuška et al [48], Zienkiewicz and Zhu [593], see also [6], [41], [80], [294], [296], [461], [462], [548], [570], [576]. Several asymptotically exact estimators are presented in Durán et al [203]–[205].

3   A BRIEF CLASSIFICATION OF SUPERCONVERGENCE PHENOMENA FOR OTHER METHODS AND PROBLEMS

Some references in the bibliography have nothing to do with finite elements and the Galerkin method, but they deal with superconvergence in other contexts. For instance, superconvergence phenomena for the *finite difference method* have been observed in solving quasilinear partial differential equations by Dautov et al [170],

[171], [173], [175], and variational inequalities by Lapin [332], [333]. There are other papers dealing with superconvergence of the finite difference method [158], [334], [397], [428], [429], [479], [508], [560], [600], the *boundary element method* [472], [486], [528], [532], the *least squares method* [27], [272], [386]–[388], [449], the *collocation method* [63], [67], [176], [181], [184], [260], [283], [287], [346], [446], the *box method* [599], the *α-method* [64], the *analytical method* [271], the *split iteration methods* [553], the *stream function method* [572], the *stream line diffusion method* [574], various *projection methods* [105], [177], [206], [208], [259], [411], [455], [487], [567], and the *infinite element method* [555].

Results concerning superconvergence for *initial value problems for ordinary differential equations* are presented in [235], [350], [515], [563], for *pseudodifferential equations* see [511], for *differential-algebraic equations* see [29], [30], [51], [52], [392] and for *integro-differential equations* see [91], [96], [137], [144], [285], [374], [395], [445], [517], [519], [520], [554].

Superconvergence phenomena for the *integral equations of the first kind* are obtained in Locker and Prenter [387] for the *least squares method*, in Sloan and Spence [488] for the *Galerkin method*, and in Shu [481] for the *Galerkin collocation method*.

There are many superconvergence techniques for the *Fredholm integral equations of the second kind* that usually arise form elliptic and other similar problems. Sloan and Thomée [489] propose a simple postprocessing of the *Galerkin approximation* based on polynomials of degree at most k. They shown the $\mathcal{O}(h^{2k+2})$-superconvergence in W_∞^1-norm, whereas $\mathcal{O}(h^{k+1})$ is the optimal order in L^∞-norm. Other related works corresponding to the Galerkin scheme include, e.g., Atkinson and Bogomolny [33], Bourlard [74], Chandler [110]–[115], Chatelin and Lebbar [119], [120], Elschner [219], Graham [257], Hebeker et al [269], Hsiao and Wendland [284], Omari [442], Richter [460], Sloan [482], [485], and Spence and Thomas [492]. For *collocation methods*, where superconvergence at collocation points is obtained, we refer to Chatelin [118]. Superapproximation results for collocation methods were achieved by Arnold and Wendland [25]. For superconvergence in solving the integral equations of the second kind, see also [10], [116]–[120], [391], [412], [460], [480], [482], [483], [533].

Next, let us mention the works of Brunner [87], [89], [90], [92], [93], [97] for superconvergence of *collocation methods* for the *Volterra integral equations* of the first and second kind, see also [50], [70], [97]–[100], [209]–[211], [310], [499], [552]. Other superconvergence results of collocation methods for integral equations can be found in [10], [11], [25], [34], [217], [218], [258], [484], [514]. For the finite difference method, see [9].

Superconvergence phenomena has also been found in solving various special equations, e.g., the *Abel integral equations* [88], the *airfoil equation* [243], [245], [246], the *Hammerstein equations* [94], [268], [305], [306], [329], [330], the *Korteweg-de-Vries equation* [26], *Symm's integral equation* [342], [556], and several other nonlinear integral equations [35], [95], [491].

Superconvergence properties can be used in optimal control, see Reddien [454], or in sensitivity analysis of optimal shape design problems, see Hlaváček and Chleboun [273], [274].

In Introduction, we mentioned six monographs solely devoted to superconvergence of the finite element method. But there are still other books [36], [118], [164], [165], [297], [316], [326], [401], [436], [441], [463], [497], [504], [562], [590],

[**591**], where the notion superconvergence is discussed. For conference proceedings, see [**4**], [**7**], [**11**], [**38**], [**53**], [**86**], [**228**], [**234**], [**241**], [**244**], [**248**], [**270**], [**282**], [**327**], [**328**], [**405**], [**406**], [**475**], [**510**], [**512**], [**561**].

Superconvergence in connection with wavelet functions can be found in [**247**], [**299**], [**300**]. Higher order accuracy of spline approximations has been observed in [**60**]–[**62**], [**68**], [**69**], [**73**], [**232**], [**233**], [**262**]–[**267**], [**289**], [**311**], [**312**], [**389**], [**435**], [**513**]. However, these phenomena should rather be called superapproximation than superconvergence.

Finally, note that many results on superconvergence are overlapping. It is difficult to give exhaustive annotations, since nowadays the term *superconvergence* is too wide. For example, Finn [**229**] presents superconvergent Kolmogorov's algorithm in the theory of Lie transforms, Sablonnière [**465**] investigates a convergence acceleration of real sequences, Volle [**518**] studies a superconvergence of level sets, and Leseberg [**340**] examines superconvergence spaces. We meet the notion superconvergence also in solving a bifurcation problem [**66**], in various problems of mathematical physics [**101**], [**221**], [**240**], [**398**], [**438**], [**439**], in the theory of series [**404**], [**478**], in the theory of random free variables [**65**], in the theory of extrapolation [**141**], in the operator theory [**179**], [**220**], [**402**], [**473**], etc.

REFERENCES

1. Adeboye, K.R., *Superconvergence results for Galerkin method for parabolic initial value problems via Laplace transform*, Bull. Math. Soc. Sci. Math. R. S. Roumanie (N.S.) **26** (1982), 115–127. MR 84h:65095.

2. Adjerid, S., Flaherty, J.E., *A moving-mesh finite element method with local refinement for parabolic partial differential equations*, Comput. Methods Appl. Mech. Engrg. **55** (1986), 3–26. MR 87h:65187.

3. Adjerid, S., Flaherty, J.E., *A local refinement finite-element method for two-dimensional parabolic systems*, SIAM J. Sci. Statist. Comput. **9** (1988), 792–811. MR 89g:65136.

4. Agarwal, R.P., Chow, Y.M., Wilson, S.J. (eds), "Numerical mathematics," Proc. of the Internat. Conf., National Univ. of Singapore, Singapore 1988, Internat. Series of Numer. Math. 86, Birkhäuser, Basel, 1988, pp. 1–526. MR 90i:65007.

5. Ainsworth, M., Craig, A., *A posteriori error estimators in the finite element method*, Numer. Math. **60** (1992), 429–463. MR 92j:65163.

6. Ainsworth, M., Oden, J.T, *A posteriori error estimation in finite element analysis*, Preprint no. 14, Univ. of Leicester, 1995, 1–138.

7. Albrecht, J., Collatz, L. (eds), "Numerical treatment of integral equations," Lectures delivered at the Workshop held at the Math. Research Inst., Oberwolfach, 1979, Internat. Series of Numer. Math., 53, Birkhäuser, Basel, 1980, pp. 1–275. MR 81j:65009.

8. Allen, G.D., *Smoothness and superconvergence for approximate solutions to the one-dimensional monoenergetic transport equation*, in "Advances in optimization and numerical analysis," (Oaxaca, 1992), Math. Appl., 275, Kluwer Acad. Publ., Dordrecht, 1994, pp. 219–232. MR 95a:00010.

9. Allen, G.D., Nelson, P., *On generalized finite difference methods for approximating solutions to integral equations*, in "Advances in numerical partial differential equations and optimization," (Mérida, 1989), SIAM, Philadelphia, PA, 1991, pp. 112–140. MR 92i:65197.

10. Amini, S., Sloan, I.H., *Collocation methods for second kind integral equations with noncompact operators*, J. Integral Equations Appl. **2** (1989), 1–30. MR 91c:45015.

11. Anderssen, R.S., de Hoog, F.R., Lukas, M.A. (eds), "The application and numerical solution of integral equations," Proc. of a Seminar held at the Australian National Univ., Canberra, 1978, Monographs and Textbooks on Mechanics of Solids and Fluids: Mechanics and Analysis 6, Martinus Nijhoff Publishers, The Hague, 1980, pp. 1–259. MR 81f:45001.

12. Andreasyan, G.D., *Superconvergence of the finite element method for high-order one-dimensional boundary value problems*, Akad. Nauk Armyan. SSR Dokl. **84** (1987), 3–8. MR 88e:65131. (in Russian)

13. Andreasyan, G.D., *Superconvergence of averaged gradients on an irregular mesh in the finite element method*, Vestnik Moskov. Univ. Ser. XV Vychisl. Mat. Kibernet. (1987), 11–16, 72. MR 88j:65246. (in Russian)

14. Andreasyan, G.D., *Superconvergence of averaged gradients in the finite element method on rectangles*, Applied mathematics, No. 5, 130, Erevan. Univ., Erevan (1987), 5–10. MR 90e:65149. (in Russian)

15. Andreasyan, G.D., *Superconvergence of the finite element method for higher order one-dimensional boundary value problems*, Akad. Nauk Armyan. SSR Dokl. **84** (1987), 3–8. MR 88e:65131. (in Russian)

16. Andreev, A.B., *Superconvergence of the gradient for linear triangle elements for elliptic and parabolic equations*, C. R. Acad. Bulgare Sci. **37** (1984), 293–296. MR 85h:65230.

17. Andreev, A.B., *Error estimate of type superconvergence of the gradient for quadratic triangular elements*, C. R. Acad. Bulgare Sci. **37** (1984), 1179–1182. MR 86f:65186.

18. Andreev, A.B., *Superconvergence of the gradient of approximate solutions of parabolic equations*, Serdica **11** (1985), 359–368. MR 87j:65142. (in French)

19. Andreev, A.B., *Superconvergence of the gradient of finite element eigenfunctions*, C. R. Acad. Bulgare Sci. **43** (1990), 9–11. MR 92d:65185.

20. Andreev, A.B., Lazarov, R.D., *Superconvergence of the gradient for quadratic triangular finite elements*, Numer. Methods Partial Differential Equations **4** (1988), 15–32. MR 90m:65190.

21. Andreev, V.B., Andreasyan, G.D., *Superconvergence of derivatives and their averagings in the finite element method for ordinary 2mth-order differential equations*, in "Computational processes and systems," No. 6, "Nauka", Moscow, 1988, pp. 31–39. MR 90f:65147. (in Russian)

22. Arnold, D.N., Brezzi, F., *Mixed and nonconforming finite element methods: implementation, postprocessing and error estimates*, RAIRO Modél. Math. Anal. Numér **19** (1985), 7–32. MR 87g:65126.

23. Arnold, D.N., Douglas, J., *Superconvergence of the Galerkin approximation of a quasilinear parabolic equation in a single space variable*, Calcolo **16** (1979), 345–369. MR 83b:65118.

24. Arnold, D.N., Douglas, J., Thomée, V., *Superconvergence of a finite element approximation to the solution of a Sobolev equation in a single space variable*, Math. Comp. **36** (1981), 53–63. MR 82f:65108.

25. Arnold, D.N., Wendland, W.L., *On the asymptotic convergence of collocation methods*, Math. Comp. **41** (1983), 349–381. MR 85h:65254.

26. Arnold, D.N., Winther, R., *A superconvergent finite element method for the Korteweg-de Vries equation*, Math. Comp. **38** (1982), 23–36. MR 82m:65087.

27. Ascher, U., *Discrete least squares approximations for ordinary differential equations*, SIAM J. Numer. Anal. **15** (1978), 478–496. MR 81e:65043.

28. Ascher, U., *Two families of symmetric difference schemes for singular perturbation problems*, in "Numerical boundary value ODEs," (Vancouver, B.C., 1984), Progr. Sci. Comput., 5, Boston, Mass., 1985, pp. 173–191. MR 87h:65125.

29. Ascher, U.M., Petzold, L.R., *Projected implicit Runge–Kutta methods for differential-algebraic equations*, SIAM J. Numer. Anal. **28** (1991), 1097–1120. MR 92f:65082.

30. Ascher, U.M., Petzold, L.R., *Projected collocation for higher-order higher-index differential-algebraic equations. Orthogonal polynomials and numerical methods*, J. Comput. Appl. Math. **43** (1992), 243–259. MR 93i:65080.

31. Ascher, U.M., Weiss, R., *Collocation for singular perturbation problems. I. First order systems with constant coefficients*, SIAM J. Numer. Anal. **20** (1983), 537–557. MR 85a:65113.

32. Ascher, U.M., Weiss, R., *Collocation for singular perturbation problems. III. Nonlinear problems without turning points*, SIAM J. Sci. Statist. Comput. **5** (1984), 811–829. MR 86g:65138b.

33. Atkinson, K.E., Bogomolny, A., *The discrete Galerkin method for integral equations*, Math. Comp. **48** (1987), 595–616, S11–S15. MR 88k:65125.

34. Atkinson, K.E., Chien, D., *Piecewise polynomial collocation for boundary integral equations*,

SIAM J. Sci. Comput. **16** (1995), 651–681. MR 95m:65200.

35. Atkinson, K.E., Potra, F.A., *Projection and iterated projection methods for nonlinear integral equations.*, SIAM J. Numer. Anal. **24** (1987), 1352–1373. MR 88j:65289.

36. Axelsson, O., Barker, V.A., "Finite element solution of boundary value problems. Theory and computation," Academic Press, New York, 1984. MR 85m:65116.

37. Aziz, A.K., Monk, P., *Continuous finite elements in space and time for the heat equation,* Math. Comp. **52** (1989), 255–274. MR 90d:65189.

38. Babuška, I., Flaherty, J.F., Henshaw, W.D, Hopcroft, J.E., Oliger, J.E., Tezduyar, T. (eds), "Modeling, mesh generation, and adaptive numerical methods for partial differential equations," Proc. of the IMA Summer Program held at the Univ. of Minnesota, Minneapolis, 1993, IMA Volumes in Mathematics and its Applications 75, Springer, New York, 1995. MR 96g:65002.

39. Babuška, I., Izadpanah, K., Szabó, B., *The postprocessing technique in the finite element method. The theory & experience,* in "Unification of finite element methods," Math. Stud., 94, North-Holland, Amsterdam, 1984, pp. 97–121. MR 87g:65127.

40. Babuška, I., Miller, A., *The post-processing in the finite element, Parts I–II,* Internat. J. Numer. Methods Engrg. **20** (1984), 1085–1109, 1111–1129.

41. Babuška, I., Planck, L., Rodríguez, R., *Basic problems of a posteriori error estimation,* Comput. Methods Appl. Mech. Engrg. **101** (1992), 97–112. MR 93h:73036.

42. Babuška, I., Strouboulis, T., Gangaraj, S.K., Upadhyay, C.S., *Validation of recipes for the recovery of stresses and derivatives by a computer-based approach,* Math. Comput. Modelling **20** (1994), 45–89. MR 95g:73044.

43. Babuška, I., Strouboulis, T., Gangaraj, S.K., Upadhyay, C.S., *$\eta\%$-superconvergence in the interior of locally refined meshes of quadrilaterals: superconvergence of the gradient in finite element solutions of Laplace's and Poisson's equations,* Appl. Numer. Math. **16** (1994), 3–49. MR 95m:65186.

44. Babuška, I., Strouboulis, T., Upadhyay, C.S., *A model study of the quality of a posteriori error estimators for linear elliptic problems. Error estimation in the interior of patchwise uniform grids of triangles,* Comput. Methods Appl. Mech. Engrg. **114** (1994), 307–378. MR 95d:65093.

45. Babuška, I., Strouboulis, T., Upadhyay, C.S., *$\eta\%$-superconvergence of finite element approximations in the interior of general meshes of triangles,* Comput. Methods Appl. Mech. Engrg. **122** (1995), 273–305. MR 96e:65066.

46. Babuška, I., Strouboulis, T., Upadhyay, C.S., Gangaraj, S.K., *Study of superconvergence by a computer-based approach. Superconvergence of the gradient in finite element solutions of Laplace's and Poisson's equations,* CMC Report No. 93-07, Texas A&M Univ., 1993, 1–59.

47. Babuška, I., Strouboulis, T., Upadhyay, C.S., Gangaraj, S.K., *Computer-based proof of the existence of superconvergence points in the finite element method; superconvergence of the derivatives in finite element solutions of Laplace's, Poisson's, and the elasticity equations,* Numer. Methods Partial Differential Equations **12** (1996), 347–392. MR 97c:65160.

48. Babuška, I., Strouboulis, T., Upadhyay, C.S., Gangaraj, S.K., Copps, K., *Validation of a-posteriori error estimators by numerical approach,* Internat. J. Numer. Methods Engrg. **37** (1994), 1073–1123. MR 95e:65096.

49. Babuška, I., Szabó, B., *Trends and new problems in finite element methods,* TICAM Report 96-37, Univ. of Texas at Austin, 1996, 1–33.

50. Baddour, N., Brunner, H., *Continuous Volterra–Runge–Kutta methods for integral equations with pure delay,* Computing **50** (1993), 213–227. MR 94m:65199.

51. Bai, Y., *A perturbed collocation method for boundary-value problems in differential-algebraic equations,* Appl. Math. Comput. **45** (1991), 269–291. MR 92i:65123.

52. Bai, Y., *A modified Lobatto collocation for linear boundary value problems of differential-algebraic equations,* Computing **49** (1992), 139–150. MR 93m:65109.

53. Baker, C.T.H., Miller, G.F. (eds), "Treatment of integral equations by numerical methods," Proc. of the symposium, Durham Univ., 1982, Academic Press [Harcourt Brace Jovanovich, Publishers], London, 1982, pp. 1–493. MR 85e:65005.

54. Bakker, M., *On the numerical solution of parabolic equations in a single space variable by the continuous time Galerkin method,* SIAM J. Numer. Anal. **17** (1980), 162–177. MR 81m:65167.

55. Bakker, M., *A note on C^0 Galerkin methods for two-point boundary problems*, Numer. Math. **38** (1981/82), 447–453. MR 83e:65140.

56. Bakker, M., *One-dimensional Galerkin methods and superconvergence at interior nodal points*, SIAM J. Numer. Anal. **21** (1984), 101–110. MR 85f:65080.

57. Barlow, J., *Optimal stress location in finite-element models*, Internat. J. Numer. Methods Engrg. **10** (1976), 243–251.

58. Barlow, J., *More on optimal stress points – reduced integration, element distortions and error estimation*, Internat. J. Numer. Methods Engrg. **28** (1989), 1487–1504. MR 90h:73022a,b.

59. Barrett, J.W., Moore, G., Morton, K.W., *Optimal recovery in the finite-element method. I. Recovery from weighted L^2 fits*, IMA J. Numer. Anal. **8** (1988), 149–184. MR 90b:65167.

60. Behforooz, G.H., Papamichael, N., *Improved orders of approximation derived from interpolatory cubic splines*, BIT **19** (1979), 19–26. MR 80e:41004.

61. Behforooz, G.H., *A comparison of the $E(3)$ and not-a-knot cubic splines*, Appl. Math. Comput. **72** (1995), 219–223. MR 96d:65019.

62. Behforooz, G.H, Papamichael, N., *Overconvergence properties of quintic interpolatory splines*, J. Comput. Appl. Math. **24** (1988), 337–347. MR 90a:65019.

63. Bellen, A., *One-step collocation for delay differential equations*, J. Comput. Appl. Math. **10** (1984), 275–283. MR 85m:65061.

64. Bensebah, A., Dubeau, F., Gélinas, J., *Existence and superconvergence results for the α-method*, Ann. Sci. Math. Québec **17** (1993), 115–138. MR 94m:65113. (in French)

65. Bercovici, H., Voiculescu, D., *Superconvergence to the central limit and failure of the Cramér theorem for free random variables*, Probab. Theory Related Fields **103** (1995), 215–222. MR 96k:46115.

66. Beyn, W.J., *Global bifurcations and their numerical computation*, in "Continuation and bifurcations: numerical techniques and applications," (Leuven, 1989), NATO Adv. Sci. Inst. Ser. C: Math. Phys. Sci., 313, Kluwer Acad. Publ., Dordrecht, 1990, pp. 169–181. MR 91k:58092.

67. Bialecki, B., Cai, X.C., *H^1-norm error bounds for piecewise Hermite bicubic orthogonal spline collocation schemes for elliptic boundary value problems*, SIAM J. Numer. Anal. **31** (1994), 1128–1146. MR 96e:65071.

68. Binev, P.G., *Superconvergence in spline-interpolation*, C. R. Acad. Bulgare Sci. **37** (1984), 1613–1616. MR 87c:41004.

69. Binev, P.G., *Convergence and superconvergence in Hermite spline interpolation*, in "Numerical Methods and Applications," Proc. Internat. Conf., Sofia, 1984, Izd. Bulg. Akad. Nauk, Sofia, 1985, pp. 179–184.

70. Blom, J.G., Brunner, H., *The numerical solution of nonlinear Volterra integral equations of the second kind by collocation and iterated collocation methods*, SIAM J. Sci. Statist. Comput. **8** (1987), 806–830. MR 89f:65138.

71. Blum, H, Lin, Q., Rannacher, R., *Asymptotic error expansion and Richardson extrapolation for linear finite elements*, Numer. Math. **49** (1986), 11–38. MR 87m:65172.

72. Blum, H, Rannacher, R., *Extrapolation techniques for reducing the pollution effect of reentrant corners in the finite element method*, Numer. Math. **52** (1988), 539–564. MR 89e:65117.

73. Bokhari, M. A., *A note on certain next-to-interpolatory rational functions*, Acta Math. Hungar. **62** (1993), 49–55. MR 95a:41023.

74. Bourlard, M., *Superconvergence for adapted Galerkin methods solving Neumann and Dirichlet problems on polygonal domains*, C. R. Acad. Sci. Paris Sér. I Math. **306** (1988), 211–216. MR 88m:65171. (in French)

75. Bramble, J.H. Schatz, A.H., *Estimates for spline projection*, RAIRO Anal. Numér. **10** (1976), 5–37. MR 55#9563.

76. Bramble, J.H., Schatz, A.H., *Higher order local accuracy by averaging in the finite element method*, Math. Comp. **31** (1977), 94–111. MR 55#4739.

77. Bramble, J.H., Thomée, V., *Interior maximum norm estimates for some simple finite element methods*, RAIRO Anal. Numér. **8** (1974), 5–18. MR 50#11808.

78. Bramble, J.H., Xu, J., *A local post-processing technique for improving the accuracy in mixed finite element approximations*, SIAM J. Numer. Anal. **26** (1989), 1267–1275. MR 90m:65193.

79. Brandts, J.H., *A note on uniform superconvergence for the Timoshenko beam using mixed finite elements*, Math. Models Methods Appl. Sci. **4** (1994), 795–806. MR 95h:73064.

80. Brandts, J.H., *Superconvergence and a posteriori error estimation for triangular mixed finite elements*, Numer. Math. **68** (1994), 311–324. MR 96a:65162.

81. Brandts, J.H., *Superconvergence phenomena in finite element methods*, PhD thesis, Utrecht Univ., 1994, 1–115.

82. Brandts, J.H., *Superconvergence for second order triangular mixed and standard finite elements*, Report 9/1996, Dept. of Math., Univ. of Jyväskylä, 1996, 1–20.

83. Brandts, J.H., *Superconvergence similarities in standard and mixed finite element methods*, in [328], 1997.

84. Brezzi, F., Douglas, J., Fortin, M., Marini, L.D., *Efficient rectangular mixed finite elements in two and three space variables*, RAIRO Modél. Math. Anal. Numér. **21** (1987), 581–604. MR 88j:65249.

85. Brezzi, F., Douglas, J., Marini, L.D., *Two families of mixed finite elements for second order elliptic problems*, Inst. of Numer. Anal. of the National Research Council, 435, Pavia, 1984, 1–32. MR 86c:65128.

86. Broeckx, F., Goovaerts, M.J., Piessens, R., Wuytack, L. (eds), "Proceedings of the 2nd internatational conference on computational and applied mathematics," Univ. of Leuven, 1986, J. Comput. Appl. Math. **20** (1987), Special Issue, North-Holland, Amsterdam, 1987, pp. 1–424. MR 88h:65012.

87. Brunner, H., *Superconvergence of collocation methods for Volterra integral equations of the first kind*, Computing **21** (1978), 151–157. MR 83a:65125.

88. Brunner, H., *On superconvergence in collocation methods for Abel integral equations*, in "Proc. of the Eighth Manitoba Conf. on Numer. Math. and Computing," (Univ. Manitoba, Winnipeg, Man., 1978), Congress. Numer., XXII, Utilitas Math., Winnipeg, Man., 1979, pp. 117–128. MR 80m:45023.

89. Brunner, H., *A note on collocation methods for Volterra integral equations of the first kind*, Computing **23** (1979), 179–187. MR 83b:65145.

90. Brunner, H., *Superconvergence in collocation and implicit Runge–Kutta methods for Volterra-type integral equations of the second kind*, in "Numerical treatment of integral equations," (Workshop, Math. Res. Inst., Oberwolfach, 1979), Internat. Ser. Numer. Math., 53, Birkhäuser, Basel, 1980, pp. 54–72. MR 81m:65194.

91. Brunner, H., *The application of the variation of constants formulas in the numerical analysis of integral and integro-differential equations*, Utilitas Math. **19** (1981), 255–290. MR 83b:65146.

92. Brunner, H., *Iterated collocation methods and their discretizations for Volterra integral equations*, SIAM J. Numer. Anal. **21** (1984), 1132–1145. MR 86d:65160.

93. Brunner, H., *Collocation methods for one-dimensional Fredholm and Volterra integral equations*, in "The state of the art in numerical analysis," (Birmingham, 1986), Inst. Math. Appl. Conf. Ser. New Ser., 9, Oxford University Press, New York, 1987, pp. 563–600. MR 89m:65112.

94. Brunner, H., *On implicitly linear and iterated collocation methods for Hammerstein integral equations*, J. Integral Equations Appl. **3** (1991), 475–488. MR 93b:65206.

95. Brunner, H., *On discrete superconvergence properties of spline collocation methods for nonlinear Volterra integral equations*, J. Comput. Math. **10** (1992), 348–357. MR 93k:65107.

96. Brunner, H., *The numerical solution of neutral Volterra integro-differential equations with delay arguments*, Scientific computation and differential equations (Auckland, 1993), Ann. Numer. Math. **1** (1994), 309–322. MR 96m:45014.

97. Brunner, H., *Iterated collocation methods for Volterra integral equations with delay arguments*, Math. Comp. **62** (1994), 581–599. MR 94g:65143.

98. Brunner, H., Kauthen, J.P., *The numerical solution of two-dimensional Volterra integral equations by collocation and iterated collocation*, IMA J. Numer. Anal. **9** (1989), 47–59. MR 90g:65176.

99. Brunner, H., Nørsett, S.P., *Superconvergence of collocation methods for Volterra and Abel integral equations of the second kind*, Numer. Math. **36** (1980/81), 347–358. MR 83e:65202.

100. Brunner, H., Yan, N., *On global superconvergence of iterated collocation solutions to linear second-kind Volterra integral equations*, J. Comput. Appl. Math. **67** (1996), 185–189. MR 96m:65120.

101. Calucci, G., Jengo, R., Furlan, G., Rebbi, C., *Smoothness near the light-cone and supercon-*

vergence, Phys. Lett. B **37** (1971), 416–419. MR 47#2936.

102. Cao, L.Q., *A one-dimensional hierarchical fast finite element method with high accuracy*, Natur. Sci. J. Xiangtan Univ. **15** (1993), 1–7. MR 94m:65174. (in Chinese)

103. Cao, L.Q., Zhu, Q.D., *A new kind of finite element and its superconvergence estimate*, Natur. Sci. J. Xiangtan Univ. **12** (1990), 1–8. MR 92j:65166. (in Chinese)

104. Carey, G.F., *Derivative calculation form finite element solutions*, Comput. Methods Appl. Mech. Engrg. **35** (1982), 1–14. MR 83m:73067.

105. Carey, G.F., *Projections in finite element analysis and application*, in [328], 1997.

106. Carey, G.F., Chow, S.S., Seager, M., *Approximate boundary-flux calculations*, Comput. Methods Appl. Mech. Engrg. **50** (1985), 107–120. MR 87h:65191.

107. Carey, G.F., Humphrey, D., Wheeler, M.F., *Galerkin and collocation-Galerkin methods with superconvergence and optimal fluxes*, Internat. J. Numer. Methods Engrg. **17** (1981), 939–950. MR 82g:80008.

108. Carey, G.F., Seager, M., *Projection and iteration in adaptive finite element refinement*, Internat. J. Numer. Methods Engrg. **21** (1985), 1681–1695. MR 87b:65118.

109. Carey, G.F., Wheeler, M.F., C^0-*collocation-Galerkin methods*, in "Codes for Boundary-Value Problems in Ordinary Differential Equations," Proc. Conf. Houston, 1978, Lecture Notes in Comp. Sci., 76,, Springer, Berlin, 1979, pp. 250–256. MR 82b:65074.

110. Chandler, G.A., *Global superconvergence of iterated Galerkin solutions for second kind integral equations*, Technical Report, Australian Nat. Univ. Canberra, 1978.

111. Chandler, G.A., *Superconvergence of numerical solutions to second kind integral equations*, PhD thesis, Australian Nat. Univ. Canberra, 1979.

112. Chandler, G.A., *Superconvergence for second kind integral equations*, in "Application and numerical solution of integral equations," Proc. Sem., Australian Nat. Univ., Canberra, 1978, Monographs Textbooks Mech. Solids Fluids: Mech. Anal., 6, Nijhoff, The Hague, 1980, pp. 103–117. MR 81h:45027.

113. Chandler, G.A., *Mesh grading for boundary integral equations*, in "Computational techniques and applications: CTAC-83," (Sydney, 1983), North-Holland, Amsterdam, 1984, pp. 289–296. MR 86j:65149.

114. Chandler, G.A., *Galerkin's method for boundary integral equations on polygonal domains*, J. Austral. Math. Soc. Ser. B **26** (1984), 1–13. MR 86a:65116.

115. Chandler, G.A., *Superconvergent approximations to the solution of a boundary integral equation on polygonal domains*, SIAM J. Numer. Anal. **23** (1986), 1214–1229. MR 88d:45014.

116. Chatelin, F., *Des resultats de superconvergence*, Seminaire d'analyse numerique, no. 331, Univ. Scientifique et Medicale de Grenoble, Laboratoire IMAG, 1980.

117. Chatelin, F., *The iterated projection solution for the Fredholm integral equation of second kind*, Research Report IMS, 2. Academia Sinica, Chengdu Branch, Chengdu, 1981, 1–19. MR 83i:45004.

118. Chatelin, F., "Spectral approximation of linear operators," Computer Science and Appl. Math., Academic Press [Harcourt Brace Jovanovich, Publishers], New York–London, 1983, pp. 1–458. MR 86d:65071.

119. Chatelin, F., Lebbar, R., *The iterated projection solution for the Fredholm integral equation of second kind*, J. Austral. Math. Soc. Ser. B **22** (1980/81), 439–451. MR 82h:65096.

120. Chatelin, F., Lebbar, R., *Superconvergence results for the iterated projection method applied to a Fredholm integral equation of the second kind and the corresponding eigenvalue problem*, J. Integral Equations **6** (1984), 71–91. MR 85i:65167.

121. Chen, C.M., *Superconvergent points of Galerkin's method for two-point boundary value problems*, Numer. Math. J. Chinese Univ. **1** (1979), 73–79. MR 82e:65085.

122. Chen, C.M., *Optimal points of the stresses for triangular linear element*, Numer. Math. J. Chinese Univ. **2** (1980), 12–20. MR 83d:65279.

123. Chen, C.M., *Optimal points of stresses for tetrahedron linear element*, Natur. Sci. J. Xiangtan Univ. **3** (1980), 16–24. MR 83d:65279.

124. Chen, C.M., *Superconvergence of finite element solutions and their derivatives*, Numer. Math. J. Chinese Univ. **3** (1981), 118–125. MR 82m:65100. (in Chinese)

125. Chen, C.M., *Superconvergence of finite elements for nonlinear problems*, Numer. Math. J. Chinese Univ. **4** (1982), 222–228. MR 83m:65083. (in Chinese)

126. Chen, C.M., "Finite element method and its analysis in improving accuracy," Hunan Sci.

and Techn. Press, Changsha, 1982, pp. 1–242. (in Chinese)

127. Chen, C.M., *Superconvergence of finite element approximations to nonlinear elliptic problems*, in "Proc. of the China–France symposium on finite element methods," Beijing, 1982, Science Press, Beijing, 1983, pp. 622–640. MR 85h:65235.

128. Chen, C.M., *An estimate for elliptic boundary value problems and its applications to finite element methods*, Numer. Math. J. Chinese Univ. **5** (1983), 215–223. MR 86b:65120. (in Chinese)

129. Chen, C.M., *Convergence and superconvergence of finite element method for axisymmetric problems*, Hunan Ann. Math. **3** (1983), 81–88.

130. Chen, C.M., *Superconvergence of Galerkin solutions for singular nonlinear two-point boundary value problems*, Math. Numer. Sinica **7** (1985), 113–123. MR 87b:65119. (in Chinese)

131. Chen, C.M., *Superconvergence of finite element methods*, Adv. in Math. (Beijing) **14** (1985), 39–51. MR 87g:65134. (in Chinese)

132. Chen, C.M., $W^{1,\infty}$-*interior estimates for finite element method on regular mesh*, J. Comput. Math. **3** (1985), 1–7. MR 87b:65194.

133. Chen, C.M., *Superconvergence and extrapolation of the finite element approximations to quasilinear elliptic problems*, Dongbei Shuxue **2** (1986), 228–236. MR 88g:65104.

134. Chen, C.M., *High accuracy analysis of finite element methods–superconvergence and postprocessing*, Natur. Sci. J. Xiangtan Univ. **10** (1988), 114–123.

135. Chen, C.M., *Superconvergence of finite element in domain with reentrant corner*, Natur. Sci. J. Xiangtan Univ. **12** (1990), 134–141. MR 92c:65124.

136. Chen, C.M., *Correction of bilinear finite element*, Acta Math. Sci. (English Ed.) **11** (1991), 13–19. MR 92f:65132.

137. Chen, C.M., *Finite element methods of the parabolic integrodifferential equations*, Third Conf. on the FEM, Dayong, 1992, 93–96.

138. Chen, C.M., *Element analysis method and superconvergence*, in [328], 1997.

139. Chen, C.M., Huang, Y.Q., "High accuracy theory of finite element methods," Hunan Science and Technology Press, Changsha, 1995, pp. 1–638. (in Chinese)

140. Chen, C.M., Larsson, S., Zhang, N.Y., *Error estimates of optimal order for finite element methods with interpolated coefficients for the nonlinear heat equation*, IMA J. Numer. Anal. **9** (1989), 507–524. MR 91k:65141.

141. Chen, C.M., Lin, Q., *Extrapolation of finite element approximation in a rectangular domain*, J. Comput. Math. **7** (1989), 227–233. MR 90i:65190.

142. Chen, C.M., Liu, J.G., *Superconvergence of gradient of triangular linear element in general domain*, Natur. Sci. J. Xiangtan Univ. **9** (1987), 114–127. MR 88f:65193.

143. Chen, C.M., Thomée, V., *The lumped mass finite element method for a parabolic problem*, J. Austral. Math. Soc. Ser. B **26** (1985), 329–354. MR 86m:65117.

144. Chen, C.M., Wahlbin, L.B., *Superconvergence for the gradient in finite element method for an integrodifferential problem*, Preprint, 1997.

145. Chen, C.M., Zhu, Q.D., *A new estimate for finite element method and optimal point theorem for stresses*, Natur. Sci. J. Xiangtan Univ. **1** (1978), 10–20. (in Chinese)

146. Chen, H.S., *A mixed finite element and superconvergence*, Natur. Sci. J. Xiangtan Univ. **10** (1988), 7–10. MR 90g:65140. (in Chinese)

147. Chen, H.S., *A rectangular finite element analysis*, Natur. Sci. J. Xiangtan Univ. **11** (1989), 1–11. MR 92k:65160. (in Chinese)

148. Chen, H.S., *A superconvergence estimate for mixed finite element methods for plate bending problems*, Math. Numer. Sinica **12** (1990), 28–32. MR 91e:73075. (in Chinese)

149. Chen, H.S., *An L^2- and L^∞-error analysis for parabolic finite element equations with applications to superconvergence and error expansions*, Thesis, Preprint 93-11 (SFB 359), Univ. Heidelberg, 1993, 1–134.

150. Chen, H.S., Li, B., *Superconvergence analysis and error expansion for the Wilson nonconforming finite element*, Numer. Math. **69** (1994), 125–140. MR 95k:65105.

151. Chen, H.S., Lin, Q., *Finite element approximation using a graded mesh on domains with re-entrant corners*, Systems Sci. Math. Sci. **5** (1992), 127–140. MR 94c:65115.

152. Chen, H.S., Lin, Q., *A high accuracy analysis of the mixed finite element method for biharmonic equations*, Systems Sci. Math. Sci. **5** (1992), 352–364. MR 93k:65091.

153. Chen, H.S., Rannacher, R., *Superconvergence properties of finite element schemes for the*

Navier–Stokes problem, Preprint 93-37, IWR, Univ. Heidelberg, 1993, 1–27.

154. Chen, Y., *Superconvergent recovery of gradients of piecewise linear element approximations on non-uniform mesh partitions*, submitted to Numer. Methods Partial Differential Equations.

155. Chen, Y.P., Huang, Y.Q., *Global high accuracy of finite element solutions to singular non-symmetric two-point boundary value problems*, Numer. Math. J. Chinese Univ. **16** (1994), 271–278. MR 95i:65116. (in Chinese)

156. Chen, Z., *A quasi-projection analysis for elastic wave propagation in fluid-saturated porous media*, J. Comput. Math. **10** (1992), 366–375. MR 93g:65121.

157. Chen, Z., *Superconvergence results for Galerkin methods for wave propagation in various porous media*, Numer. Methods Partial Differential Equations **12** (1996), 99–122. MR 97f:65058.

158. Chen, Z.Y., *Superconvergence of generalized difference method for elliptic boundary value problem*, Numer. Math. J. Chinese Univ. (English Ser.) **3** (1994), 163–171. MR 96b:65106.

159. Cheng, S.J., *Superconvergence of finite element approximation for Navier–Stokes equations*, in "Extrapolation procedures in the finite element method," (Bonn, 1983), Bonner Math. Schriften, 158, Univ. Bonn, Bonn, 1984, pp. 31–45. MR 87b:65195.

160. Chou, S.H., Li, Q., *Convergence of the nonconforming Wilson element for a class of nonlinear parabolic problems*, Math. Comp. **54** (1990), 509–524. MR 90i:65192.

161. Chow, S.S., Carey, G.F., *Superconvergence phenomena in nonlinear two-point boundary value problems*, Numer. Methods Partial Differential Equations **9** (1993), 561–577. MR 94h:65112.

162. Chow, S.S., Carey, G.F., Lazarov, R.D., *Natural and postprocessed superconvergence in semilinear problems*, Numer. Methods Partial Differential Equations **7** (1991), 245–259. MR 92m:65131.

163. Chow, S.S., Lazarov, R.D., *Superconvergence analysis of flux computations for nonlinear problems*, Bull. Austral. Math. Soc. **40** (1989), 465–479. MR 91d:65149.

164. Ciarlet, P.G., "The finite element method for elliptic problems," North-Holland, Amsterdam, 1978, pp. 1–530. MR 58#25001.

165. Ciarlet, P.G., *Basic error estimates for elliptic problems*, in "Handbook of Numer. Anal.," Vol. II, P.G. Ciarlet, J.L. Lions (eds), North-Holland, Amsterdam, 1991, pp. 17–351. MR 92f:65001.

166. Ciavaldini, J.F., Crouzeix, M., *A finite element method scheme for one-dimensional elliptic equations with high superconvergence of nodes*, Numer. Math. **46** (1985), 417–427. MR 86g:65140.

167. Codina, R., *A finite element formulation for the numerical solution of the convection-diffusion equation*, Monograph, 14, Centro Internacional de Métodos Numéricos en Ingeniería, Barcelona, 1993. MR 96c:65160.

168. Cohen, G., Joly, P., Tordjman, N., *Construction and analysis of higher order finite elements with mass lumping for the wave equation*, in "Second International Conference on Mathematical and Numerical Aspects of Wave Propagation," (Newark, DE, 1993), SIAM, Philadelphia, PA, 1993, pp. 152–160. MR 94d:65058.

169. Comodi, M.I., *The Hellan–Herrmann–Johnson method: some new error estimates and postprocessing*, Math. Comp. **52** (1989), 17–29. MR 89f:65120.

170. Dautov, R.Z., *Superconvergence of difference schemes for the third boundary value problem for quasilinear elliptic equations*, Chisl. Metody Mekh. Sploshn. Sredy **11** (1980), 62–80. MR 84b:65102. (in Russian)

171. Dautov, R.Z., *Superconvergence of finite-element schemes with numerical integration for fourth-order quasilinear elliptic equations*, Differential Equations **18** (1982), 818–824, 1285. MR 83k:65080.

172. Dautov, R.Z., *Accuracy estimation of schemes of the finite element method in mesh norms*, in "Variational-Difference Methods in Mathematical Physics," Proc. Conf., Moscow, 1984, Viniti, Moscow, 1985, pp. 92–97.

173. Dautov, R.Z., Lapin, A.V., *Difference schemes of an arbitrary order of accuracy for quasilinear elliptic equations*, Izv. Vysš. Učebn. Zaved. Matematika **209** (1979), 24–37. MR 81g:65128.

174. Dautov, R.Z., Lapin, A.V., *Investigation of the convergence, in mesh norms, of finite-element-method schemes with numerical integration for fourth-order elliptic equations*, Differ-

ential Equations **17** (1981), 807–817. MR 83h:65122.

175. Dautov, R.Z., Lapin, A.V., Lyashko, A.D., *Some mesh schemes for quasilinear elliptic equations*, U.S.S.R Comput. Math. and Math. Phys. **20** (1980), 62–78. MR 81j:65121.

176. De Boor, C., Swartz, B., *Collocation at Gaussian points*, SIAM J. Numer. Anal. **10** (1973), 582–606. MR 51#9528.

177. De Boor, C., Swartz, B., *Local piecewise polynomial projection methods for an O.D.E. which give high-order convergence at knots*, Math. Comp. **36** (1981), 21–33. MR 82f:65091.

178. De Groen, P.P.N., *A finite element method with a large mesh-width for a stiff two-point boundary value problem*, J. Comput. Appl. Math. **7** (1981), 3–15. MR 82d:65059.

179. Del Prete, I., Lignola, M.B., *On convergence of closed-valued multifunctions*, Boll. Un. Mat. Ital. B (6) **2** (1983), 819–834. MR 85i:54017.

180. Dieci, L., Osborne, M.R., Russell, R.D.A, *Riccati transformation method for solving linear BVPs. II. Computational aspects*, SIAM J. Numer. Anal. **25** (1988), 1074–1092. MR 90b:65154.

181. Doucette, R.L., *A collocation method for the numerical solution of Laplace's equation with nonlinear boundary conditions on a polygon*, SIAM J. Numer. Anal. **30** (1993), 717–732. MR 94f:65113.

182. Dougalis, V.A., *Multistep-Galerkin methods for hyperbolic equations*, Math. Comp. **33** (1979), 563–584. MR 81b:65081.

183. Dougalis, V.A., Serbin, S.M., *On the superconvergence of Galerkin approximations to second-order hyperbolic equations*, SIAM J. Numer. Anal. **17** (1980), 431–446. MR 81j:65106.

184. Douglas, J., *A superconvergence result for the approximate solution of the heat equation by a collocation method*, in "The Mathematical Foundations of the Finite Element Method with Applications to Partial Differential Equations," (Proc. Sympos., Univ. Maryland, Baltimore, Md., 1972), Academic Press, New York, 1972, pp. 475–490. MR 53#7063.

185. Douglas, J., *Improved accuracy through superconvergence in the pressure in the simulation of miscible displacement*, in "Computing methods in applied sciences and engineering, VI," (Versailles, 1983), North-Holland, Amsterdam, 1984, pp. 633–638. MR 86f:65009.

186. Douglas, J., *Superconvergence in the pressure in the simulation of miscible displacement*, SIAM J. Numer. Anal. **22** (1985), 962–969. MR 86j:65129.

187. Douglas, J., Dupont, T., *Some superconvergence results for Galerkin methods for the approximate solution of two-point boundary problems*, in "Topics in numerical analysis," (Proc. Roy. Irish Acad. Conf., Univ. Coll., Dublin, 1972), Academic Press, London, 1973, pp. 89–92. MR 51#2295.

188. Douglas, J., Dupont, T., *Superconvergence for Galerkin methods for the two point boundary problem via local projections*, Numer. Math. **21** (1973), 270–278. MR 48#10130.

189. Douglas, J., Dupont, T., Wahlbin, L.B., *Optimal L_∞ error estimates for Galerkin approximations to solutions of two-point boundary value problems*, Math. Comp. **29** (1975), 475–483. MR 51#7298.

190. Douglas, J., Dupont, T., Wheeler, M.F., *Some superconvergence results for an H^1-Galerkin procedure for the heat equation*, in "Computing Meth. in Appl. Sciences and Engineering," (Proc. Internat. Sympos., Versailles, 1973), Part 1, Lecture Notes in Comput. Sci., Vol. 10, Springer, Berlin, 1974, pp. 288–311. MR 56#10056.

191. Douglas, J., Dupont, T., Wheeler, M.F., *An L^∞ estimate and a superconvergence result for a Galerkin method for elliptic equations based on tensor products of piecewise polynomials*, RAIRO Anal. Numér. **8** (1974), 61–66. MR 50#11812.

192. Douglas, J., Dupont, T., Wheeler, M.F., *A Galerkin procedure for approximating the flux on the boundary for elliptic and parabolic boundary value problems*, RAIRO Anal. Numér. **8** (1974), 47–59. MR 50#11811.

193. Douglas, J., Gupta, C.P., *Superconvergence for a mixed finite element method for elastic wave propagation in a plane domain*, Numer. Math. **49** (1986), 189–202. MR 88c:65095.

194. Douglas, J., Gupta, C.P., *Superconvergence and interior estimates for a mixed finite element method for elastic waves in a planar domain*, Mat. Apl. Comput. **7** (1988), 75–99. MR 90k:65172.

195. Douglas, J., Gupta, C.P., Li, G.Y., *Interior and superconvergence estimates for a primal hybrid finite element method for second order elliptic problems*, Calcolo **22** (1985), 187–207. MR 87g:65137.

196. Douglas, J., Milner, F.A., *Interior and superconvergence estimates for mixed methods for second order elliptic problems*, RAIRO Modél. Math. Anal. Numér. **19** (1985), 397–428. MR 88a:65129.

197. Douglas, J., Wang, J., *Superconvergence of mixed finite element methods on rectangular domains*, Calcolo **26** (1989), 121–133. MR 92e:65147.

198. Douglas, J., Wang, J.P., *A new family of mixed finite element spaces over rectangles.*, Mat. Apl. Comput. **12** (1993), 183–197. MR 95g:65150.

199. Druskin, V., Knizhnerman, L., *Krylov subspace approximation of eigenpairs and matrix functions in exact and computer arithmetic*, Numer. Linear Algebra Appl. **2** (1995), 205–217. MR 96b:65037.

200. Du, M.S., Liu, C.F., *Stability and convergence for the discrete-discontinuous finite element method*, J. Numer. Methods Comput. Appl. **5** (1984), 219–231. MR 86j:65131. (in Chinese)

201. Dupont, T., *A unified theory of superconvergence for Galerkin methods for two-point boundary problems*, SIAM J. Numer. Anal. **13** (1976), 362–368. MR 53#12021.

202. Durán, R., *Superconvergence for rectangular mixed finite elements*, Numer. Math. **58** (1990), 287–298. MR 92a:65289.

203. Durán, R., Rodríguez, R., *On the asymptotic exactness of Bank-Weiser's estimator*, Numer. Math. **62** (1992), 297–303. MR 93e:65135.

204. Durán, R., Muschietti, M.A., Rodríguez, R., *On the asymptotic exactness of error estimators for linear triangular finite elements*, Numer. Math. **59** (1991), 107–127. MR 92b:65086.

205. Durán, R., Muschietti, M.A., Rodríguez, R., *Asymptotically exact error estimators for rectangular finite elements*, SIAM J. Numer. Anal. **29** (1992), 78–88. MR 93a:65148.

206. Dzhishkariani, A.V., *Superconvergence of an approximate projection method with one iteration*, in "Current problems in mathematical physics, Vol. I," Tbilis. Gos. Univ., Tbilisi, 1987, pp. 231–238, 483. MR 90c:65078. (in Russian)

207. Dzhishkariani, A. V., *On the convergence of the finite element method with one iteration*, Zh. Vychisl. Mat. i Mat. Fiz. **30** (1990), 791–796, 800. MR 91h:65171. (in Russian)

208. Dzhishkariani, A.V., Manveliani, G.S., *On the superconvergence of an approximate projection method with one iteration*, Proc. A. Razmadze Math. Inst. **100** (1992), 89–95. MR 95b:65135. (in Russian)

209. Eggermont, P.P.B., *Collocation as a projection method and superconvergence for Volterra integral equations of the first kind*, in "Treatment of integral equations by numerical methods," (Durham, 1982), Academic Press, London, 1982, pp. 131–138.

210. Eggermont, P.P.B., *Beyond superconvergence of collocation methods for Volterra integral equations of the first kind*, in "Constructive methods for the practical treatment of integral equations," (Oberwolfach, 1984), Internat. Schriftenreihe Numer. Math., 73, Birkhäuser, Basel, 1985, pp. 110–119.

211. Eggermont, P.P.B., *Improving the accuracy of collocation solutions of Volterra integral equations of the first kind by local interpolation*, Numer. Math. **48** (1986), 263–279. MR 87f:65150.

212. El Hatri, M., *Superconvergence of the lumped mass approximation of parabolic equations*, C. R. Acad. Bulgare Sci. **36** (1983), 575–578. MR 85h:65187.

213. El Hatri, M., *Superconvergence of axisymmetrical boundary value problem*, C. R. Acad. Bulgare Sci. **36** (1983), 1499–1502. MR 86e:65153.

214. El Hatri, M., *Superconvergence in finite element method for a degenerated boundary value problem*, in "Proc. Conf. Constructive Theory of Functions 84," Izd. Bulg. Acad. Sci., Sofia, 1984, pp. 328–333.

215. El Hatri, M., *Estimation de l'erreur de la méthode des éléments finis avec integration numérique pour un problème aux limites dégénéré*, Thesis, Fac. de Math. et Mécanique, Univ. Sofia, 1985.

216. El Hatri, M., *Optimal and superconvergence-type error estimation of the finite-element method for a degenerate boundary value problem*, RAIRO Modél. Math. Anal. Numér. **21** (1987), 27–61. MR 89a:65168. (in French)

217. Elschner, J., *On spline approximation for a class of integral equations. I. Galerkin and collocation methods with piecewise polynomials*, Math. Methods Appl. Sci. **10** (1988), 543–559. MR 89m:65113.

218. Elschner, J., *On spline collocation for convolution equations*, Integral Equations Operator

Theory **12** (1989), 486–510. MR 91b:65149.

219. Elschner, J., *On spline approximation for a class of noncompact integral equations*, Math. Nachr. **146** (1990), 271–321. MR 91m:65326.

220. Engl, H.W., Neubauer, A., *A parameter choice strategy for (iterated) Tikhonov regularization of ill-posed problems leading to superconvergence with optimal rates*, Appl. Anal. **27** (1988), 5–18. MR 89b:65135.

221. Englert, F., Brout, R., Stern, H., *Mass quantization conditions from infinite superconvergence and chiral symmetry*, Nuovo Cimento A (10) **66** (1970), 845–855. MR 41#7959.

222. Eriksson, K., Nie, Y.Y., *Convergence analysis for a nonsymmetric Galerkin method for a class of singular boundary value problems in one space dimension*, Math. Comp. **49** (1987), 167–186. MR 88h:65182.

223. Ewing, R.E., Lazarov, R.D., *Superconvergence of the mixed finite element approximations of parabolic problems using rectangular finite elements*, East-West J. Numer. Math. **1** (1993), 199–212. MR 94m:65158.

224. Ewing, R.E., Lazarov, R.D., Russell, T.F., Vassilevski, P.S., *Local refinement via domain decomposition techniques for mixed finite element methods with rectangular Raviart–Thomas elements*, in "Third Internat. Symposium on Domain Decomposition Methods for Partial Differential Equations," (Houston, TX, 1989), SIAM, Philadelphia, PA, 1990, pp. 98–114. MR 91f:65191.

225. Ewing, R.E., Lazarov, R.D., Wang, J., *Superconvergence of the velocity along the Gauss lines in mixed finite element methods*, SIAM J. Numer. Anal. **28** (1991), 1015–1029. MR 92e:65149.

226. Ewing, R.E., Shen, J., Wang, J., *Application of superconvergence to problems in the simulation of miscible displacement*, Second World Congress on Computational Mechanics, Part I (Stuttgart, 1990), Comput. Methods Appl. Mech. Engrg. **89** (1991), 73–84. MR 92k:76078.

227. Ewing, R.E., Wang, J.P., *Analysis of mixed finite element methods on locally refined grids*, Numer. Math. **63** (1992), 183–194. MR 93h:65137.

228. Feng, K., Lions, J.L. (eds), "Proceedings of the China–France symposium on finite element methods," Beijing, 1982. Science Press, Beijing; Gordon & Breach Science Publishers, New York, 1983, pp. 1–1036. MR 85g:65010.

229. Finn, J.M., *Lie transforms: a perspective*, in "Local and global methods of nonlinear dynamics," (Silver Spring, Md., 1984), Lecture Notes in Phys., 252, Springer, Berlin, 1986, pp. 63–86. MR 87m:58055.

230. Franca, L.P., *On the superconvergence of the satisfying-Babuška-Brezzi method*, Internat. J. Numer. Methods Engrg. **29** (1990), 1715–1726. MR 91g:65182.

231. French, D.A., Peterson, T.E., *A continuous space-time finite element method for the wave equation*, Math. Comp. **65** (1996), 491–506. MR 96g:65098.

232. Fu, K.S., *A superconvergence consequence of even spline interpolation*, Natur. Sci. J. Xiangtan Univ. (1984), 22–31. MR 86m:65015. (in Chinese)

233. Fu, K.X., Shen, X.Y., *Asymptotic expansion and results on superconvergence of single knot splines*, Hunan Ann. Math. **6** (1986), 105–113. MR 92d:65022. (in Chinese)

234. Gähler, W., Herrlich, H., Preuss, G. (eds), "Recent developments of general topology and its applications," Proc. of the Internat. Conf. in Memory of Felix Hausdorff (1868–1942), the Free Univ. Berlin, 1992, Math. Research, 67, Berlin, 1992, pp. 1–319. MR 93m:54004.

235. Gärtner, S., *A superconvergent time integration method for dynamical systems and other problems of mth order*, Z. Angew. Math. Mech. **66** (1986), 385–393. MR 88h:65135. (in German)

236. Gastaldi, L., Nochetto, R., *Optimal L^∞-error estimates for nonconforming and mixed finite element methods of lowest order*, Numer. Math. **50** (1987), 587–611. MR 88f:65196.

237. Gastaldi, L., Nochetto, R.H., *Erratum: On L^∞-accuracy of mixed finite element methods for second order elliptic problems*, Mat. Apl. Comput. **7** (1988), i. MR 90d:65192b.

238. Gastaldi, L., Nochetto, R.H., *Sharp maximum norm error estimates for general mixed finite element approximations to second order elliptic equations*, RAIRO Modél. Math. Anal. Numér. **23** (1989), 103–128. MR 91b:65125.

239. Genis, A.M., *On finite element methods for the Euler-Poisson-Darboux equation*, SIAM J. Numer. Anal. **21** (1984), 1080–1106. MR 86f:65171.

240. Gladwell, I., Reddien, G., Wang, J., *Energy superconvergence of one-step methods for sepa-*

rable Hamiltonian systems, Phys. Lett. A **209** (1995), 31–38.

241. Glowinski, R., Lions, J.L. (eds), "Computing methods in applied sciences and engineering, VI," Proc of the sixth internat. symposium, Versailles, 1983, North-Holland Publishing Co., Amsterdam, 1984, pp. 1–728. MR 86f:65009.

242. Golberg, M.A., *A survey of numerical methods for integral equations*, in "Solution Methods for Integral Equations," Math. Concepts and Methods in Sci. and Engrg., 18, Plenum Press, New York, 1979, pp. 1–58. MR 80m:65090.

243. Golberg, M.A., *A note on a superconvergence result for the generalized airfoil equation*, Appl. Math. Comput. **26** (1988), part I, 105–117. MR 89f:65141.

244. Golberg, M.A. (ed.), "Numerical solution of integral equations," Mathematical Concepts and Methods in Science and Engineering, 42, Plenum Press, New York, 1990, pp. 1–417. MR 91e:65007.

245. Golberg, M.A., *The discrete Sloan iterate for the generalized airfoil equation*, Appl. Math. Comput. **57** (1993), 103–115. MR 94e:76071.

246. Golberg, M.A., Lea, M., Miel, G., *A superconvergence result for the generalized airfoil equation with application to the flap problem*, J. Integral Equations **5** (1983), 175–186. MR 84e:65127.

247. Gomes, S.M., *Convergence estimates for the wavelet-Galerkin method: superconvergence at the node points*, Adv. Comput. Math. **4** (1995), 261–282. MR 96j:65101.

248. Gómez, S., Hennart, J.P. (eds), "Advances in optimization and numerical analysis," Proceedings of the Sixth Workshop on Optimization and Numer. Anal., Oaxaca, 1992, Math. and its Applications, 275, Kluwer Acad. Publ., Dordrecht, 1994, pp. 1–285. MR 95a:00010.

249. Goodsell, G., *Gradient superconvergence properties for finite element approximations to two-dimensional Poisson problems*, MSc Dissertation, Brunel Univ., 1985.

250. Goodsell, G., *Pointwise superconvergence of the gradient for the linear tetrahedral element*, Numer. Methods Partial Differential Equations **10** (1994), 651–666. MR 95e:65101.

251. Goodsell, G., Whiteman, J.R., *Superconvergent recovered gradient functions for piecewise linear finite element approximations, with extensions to subdomains*, in "Proc. Conf. MAFELAP VI," J.R. Whiteman (ed.), 1987, Academic Press, London, 1988, pp. 582–583.

252. Goodsell, G., Whiteman, J.R., *A unified treatment of superconvergent recovered gradient functions for piecewise linear finite element approximations*, Internat. J. Numer. Methods Engrg. **27** (1989), 469–481. MR 91k:65144.

253. Goodsell, G., Whiteman, J.R., *Pointwise superconvergence of recovered gradients for piecewise linear finite element approximations to problems of planar linear elasticity*, Numer. Methods Partial Differential Equations 6 (1990), 59–74. MR 91c:73081.

254. Goodsell, G., Whiteman, J.R., *Superconvergence of recovered gradients of piecewise quadratic finite element approximations. I. L^2-error estimates*, Numer. Methods Partial Differential Equations 7 (1991), 61–83. MR 92e:65151a.

255. Goodsell, G., Whiteman, J.R., *Superconvergence of recovered gradients of piecewise quadratic finite element approximations. II. L^∞-error estimates*, Numer. Methods Partial Differential Equations 7 (1991), 85–99. MR 92e:65151b.

256. Gottlieb, D., Gustafsson, B., Olsson, P., Strand, B., *On the superconvergence of Galerkin methods for hyperbolic IBVP*, SIAM J. Numer. Anal. **33** (1996), 1778–1796.

257. Graham, I.G., *Numerical methods for multidimensional integral equations*, in "Computational techniques and applications: CTAC-83," (Sydney, 1983), North-Holland, Amsterdam, 1984, pp. 335–351. MR 86h:65195.

258. Graham, I.G., Joe, S., Sloan, I.H., *Iterated Galerkin versus iterated collocation for integral equations of the second kind*, IMA J. Numer. Anal. **5** (1985), 355–369. MR 86j:65178.

259. Griewank, A., Reddien, G.W., *The approximation of generalized turning points by projection methods with superconvergence to the critical parameter*, Numer. Math. **48** (1986), 591–606. MR 87g:65072.

260. Grossmann, C., Roos, H.G., *On deriving the rate of convergence in C^1-spline collocation methods for a monotone boundary value problem*, Z. Angew. Math. Mech. **68** (1988), 59–60. MR 89b:65188. (in German)

261. Gruber, R., *Finite elements in magnetohydrodynamics: ideal linear stability problem*, in "Finite elements in physics," (Lausanne, 1986), North-Holland, Amsterdam, 1987, pp. 299–333. MR 89b:65264.

262. Hämmerlin, G., Nikolis, A., *Superconvergence in trigonometric splines and spline-on-splines*, in "Beiträge zur angewandten Analysis und Informatik," Shaker, Aachen, 1994, pp. 112–122. (in German)

263. Han, G.Q., *Asymptotic expressions and superconvergence for odd degree periodic spline interpolation*, Numer. Math. J. Chinese Univ. **9** (1987), 193–197. MR 89f:65014. (in Chinese)

264. Han, G.Q., *Asymptotic expressions for the remainder terms in fourth order operator spline interpolation, and their points of superconvergence*, Numer. Math. J. Chinese Univ. **10** (1988), 126–135. MR 90c:65023. (in Chinese)

265. Han, G.Q., *Asymptotic formulas for interpolation by a class of even splines and their superconvergence*, J. Math. (Wuhan) **11** (1991), 475–478. MR 94e:41023. (in Chinese)

266. Han, G.Q., *Superconvergence and error estimates for the multivariate optimal interpolation to scattered data in the case of discrete boundary conditions*, Math. Numer. Sinica **15** (1993), 39–48. (in Chinese)

267. Han, G.Q., *Superconvergence and error bounds for multivariate optimal interpolation of scattered data on rectangular domains I. The case of continuous boundary conditions*, Gaoxiao Yingyong Shuxue Xuebao Ser. A **9** (1994), 294–303. MR 96c:65011. (in Chinese)

268. Han, G.Q., *Extrapolation of a discrete collocation-type method of Hammerstein equations.*, J. Comput. Appl. Math. **61** (1995), 73–86. MR 96i:65120.

269. Hebeker, F.K., Mika, J., Pack, D.C., *Application of the superconvergence properties of the Galerkin approximation to the calculation of upper and lower bounds for linear functionals of solutions of integral equations*, IMA J. Appl. Math. **38** (1987), 61–70. MR 90f:65238.

270. Hennart, J.P. (ed.), "Numerical analysis," Proc. of the fourth Instit. for Research in Appl. Math. and Systems (IIMAS) workshop, Guanajuato, 1984, Lecture Notes in Math., 1230, Springer, Berlin, 1986, pp. 1–234. MR 88a:65005.

271. Hennart, J.P., *On the numerical analysis of analytical nodal methods*, Numer. Methods Partial Differential Equations **4** (1988), 233–254. MR 90i:65218.

272. Hinton, E., Campbell, J.S., *Local and global smoothing of discontinuous finite element functions using a least squares method*, Internat. J. Numer. Methods Engrg. **8** (1974), 461–480. MR 53#14859.

273. Hlaváček, I., Chleboun, J., *A recovered gradient method applied to smooth optimal shape problems*, Appl. Math. **41** (1996), 281–297.

274. Hlaváček, I., Chleboun, J., *Shape design sensitivity formulae approximated by means of a recovered gradient method*, in [328], 1997.

275. Hlaváček, I., Křížek, M., *On a superconvergent finite element scheme for elliptic systems*, in "Software a algoritmy numerické matematiky," ed. I. Marek, Doksy, 1985, VŠSE Plzeň, 1986, pp. 5–14.

276. Hlaváček, I., Křížek, M., *On a superconvergent finite element scheme for elliptic systems. I. Dirichlet boundary condition*, Apl. Mat. **32** (1987), 131–154. MR 88c:65099.

277. Hlaváček, I., Křížek, M., *On a superconvergent finite element scheme for elliptic systems. II. Boundary conditions of Newton's or Neumann's type*, Apl. Mat. **32** (1987), 200–213. MR 88j:65251.

278. Hlaváček, I., Křížek, M., *On a superconvergent finite element scheme for elliptic systems. III. Optimal interior estimates*, Apl. Mat. **32** (1987), 276–289. MR 88j:65252.

279. Hlaváček, I., Křížek, M., *Superconvergence phenomena in the finite element method*, in "Proc. Conf. Programmes and Algorithms of Numer. Math. 7, Bratříkov, 1994," MÚ AV ČR, Praha, 1994, pp. 48–63. (in Czech)

280. Hlaváček, I., Křížek, M., *Optimal interior and local error estimates of a recovered gradient of linear elements on nonuniform triangulations*, J. Comput. Math. **14** (1996), 345–362. MR 97g:65234.

281. Hlaváček, I., Křížek, M., Pištora, V., *How to recover the gradient of linear elements on nonuniform triangulations*, Appl. Math. **41** (1996), 241–267. MR 97g:65233.

282. Hogarth, W.L., Noye B.J. (eds), "Computational techniques and applications: CTAC-89," Proc. of the Fourth International Conference, Griffith University, Brisbane, 1989, Hemisphere Publishing Corp., New York, 1990, pp. 1–755. MR 92d:00025.

283. Houstis, E.N., Christara, C.C., Rice, J.R., *Quadratic-spline collocation methods for two-point boundary value problems*, Internat. J. Numer. Methods Engrg. **26** (1988), 935–952. MR 89g:65103.

284. Hsiao, G.C., Wendland, W.L., *The Aubin-Nitsche lemma for integral equations*, J. Integral Equations **3** (1981), 299–315. MR 83j:45019.

285. Hu, Q., *Stieltjes derivatives and β-polynomial spline collocation for Volterra integrodifferential equations with singularities*, SIAM J. Numer. Anal. **33** (1996), 208–220. MR 97a:65112.

286. Hu, R.Z., *Spline collocation for generalized multipoint boundary value problems*, Numer. Math. J. Chinese Univ. **5** (1983), 127–138. MR 85a:65120. (in Chinese)

287. Hu, R.Z., *A spline collocation method for generalized multipoint boundary value problems of mth-order ordinary differential equations*, Acta Sci. Natur. Univ. Sunyatseni **1984**, no. 2., 53–61. MR 86e:65109. (in Chinese)

288. Hu, R.Z., *Spline collocation method for the vibrating string equation with inner point constraint conditions*, Numer. Math. J. Chinese Univ. **8** (1986), 21–30. MR 87i:65196. (in Chinese)

289. Huang, D, *The error expressions and the superconvergence of the cubic interpolation spline*, Numer. Math. J. Chinese Univ. **5** (1983), 179–184. MR 84m:65023. (in Chinese)

290. Huang, D., Wu, D., *The superconvergence of the spline finite element solution and its second order derivatives for the two-point boundary value problem of a fourth order differential equation*, J. Zhejiang Univ. **3** (1982), 92–99.

291. Huang, M.Y., Wang, X.F., *A note on the L_∞ estimates of semidiscrete Galerkin approximations*, Numer. Math. J. Chinese Univ. **6** (1984), 376–379. MR 87b:65162. (in Chinese)

292. Huang, X.F., *The optimal point of the gradient of finite element solution*, Appl. Math. Mech. (English Ed.) **7** (1986), 785–794. MR 88d:65164.

293. Huang, Y.Q., *The superconvergence of finite element methods on domains with reentrant corners*, in [328], 1997.

294. Huang, Y.Q., Chen, Y.P., *Superconvergence and asymptotically exact a posteriori error estimates for finite elements on a K-mesh*, Math. Numer. Sinica **16** (1994), 278–285. (in Chinese)

295. Huang, Y.Q., Lin, Q., *Asymptotic expansions and superconvergence of finite element solutions to parabolic equations*, J. Systems Sci. Math. Sci. **11** (1991), 327–335. MR 92h:65170. (in Chinese)

296. Hugger, J., *An asymptotically exact, pointwise, a posteriori error estimator for the finite element method with superconvergence properties*, in "Modeling, mesh generation, and adaptive numerical methods for partial differential equations," (Minneapolis, MN, 1993), IMA Vol. Math. Appl., 75, Springer, New York, 1995, pp. 277–305. MR 96i:65096.

297. Hughes, T.J.R., "The finite element method. Linear static and dynamic finite element analysis," With the collaboration of Robert M. Ferencz and Arthur M. Raefsky, Prentice-Hall, Englewood Cliffs, NJ, 1987, pp. 1–803. MR 90i:65001.

298. Ioakimidis, N.I., *A superconvergence result for the natural extrapolation formula for the numerical determination of stress intensity factors*, Internat. J. Numer. Methods Engrg. **21** (1985), 1391–1401. MR 87c:73088.

299. Jameson, L., *On the spline-based wavelet differentiation matrix*, Appl. Numer. Math. **17** (1995), 53–45. MR 96a:65022.

300. Jameson, L., *The differentiation matrix for Daubechies-based wavelets on an interval*, SIAM J. Sci. Comput. **17** (1996), 498–516. MR 97b:65027.

301. Jay, L., *Convergence of a class of Runge-Kutta methods for differential-algebraic systems of index 2*, BIT **33** (1993), 137–150. MR 96a:65106.

302. Jiang, Z., Zhang, S.H., *Non-standard Galerkin methods of high accuracy for parabolic problems*, Preprint of Tianjin Univ., 1996.

303. Jin, J.C., Huang, Y.Q., *Superconvergence of derivatives of biquadratic finite element solutions*, Natur. Sci. J. Xiangtan Univ. **15** (1993), 8–14. MR 94m:65175. (in Chinese)

304. Jones, L., Pani, A.K., *On superconvergence results and negative norm estimates for a unidimensional single phase Stefan problem*, Numer. Funct. Anal. Optim. **16** (1995), 153–175. MR 96c:65161.

305. Kaneko, H., Noren, R.D., Xu, Y., *Numerical solutions for weakly singular Hammerstein equations and their superconvergence*, J. Integral Equations Appl. **4** (1992), 391–407. MR 93j:65223.

306. Kaneko, H., Xu, Y., *Superconvergence of the iterated Galerkin methods for Hammerstein equations*, SIAM J. Numer. Anal. **33** (1996), 1048–1064. MR 97g:65278.

307. Kantchev, V., *Superconvergence of the gradient for linear finite elements for nonlinear elliptic problems*, in "Proc. of the Second Internat. Symposium on Numer. Anal.," Prague, 1987, Teubner-Texte Math., 107, Teubner, Leipzig, 1988, pp. 199–204. MR 93b:65007.

308. Kantchev, V., *Superconvergent L^∞ error estimates for linear finite element approximations of quasilinear boundary value problems*, C. R. Acad. Bulgare Sci. **46** (1993), 25–28. MR 95j:65138.

309. Kantchev, V., Lazarov, R.D., *Superconvergence of the gradient of linear finite elements for 3D Poisson equation*, in "Proc. Internat. Conf. Optimal Algorithms," B. Sendov (ed.), Blagoevgrad, 1986, Izd. Bulg. Akad. Nauk, Sofia, 1986, pp. 172–182.

310. Kauthen, J.P., *Continuous time collocation methods for Volterra-Fredholm integral equations*, Numer. Math. **56** (1989), 409–424. MR 90m:65236.

311. Kindalev, B.S., *The accuracy of approximation by periodic interpolation splines of odd degree*, Vychisl. Sistemy No. 98 (1983), 67–82, 153. MR 86k:41015. (in Russian)

312. Kindalev, B.S., *Error asymptotics and superconvergence of periodic interpolation splines of even degree*, Vychisl. Sistemy **115** (1986), 3–25, 150. MR 88j:41034. (in Russian)

313. Klouček, P., Li, B., Luskin, M., *Analysis of a class of nonconforming finite elements for crystalline microstructures*, Math. Comp. **65** (1996), 1111–1135. MR 97a:73076.

314. Korneev, V.G., *Superconvergence of solutions of the finite element method in mesh norms*, Zh. Vychisl. Mat. i Mat. Fiz. **22** (1982), 1133–1148, 1277. MR 84b:65114. (in Russian)

315. Korneev, V.G., *Convergence of finite element solutions in difference norms*, Metody Vychisl. **13** (1983), 3–42. MR 85c:65138. (in Russian)

316. Korobeĭnik, Yu.F. (ed.), "Current problems in function theory," North Caucasus Regional School on Function Theory, Teberda, 1985, Rostov. Gos. Univ., Rostov-on-Don, 1987, pp. 1–191. MR 91a:00015. (in Russian)

317. Kreiss, H.O., Manteuffel, T.A., Swartz, B., Wendroff, B., White, A.B., *Supra-convergent schemes on irregular grids*, Math. Comp. **47** (1986), 537–554. MR 88b:65082.

318. Křížek, M., *Superconvergence results for linear triangular elements*, in "EQUADIFF 6," Brno, 1985, Lecture Notes in Math., 1192, Springer, Berlin, 1986, pp. 315–320. MR 88i:65132.

319. Křížek, M., *Superconvergence phenomena in the finite element method*, ICOSAHOM'92, Montpellier, 1992, Comput. Methods Appl. Mech. Engrg. **116** (1994), 157–163. MR 95b:65007.

320. Křížek, M., *Higher order global accuracy of a weighted averaged gradient of the Courant elements on irregular meshes*, in [327], 1994, pp. 267–276. MR 95j:65139.

321. Křížek, M., Lin, Q., Huang, Y.Q., *A nodal superconvergence arising from combination of linear and bilinear elements*, Systems Sci. Math. Sci. **1** (1988), 191–197. MR 91f:65170.

322. Křížek, M., Neittaanmäki, P., *Superconvergence phenomenon in the finite element method arising from averaging gradients*, Numer. Math. **45** (1984), 105–116. MR 86c:65135.

323. Křížek, M., Neittaanmäki, P., *On a global superconvergence of the gradient of linear triangular elements*, J. Comput. Appl. Math. **18** (1987), 221–233. MR 88h:65209.

324. Křížek, M., Neittaanmäki, P., *On superconvergence techniques*, Acta Appl. Math. **9** (1987), 175–198. MR 88h:65208.

325. Křížek, M., Neittaanmäki, P., *On $O(h^4)$-superconvergence of piecewise bilinear FE-approximations*, Mat. Apl. Comput. **8** (1989), 49–61. MR 91g:65242.

326. Křížek, M., Neittaanmäki, P., "Finite element approximation of variational problems and applications," Pitman Monographs and Surveys in Pure and Applied Mathematics, 50, Longman Scientific & Technical, Harlow; copublished in the United States with John Wiley & Sons, New York, 1990, pp. 1–239. MR 91h:65174.

327. Křížek, M., Neittaanmäki, P., Stenberg, R. (eds), "Finite Element Methods: Fifty Years of the Courant Element," Proc. Conf., Univ. of Jyväskylä, 1993, M. Křížek, P. Neittaanmäki, R. Stenberg (eds), Lecture Notes in Pure and Appl. Math., 164, Marcel Dekker, New York, 1994, pp. 1–504. MR 95d:65008.

328. Křížek, M., Neittaanmäki and P., Stenberg, R. (eds), "Finite Element Methods: Superconvergence, Post-processing and A Posteriori Estimates," Proc. Conf., Univ. of Jyväskylä, 1996, M. Křížek, P. Neittaanmäki, R. Stenberg (eds), Marcel Dekker, New York, 1997.

329. Kumar, S., *Superconvergence of a collocation-type method for Hammerstein equations*, IMA J. Numer. Anal. **7** (1987), 313–325. MR 90a:65275.

330. Kumar, S., *Superconvergence of a collocation-type method for simple turning points of Ham-*

merstein equations, Math. Comp. **50** (1988), 385–398. MR 89c:65139.

331. Lakhany, A.M., Whiteman, J.R., *Superconvergent recovery operators: derivative recovery techniques*, in [328], 1997.

332. Lapin, A.V., *Difference schemes of a high order of accuracy for some classes of variational inequalities*, Preprint 86-114, Akad. Nauk SSSR, Otdel Vychisl. Mat., Moscow, 1986, 1–30. MR 88e:65141. (in Russian)

333. Lapin, A.V., *Mesh schemes of high order of accuracy for some classes of variational inequalities*, Soviet J. Numer. Anal. Math. Modelling **2** (1987), 37–55. MR 88m:65176.

334. Larsen, E.W., Nelson, P., *Finite-difference approximations and superconvergence for the discrete-ordinate equations in slab geometry*, SIAM J. Numer. Anal. **19** (1982), 334–348. MR 83h:82050.

335. Laydi, M.R., Zealouk, L., *A totally explicit iterative method for solving the Stokes problem*, C. R. Acad. Sci. Paris Sér. I Math. **321** (1995), 641–644. MR 96e:65064. (in French)

336. Layton, W.J. 2, *The finite element method for a degenerate hyperbolic partial differential equation*, BIT **23** (1983), 231–238. MR 84f:65073.

337. Lazarov, R. D., *Superconvergence of the gradient for triangular and tetrahedral finite elements of a solution of linear problems in elasticity theory*, in "Computational processes and systems," No. 6, "Nauka", Moscow, 1988, pp. 180–191. MR 90f:65207. (in Russian)

338. Lazarov, R.D., Pehlivanov, A.I., *Local superconvergence analysis of 3-D linear finite elements for boundary-flux calculations*, in "The mathematics of finite elements and applications, VII," Uxbridge, 1990, Academic Press, London, 1991, pp. 75–83. MR 92d:65007.

339. Lesaint, P., Zlámal, M., *Superconvergence of the gradient of finite element solutions*, RAIRO Anal. Numér. **13** (1979), 139–166. MR 80g:65112.

340. Leseberg, D., *Superconvergence spaces and related topics*, in "Recent developments of general topology and its applications," Berlin, 1992, Math. Res., 67, Akademie-Verlag, Berlin, 1992, pp. 187–196. MR 94c:54020.

341. Levermore, C.D., Manteuffel, T.A., White, A.B., *Numerical solution of partial differential equations on irregular grids*, in "Computational techniques and applications: CTAC-87," (Sydney, 1987), North-Holland, Amsterdam, 1988, pp. 417–426. MR 90f:65148.

342. Levesley, J., Hough, D.M., Chandler-Wilde, S.N., *A Chebyshev collocation method for solving Symm's integral equation for conformal mapping: a partial error analysis*, IMA J. Numer. Anal. **14** (1994), 57–79. MR 95a:45021.

343. Levine, N., *Superconvergent recovery of the gradient from piecewise linear finite-element approximations*, IMA J. Numer. Anal. **5** (1985), 407–427. MR 87b:65206.

344. Levine, N., *Superconvergent estimation of the gradient from linear finite element approximations on triangular elements*, Numer. Anal. Report No. 3/85, Univ. of Reading, 1985, 1–202.

345. Leyk, Z., *Superconvergence in the finite element method*, Mat. Stos. **20** (1982), 93–107. MR 85k:65090. (in Polish)

346. Leyk, Z., *A C^0-collocation-like method for two-point boundary value problems*, Numer. Math. **49** (1986), 39–53. MR 88d:65117.

347. Li, B., *Superconvergence for higher-order triangular finite elements*, Chinese J. Numer. Math. Appl. **12** (1990), 75–79. MR 92d:65196.

348. Li, Q., *Global superconvergence of Galerkin approximations to nonlinear parabolic problems*, Numer. Math. J. Chinese Univ. **16** (1994), 288–292. MR 95m:65168. (in Chinese)

349. Li, Z.C., *Superconvergence of coupling techniques in combined methods for elliptic equations with singularities*, Preprint National Sun Yat-sen Univ., Taiwan, 1995, 1–29.

350. Lie, I., Nørsett, S.P., *Superconvergence for multistep collocation*, Math. Comp. **52** (1989), 65–79. MR 89m:65069.

351. Lin, Q., *High accuracy from linear elements*, in "Proc. of the Beijing Sympos. on Differential Geom. and Differential Equations," Feng Kang (ed.), Beijing, 1984, pp. 258–262.

352. Lin, Q., *Finite element error expansion for nonuniform quadrilateral meshes*, J. Systems Sci. Math. Sci. (1989), 275–282. MR 92c:65137.

353. Lin, Q., *An integral identity and interpolated postprocess in superconvergence*, Research Report 90-07, Inst. Systems Sci., Academia Sinica, Beijing, 1990, 1–6.

354. Lin, Q., *Superconvergence of FEM for singular solution*, Proc. Conf. MAFELAP, 1990, J. Comput. Math. **9** (1991), 111–114. MR 93b:65171.

355. Lin, Q., *A new observation of FEM*, in "Proc. Syst. Sci. & Syst. Engrg.," Hongkong, 1991, pp. 389–391.

356. Lin, Q., *Interpolated finite elements and global superconvergence*, in "Numerical methods for partial differential equations," (Tianjin, 1991), World Sci. Publishing, River Edge, NJ, 1992, pp. 91–95.

357. Lin, Q., *Global error expansion and superconvergence for higher order interpolation of finite elements*, J. Comput. Math., Suppl. Issue (1992), 286–289.

358. Lin, Q., *How the superconvergence theory reduces from thick to thin*, Preprint Inst. of Systems Sci., Academia Sinica, 1993.

359. Lin, Q., *Interpolated finite elements and global error recovery*, in "Computational mathematics in China," Contemp. Math., 163, Amer. Math. Soc., Providence, RI, 1994, pp. 93–109. MR 95e:65002.

360. Lin, Q., *Superclose FE-theory becomes a table of integrals*, in [328], 1997.

361. Lin, Q., Lu, T., *Asymptotic expansions for finite element approximation of elliptic problem on polygonal domains*, in "Proc. Internat. Conf. Comput. Methods in Appl. Sci. and Engrg. VI," Versailles, 1983, R. Glowinski and J.L. Lions (eds), Elsevier, 1984, pp. 317–321. MR 86f:65009.

362. Lin, Q., Lu, T., Shen, S.M., *Maximum norm estimates, extrapolation and optimal points of stresses for the finite element methods on the strongly regular triangulation*, J. Comput. Math. **1** (1983), 376–383.

363. Lin, Q., Wang, H., Lin, T., *Interpolated finite element methods for second order hyperbolic equations and their global superconvergence*, Systems Sci. Math. Sci. **6** (1993), 331–340. MR 94i:65103.

364. Lin, Q., Wang, J.P., *Some expansions of the finite element approximation*, Shuli Kexue, Mathematical Sciences. Research Reports IMS 15, Academia Sinica, Inst. of Math. Sciences, Chengdu, 1984, 1–11. MR 86d:65148.

365. Lin, Q., Whiteman, J.R., *Superconvergence of recovered gradients of finite element approximations on nonuniform rectangular and quadrilateral meshes*, in "The mathematics of finite elements and applications, VII," (Uxbridge, 1990), Academic Press, London, 1991, pp. 563–571. MR 92d:65007.

366. Lin, Q., Xie, R.F., *Error expansions for finite element approximations and their applications*, in "Numerical methods for partial differential equations," (Shanghai, 1987), Lecture Notes in Math., 1297, Springer, Berlin, 1987, pp. 98–112. MR 89b:65272.

367. Lin, Q., Xie, R.F., *How to recover the convergent rate for Richardson extrapolation on bounded domains*, J. Comput. Math. **6** (1988), 68–79. MR 90b:65206.

368. Lin, Q., Xie, R.F., *Extrapolation for the approximations to the solution of a boundary integral equation on polygonal domains*, China-US Seminar on Boundary Integral and Boundary Element Methods in Physics and Engineering (Xi'an, 1987–88), J. Comput. Math. **7** (1989), 174–181. MR 90k:65183.

369. Lin, Q., Xie, R.F., *Error expansion for FEM and superconvergence under natural assumption*, J. Comput. Math. **7** (1989), 402–411.

370. Lin, Q., Xu, J.C., *Linear finite elements with high accuracy*, J. Comput. Math. **3** (1985), 115–133. MR 87k:65141.

371. Lin, Q., Yan, N.N., "The construction and analysis for efficient finite elements," Hebei Univ. Publ. House, 1996, pp. 1–304. (in Chinese)

372. Lin, Q., Yan, N.N., Zhou, A., *A rectangle test for interpolated finite elements*, in "Proc. of Systems Sci. & Systems Engrg.," Great Wall (H. K.), Culture Publ. Co., 1991, pp. 217–229.

373. Lin, Q., Yan, N.N., Zhou, A., *Interpolated finite elements with rectangular meshes*, Academia Sinica Inst. of Systems Science, Research Report No. 91/04, 1991, 1–14.

374. Lin, Q., Zhang, S.H., *An immediate analysis for global superconvergence for integrodifferential equations*, Appl. Math. **42** (1997), 1–21.

375. Lin, Q., Zhang, S.H., *A direct global superconvergence analysis for Sobolev and viscoelasticity type equations*, Appl. Math. **42** (1997), 23–34.

376. Lin, Q., Zhou, A., *Some arguments for recovering the finite element*, Acta Math. Sci. (English Ed.) **11** (1991), 290–297. MR 94h:65100.

377. Lin, Q., Zhou, A., *Defect correction for finite element gradient*, Systems Sci. Math. Sci. **5** (1992), 287–288. MR 93h:65039.

378. Lin, Q., Zhou, A., *Notes on superconvergence and its related topics*, J. Comput. Math. **11** (1993), 211–214.

379. Lin, Q., Zhu, Q.D., *Asymptotic expansion for the derivative of finite elements*, J. Comput. Math. **2** (1984), 361–363. MR 87m:65185.

380. Lin, Q., Zhu, Q.D., *Local asymptotic expansion and extrapolation for finite elements*, J. Comput. Math. **4** (1986), 263–265. MR 87k:65142.

381. Lin, Q., Zhu, Q.D., "The preprocessing and postprocessing for the finite element method," Shanghai Scientific & Technical Publishers, 1994, pp. 1–217. (in Chinese)

382. Lin, T., Wang, H., *Recovering the gradients of the solutions of second-order hyperbolic equations by interpolating the finite element solutions*, IMA Preprint Series #1157, Univ. of Minnesota, 1993, 1–38.

383. Lin, T., Wang, H., *A class of globally super-convergent post-processing techniques for approximating the gradients of solutions to elliptic boundary value problems*, ICAM Report 94-01-01, Univ. of South Carolina, 1994, 1–34.

384. Lin, T., Wang, H., *A class of error estimators based on interpolating the finite element solutions for reaction-diffusion equations*, in "Modeling, mesh generation, and adaptive numerical methods for partial differential equations," (Minneapolis, MN, 1993), IMA Vol. Math. Appl., 75, Springer, New York, 1995, pp. 129–151. MR 97b:65107.

385. Liu, X.Q., Peng, L., *The superconvergence problem for a nonconforming finite element method*, J. Southeast Univ. **26** (1996), 81–86. (in Chinese)

386. Locker, J., Prenter, P.M., *On least squares methods for linear two-point boundary value problems*, in "Functional analysis methods in numerical analysis," Proc. Special Session, Annual Meeting, Amer. Math. Soc., St. Louis, Mo., 1977, Lecture Notes in Math., 701, Springer, Berlin, 1979, pp. 149–168. MR 80j:65029.

387. Locker, J., Prenter, P.M., *Regularization with differential operators. II. Weak least squares finite element solutions to first kind integral equations*, SIAM J. Numer. Anal. **17** (1980), 247–267. MR 83j:65062b.

388. Locker, J., Prenter, P.M., *Splitting least squares and collocation procedures for two-point boundary value problems*, J. Austral. Math. Soc. Ser. B **27** (1985), 167–193. MR 86m:65088.

389. López L.G., *On a superconvergence theorem of J. L. Walsh*, Cienc. Mat. (Havana) **4** (1983), 67–78. MR 86c:30072. (in Spanish)

390. Louis, A., *Acceleration of convergence for finite element solutions of the Poisson equation*, Numer. Math. **33** (1979), 43–53. MR 81i:65091.

391. Lü, T., *Extrapolation and superconvergence of allocation solutions to weak singular boundary integral equations of the second kind*, Acta Math. Sci. **11** (1991), 16–22. MR 92f:65157. (in Chinese)

392. Lubich, C., *On projected Runge–Kutta methods for differential-algebraic equations*, BIT **31** (1991), 545–550. MR 92h:65109.

393. Luo, P., *The superconvergence problem for the average gradient in P^1 nonconforming finite element solutions*, Huaihua Shizhuan Xuebao **10** (1991), 84–88. (in Chinese)

394. Luo, P., *Superconvergence estimates for a nonconforming membrane element*, Northeast. Math. J. **8** (1992), 477–482. MR 94g:65126.

395. Luo, X.N., Hu, Q.Y., *Natural superconvergence of collocation solutions to integrodifferential equations*, J. Changsha Univ. Electr. Power Nat. Sci. Ed. **10** (1995), 234–238. (in Chinese)

396. MacKinnon, R.J., Carey, G.F., *Superconvergent derivatives: a Taylor series analysis*, Internat. J. Numer. Methods Engrg. **28** (1989), 489–509. MR 90e:73083.

397. MacKinnon, R.J., Carey, G.F., *Nodal superconvergence and solution enhancement for a class of finite-element and finite-difference methods*, SIAM J. Sci. Statist. Comput. **11** (1990), 343–353. MR 90m:65201.

398. Mahoux, G., Martin, A., *Extension of analyticity domains of functions with positivity properties and axiomatic proof of superconvergence relations*, in "Analytic methods in mathematical physics," (Sympos., Indiana Univ., Bloomington, Ind., 1968), Gordon and Breach, New York, 1970, pp. 265–277. MR 50#12037.

399. Makarenko, E.N., *Superconvergence of schemes of the finite element method for the Poisson equation on regular nets*, Vestnik Moskov. Univ. Ser. XV Vychisl. Mat. Kibernet. **1987**, no. 4, 63–66, 74. MR 89a:65169. (in Russian)

400. Mansfield, L., *On mixed finite element methods for elliptic equations*, Comput. Math. Appl.

7 (1981), 59–66. MR 82b:65133.

401. Marchuk, G.I., "Computational processes and systems," No. 6, "Nauka", Moscow, 1988, pp. 1–271. MR 89j:65009. (in Russian)

402. Marmi, S., *Diophantine conditions and K.A.M. estimates for the Siegel theorem in* $\mathbf{C}^N$, Meccanica **23** (1988), 139–146. MR 90k:58187.

403. Marshall, R.S., *An alternative interpretation of superconvergence*, Internat. J. Numer. Methods Engrg. **14** (1979), 1707–1709. MR 80m:65079.

404. Mavrodi, N.N., *Necessary and sufficient conditions for the superconvergence of power series*, in [316], 1987, 147–149, 180. MR 91b:30006. (in Russian)

405. McCarthy D., Williams, H.C. (eds), "Proc. of the Eighth Manitoba Conf. on Numer. Math. and Computing," Univ. of Manitoba, Winnipeg, Man., 1978, Congressus Numerantium, XXII, Utilitas Mathematica Publishing, Winnipeg, Man, 1979, pp. 1–487. MR 80g:65005.

406. Miller, J.J.H. (ed.), "BAIL III. Proc. of the third internat. conf. on boundary and interior layers—computational and asymptotic methods," Trinity College, Dublin, 1984, Boole Press Conference Series, 6, Boole Press, Dún Laoghaire, 1984, pp. 1–339. MR 85k:00008.

407. Milner, F.A., L^∞-*error estimates for linear elasticity problems*, J. Comput. Appl. Math. **25** (1989), 305–313. MR 90e:73023.

408. Monk, P., *Superconvergence of finite element approximations to Maxwell's equations*, Numer. Methods Partial Differential Equations **10** (1994), 793–812. MR 95h:65090.

409. Moore, P.K., Flaherty, J.E., *A local refinement finite-element method for one-dimensional parabolic systems*, SIAM J. Numer. Anal. **27** (1990), 1422–1444. MR 91h:65146.

410. Morley, M.E., *A family of mixed finite elements for linear elasticity*, Numer. Math. **55** (1989), 633–666. MR 90f:73006.

411. Nair, M.T., *On some variants of projection method for operator equations of the second kind*, in "Proc. of the Symposium on Operator Theory and Functional Analysis," (Cochin, 1989), Publication, 18, Centre Math. Sci., Trivandrum, 1989, pp. 21–40. MR 92a:65185.

412. Nair, M.T., Anderssen, R.S., *Superconvergence of modified projection method for integral equations of the second kind*, J. Integral Equations Appl. **3** (1991), 255–269. MR 92d:65235.

413. Nakao, M., *Some superconvergence estimates for a collocation-H^{-1}-Galerkin method for parabolic problems*, Mem. Fac. Sci. Kyushu Univ. Ser. A **35** (1981), 291–306. MR 83a:65111.

414. Nakao, M., *Interior estimates and superconvergence for H^{-1}-Galerkin method to elliptic equations*, Bull. Kyushu Inst. Tech. Math. Natur. Sci. **30** (1983), 19–30. MR 85i:65154.

415. Nakao, M., *Superconvergence estimates at Jacobi points of the collocation-Galerkin method for two-point boundary value problems*, J. Inform. Process. **7** (1984), 31–34. MR 85m:65079.

416. Nakao, M., *Some superconvergence estimates for a Galerkin method for elliptic problems*, Bull. Kyushu Inst. Tech. Math. Natur. Sci. No. **31** (1984), 49–58. MR 86e:65159.

417. Nakao, M.T., L^∞ *error estimates and superconvergence results for a collocation-H^{-1}-Galerkin method for elliptic equations*, Mem. Fac. Sci. Kyushu Univ. Ser. A **39** (1985), 1–25. MR 87c:65156.

418. Nakao, M.T., *Error estimates of a Galerkin method for some nonlinear Sobolev equations in one space dimension*, Numer. Math. **47** (1985), 139–157. MR 86j:65134.

419. Nakao, M.T., *Some superconvergence of Galerkin approximations for parabolic and hyperbolic problems in one space dimension*, Bull. Kyushu Inst. Tech. Math. Natur. Sci. No. 32 (1985), 1–14. MR 86j:65123.

420. Nakao, M.T., *Some superconvergence for a Galerkin method by averaging gradients in one-dimensional problems*, J. Inform. Process. **9** (1986), 130–134. MR 88e:65144.

421. Nakao, M.T., *Superconvergence of the gradient of Galerkin approximations for elliptic problems*, Proc. of the 2nd Internat. Conf. on Computational and Appl. Math. (Leuven, 1986), J. Comput. Appl. Math. **20** (1987), Special Issue, 341–348. MR 89a:65172.

422. Nakao, M.T., *Superconvergence of the gradient of Galerkin approximations for elliptic problems*, RAIRO Modél. Math. Anal. Numér. **21** (1987), 679–695. MR 88m:65180.

423. Nakata, M., Weiser, A., Wheeler, M.F., *Some superconvergence results for mixed finite element methods for elliptic problems on rectangular domains*, The mathematics of finite elements and applications, V (Uxbridge, 1984), Academic Press, London-New York, 1985, 367–389. MR 87g:65144.

424. Nakata, M., Wheeler, M.F., *Some superconvergence results for mixed finite element methods*

for linear parabolic problems, in "Numerical analysis," (Guanajuato, 1984), Lecture Notes in Math., 1230, Springer, Berlin, 1986, pp. 167–174. MR 89b:65273.

425. Neittaanmäki, P., Křížek, M., *Superconvergence of the finite element schemes arising from the use of averaged gradients*, in "Proc. of Int. Conf. on Accuracy Estimates and Adaptive Refinements in Finite Element Computations," I. Babuška, O.C. Zienkiewicz (eds), 1984, pp. 169–178.

426. Neittaanmäki, P., Křížek, M., *Postprocessing of a finite element scheme with linear elements*, in "Numerical Techniques in Continuum Mechanics," Proc. of a GAMM-Seminar, ed. W.Hackbusch and K.Witsch, Kiel, 1986, Vieweg & Sohn, Wiesbaden, 1987, pp. 69–83.

427. Neittaanmäki, P., Křížek, M., *On $O(h^4)$-superconvergence of piecewise bilinear FE-approximations*, in "Proc. of the Second Internat. Symposium on Numer. Anal.," (Prague, 1987), Teubner-Texte Math., 107, Teubner, Leipzig, 1988, pp. 250–255. MR 93b:65007.

428. Neittaanmäki, P., Lin, Q., *Acceleration of the convergence in finite difference method by predictor-corrector and splitting extrapolation methods*, J. Comput. Math. **5** (1987), 181–190. MR 91i:65172.

429. Neta, B., Victory, H.D., *A new fourth-order finite-difference method for solving discrete-ordinates slab transport equations*, SIAM J. Numer. Anal. **20** (1983), 94–105. MR 84e:65082.

430. Ni, P., *Superconvergence of generalized Galerkin method solutions and their derivatives*, Dongbei Shida Xuebao (1986), 13–17. MR 87m:65189. (in Chinese)

431. Ni, P., Wu, W., *Estimates of superconvergence for generalized Galerkin methods*, Numer. Math. J. Chinese Univ. **8** (1986), 153–158. MR 88b:65096. (in Chinese)

432. Nitsche, J.A., Schatz, A.H., *Interior estimates for Ritz-Galerkin methods*, Math. Comp. **28** (1974), 937–958. MR 51#9525.

433. Niu, Q.X., Shephard, M.S., *Superconvergent extraction techniques for finite element analysis*, Internat. J. Numer. Methods Engrg. **36** (1993), 811–836. MR 93k73080.

434. Niu, Q.X., Shephard, M.S., *Superconvergent boundary stress extraction and some experiments with adaptive pointwise error control*, Internat. J. Numer. Methods Engrg. **37** (1994), 877–891.

435. Nørsett, S.P., *Splines and collocation for ordinary initial value problems*, in "Approximation theory and spline functions," (St. John's, Nfld., 1983), NATO Adv. Sci. Inst. Ser. C: Math. Phys. Sci., 136, Reidel, Dordrecht, 1984, pp. 397–417. MR 86j:41008.

436. Oden, J.T., "Finite elements. Vol. IV. Mathematical aspects," The Texas Finite Element Series, IV, Prentice-Hall, Englewood Cliffs, NJ, 1983, pp. 1–195. MR 86m:65001d.

437. Oden, J.T., *Optimal h-p finite element methods*, Finite element methods in large-scale computational fluid dynamics (Minneapolis, MN, 1992), Comput. Methods Appl. Mech. Engrg. **112** (1994), 309–331. MR 94m:65163.

438. Oehme, R., *Renormalization group, BRST cohomology, and the problem of confinement*, Phys. Rev. D (3) **42** (1990), 4209–4221. MR 91m:81168.

439. Oehme, R., *Superconvergence, confinement and duality*, Ukrain. Fiz. Zh. **41** (1996), 311–316. MR 97g:65278.

440. Oganesjan, L.A., Ruhovec, L.A., *An investigation of the rate of convergence of variational-difference schemes for second order elliptic equations in a two-dimensional region with smooth boundary*, Ž. Vyčisl. Mat. i Mat. Fiz. **9** (1969), 1102–1120. MR 45#4665.

441. Oganesjan, L.A., Ruhovec, L.A., "Variational-difference methods for the solution of elliptic equations," Izd. Akad. Nauk Armjanskoj SSR, Jerevan, 1979. MR 82m:65105.

442. Omari, P., *On the fast convergence of a Galerkin-like method for equations of the second kind*, Math. Z. **201** (1989), 529–539. MR 90f:65083.

443. O'Riordan, E., Stynes, M., *An analysis of a superconvergence result for a singularly perturbed boundary value problem*, Math. Comp. **46** (1986), 81–92. MR 87b:65107.

444. Pani, A.K., Das, P.C., *An H^1-Galerkin method for a Stefan problem with a quasilinear parabolic equation in nondivergence form*, Internat. J. Math. Math. Sci. **10** (1987), 345–360. MR 88d:65145.

445. Pani, A.K., Sinha, R.K., *On superconvergence results and negative norm estimates for parabolic integro-differential equations*, J. Integral Equations Appl. **8** (1996), 65–98. MR 97g:65213.

446. Papini, A., *One-dimensional collocation at Gaussian points and superconvergence at interior nodal points*, Rend. Istit. Mat. Univ. Trieste **21** (1989), 224—231. MR 93a:65161.

447. Pehlivanov, A., *Superconvergence of the gradient for quadratic 3D simplex finite elements*, in "Numerical methods and applications," (Sofia, 1989), Publ. Bulgar. Acad. Sci., Sofia, 1989, pp. 362–366.

448. Pehlivanov, A.I., *Interior estimates of type superconvergence of the gradient in the finite element method*, C. R. Acad. Bulgare Sci. **42** (1989), 29–32. MR 91e:65130.

449. Pehlivanov, A.I., Carey, G.F., Lazarov, R.D., Shen, Y., *Convergence analysis of least-squares mixed finite elements*, Computing **51** (1993), 111–123. MR 95b:65096.

450. Pehlivanov, A.I., Lazarov, R.D., Carey, G.F., Chow, S.S., *Superconvergence analysis of approximate boundary-flux calculations*, Numer. Math. **63** (1992), 483–501. MR 94e:65121.

451. Peng, L., *A probabilistic algorithm for high accuracy finite elements*, J. Southeast Univ. **20** (1990), 119–126. MR 92h:65129. (in Chinese)

452. Peng, L., Wang, J.L., *Superconvergence analysis of a nonconforming model*, J. Southeast Univ. **23** (1993), 109–113. MR 96b:65108. (in Chinese)

453. Ping, L., Lin, Q., *High accuracy analysis of the Wilson element*, J. Comput. Math. (1997), 1–14 (to appear).

454. Reddien, G. W., *Collocation at Gauss points as a discretization in optimal control*, SIAM J. Control Optim. **17** (1979), 298–306. MR 80b:93171.

455. Reddien, G.W., *Projection methods for two-point boundary value problems*, SIAM Rev. **22** (1980), 156–171. MR 81e:65049.

456. Regińska, T., *Superconvergence of external approximation of eigenvalues of ordinary differential operators*, IMA J. Numer. Anal. **6** (1986), 309–323. MR 89i:65083.

457. Regińska, T., *The superconvergence effect in the finite element method for two-point boundary value problems*, Mat. Stos. **30** (1987), 97–111, 129. MR 89f:65085. (in Polish)

458. Regińska, T., *Superconvergence of external approximation for two-point boundary problems*, Apl. Mat. **32** (1987), 25–36. MR 88e:65100.

459. Regińska, T., *Error estimates for external approximation of ordinary differential equations and the superconvergence property*, Apl. Mat. **33** (1988), 277–290. MR 89f:65086.

460. Richter, G.R., *Superconvergence of piecewise polynomial Galerkin approximations, for Fredholm integral equations of the second kind*, Numer. Math. **31** (1978/79), 63–70. MR 80a:65273.

461. Rodríguez, R., *Some remarks on Zienkiewicz-Zhu estimator*, Numer. Methods Partial Differential Equations **10** (1994), 625–635. MR 95e:65103.

462. Rodríguez, R., *A-posteriori error analysis in the finite element method*, in [327], 1994, pp. 389–397. MR 95d:65008.

463. Roos, H.G., Stynes, M., Tobiska, L., "Numerical methods for singularly perturbed differential equations Convection-diffusion and flow problems," Springer, Berlin, 1996.

464. Rui, H.X., *Superconvergence of finite element solutions to a class of Sobolev equations*, Shandong Daxue Xuebao Ziran Kexue Ban **27** (1992), 301–308. MR 93i:65094. (in Chinese)

465. Sablonnière, P., *Convergence acceleration of logarithmic fixed point sequences*, J. Comput. Appl. Math. **19** (1987), 55–60. MR 88j:65008.

466. Sacco, R., *Convergence of a second-order accurate Petrov-Galerkin scheme for convection-diffusion problems in semiconductors*, Appl. Numer. Math. **11** (1993), 517–528. MR 94e:65122.

467. Sanz-Serna, J.M., Abia, L., *Interpolation of the coefficients in nonlinear elliptic Galerkin procedures*, SIAM J. Numer. Anal. **21** (1984), 77–83. MR 85c:65146.

468. Schatz, A.H., *An introduction to the analysis of the error in the finite element method for second-order elliptic boundary value problems*, in "Numerical Analysis," Lancaster, 1984, Lecture Notes in Math., 1129, Springer, Berlin, 1985, pp. 94–139. MR 86k:65114.

469. Schatz, A.H., *Pointwise error estimates, superconvergence and extrapolation*, in [328], 1997.

470. Schatz, A.H., Sloan, I.H., Wahlbin, L.B., *Superconvergence in finite element methods and meshes that are locally symmetric with respect to a point*, SIAM J. Numer. Anal. **33** (1996), 505–521.

471. Schatz, A.H., Wahlbin, L.B., *Interior maximum norm estimates for finite element methods*, Math. Comp. **31** (1977), 414–442. MR 55#4748.

472. Schmidt, G., Strese, H., *BEM for Poisson equation*, Engineering Anal. with Boundary Elements **10** (1992), 119–123.

473. Schock, E., *Approximate solution of ill-posed equations: arbitrarily slow convergence vs. su-*

perconvergence, in "Constructive methods for the practical treatment of integral equations," (Oberwolfach, 1984), Internat. Schriftenreihe Numer. Math., 73, Birkhäuser, Basel, 1985, pp. 234–243. MR 89a:65093.

474. Schock, E., *Arbitrarily slow convergence, uniform convergence and superconvergence of Galerkin-like methods*, IMA J. Numer. Anal. **5** (1985), 153–160. MR 86f:65220.

475. Schock, E. (ed.), "Contributions to applied analysis and computer science," Helmut Brakhage zu Ehren, Verlag Shaker, Aachen, 1994, pp. 1–334. MR 95b:65006.

476. Selwood, P.M., Wathen, A.J., *Convergence rates and classification for one-dimensional finite-element meshes*, IMA J. Numer. Anal. **16** (1996), 65–74. MR 96j:65072.

477. Sendov, B., Lazarov, R.D., Dimov, I. (eds), "Numerical methods and applications," Proc. of the International Conf., Sofia, 1989, Publishing House of the Bulgarian Academy of Sciences, Sofia, 1989, pp. 1–583. MR 90i:65008.

478. Shei, S.S., *Algebraic structure of current algebra, superconvergence sum rules, and the infinite-momentum method*, Phys. Rev. **188** (1969), 2274–2276. MR 41#7953.

479. Shen, J., *A block finite difference scheme for second-order elliptic problems with discontinuous coefficients*, SIAM J. Numer. Anal. **33** (1996), 686–706. MR 96m:65101.

480. Shi, J., *The extrapolation method for a class of nonsmooth Fredholm integral equations*, Systems Sci. Math. Sci. **5** (1992), 33–41. MR 94d:45009.

481. Shu, H.Z., *The superconvergence of a Galerkin collocation method for first kind boundary integral equations on smooth curves*, J. Comput. Appl. Math. **72** (1996), 1–20.

482. Sloan, I.H., *Superconvergence and the Galerkin method for integral equations of the second kind*, Treatment of integral equations by numerical methods (Durham, 1982), Academic Press, London, 1982, 197–207. MR 85e:65005.

483. Sloan, I.H., *The iterated Galerkin method for integral equations of the second kind*, in "Miniconference on operator theory and partial differential equations," (Canberra, 1983), Proc. Centre Math. Anal. Austral. Nat. Univ., 5, Canberra, 1984, pp. 153–161. MR 86e:45017.

484. Sloan, I.H., *Superconvergence in the collocation and qualocation methods*, in "Numerical mathematics," Singapore 1988, Internat. Schriftenreihe Numer. Math., 86, Birkhäuser, Basel, 1988, pp. 429–441. MR 90j:65137.

485. Sloan, I.H., *Superconvergence*, in "Numerical solution of integral equations," Math. Concepts Methods Sci. Engrg., 42, Plenum, New York, 1990, pp. 35–70. MR 91g:45011.

486. Sloan, I.H., *Boundary element methods*, in "Theory and numerics of ordinary and partial differential equations," (Leicester, 1994), Adv. Numer. Anal., IV, Oxford University Press, New York, 1995, pp. 143–180. MR 96k:65077.

487. Sloan, I.H., Spence, A., *Projection methods for integral equations on the half-line*, IMA J. Numer. Anal. **6** (1986), 153–172. MR 89h:65225.

488. Sloan, I.H., Spence, A., *The Galerkin method for integral equations of the first kind with logarithmic kernel: applications*, IMA J. Numer. Anal. **8** (1988), 123–140. MR 90d:65230b.

489. Sloan, I.H., Thomée, V., *Superconvergence of the Galerkin iterates for integral equations of the second kind*, J. Integral Equations **9** (1985), 1–23. MR 86j:65184.

490. Solov'ëv, S. I., *Superconvergence of finite-element approximations of eigenfunctions*, Differential Equations **30** (1994), 1138–1146. MR 96a:65172.

491. Song, Y.C., *On the discrete Galerkin methods for nonlinear integral equations*, J. Korean Math. Soc. **29** (1992), 297–315. MR 93j:65225.

492. Spence, A., Thomas, K.S., *On superconvergence properties of Galerkin's method for compact operator equations*, IMA J. Numer. Anal. **3** (1983), 253–271. MR 85c:65074.

493. Squeff, M.C.J., *Superconvergence of mixed finite elements for parabolic equations*, Thesis, Univ. of Chicago, 1985.

494. Squeff, M.C.J., *Superconvergence of mixed finite element methods for parabolic equations*, RAIRO Modél. Math. Anal. Numér. **21** (1987), 327–352. MR 88j:65207.

495. Stenberg, R., *Postprocessing schemes for some mixed finite elements*, Mathematical Modelling and Numerical Analysis **25** (1991), 151–168. MR 92a:65303.

496. Storti, M., Nigro, N., Idelsohn, S., *A Petrov-Galerkin formulation for the reaction-advection-diffusion equation*, Rev. Internac. Métod. Numér. Cálc. Diseñ. Ingr. **11** (1995), 247–270. MR 96f:65122. (in Spanish)

497. Strang, G., Fix, G., "An analysis of the finite element method," Prentice Hall, Englewood Cliffs, NJ, 1973. MR 65#1747.

498. Stynes, M., O'Riordan, E., *A superconvergence result for a singularly perturbed boundary value problem*, in "BAIL III," (Dublin, 1984), Boole Press Conf. Ser., 6, Boole, Dún Laoghaire, 1984, pp. 309–313. MR 86b:65084.

499. Tang, T., *Superconvergence of numerical solutions to weakly singular Volterra integro-differential equations*, Numer. Math. **61** (1992), 373–382. MR 92k:65198.

500. Taylor, R.L., Zienkiewicz, O.C., *A note on the "order of approximation"*, Internat. J. Solids and Structures **21** (1985), 793–797. MR 88a:65093.

501. Tessler, A., Riggs, H.R, Macy, S.C., *Application of a variational method for computing smooth stresses, stress gradients, and error estimation in finite element analysis*, in "The Mathematics of Finite Elements and Applications," J.R. Whiteman (ed.), John Wiley & Sons, New York, 1994, pp. 189–198.

502. Tessler, A., Riggs, H.R., Macy, S.C., *A variational method for finite element stress recovery and error estimation*, Comput. Methods Appl. Mech. Engrg. **111** (1994), 369–382. MR 94j:73062.

503. Thomée, V., *High order local approximations to derivatives in the finite element method*, Math. Comp. **31** (1977), 652–600. MR 55#11572.

504. Thomée, V., *Galerkin-finite element method for parabolic equations*, Bull. Iranian Math. Soc. No. **10** (1978/79), 3L–17L. MR 81b:65107.

505. Thomée, V., *Galerkin-finite element methods for parabolic equations*, in "Proc. of the Internat. Congress of Mathematicians," (Helsinki, 1978), Acad. Sci. Fennica, Helsinki, 1980, pp. 943–952. MR 81i:65100.

506. Thomée, V., *Negative norm estimates and superconvergence in Galerkin methods for parabolic problems*, Math. Comp. **34** (1980), 93–113. MR 81a:65092.

507. Thomée, V., *Finite element methods for parabolic problems – some steps in the evolution*, in [327], 1994, pp. 433–442, MR 95f:65184.

508. Thomée, V., Westergren, B., *Elliptic difference equations and interior regularity*, Numer. Math. **11** (1968), 196–210. MR 36#7347.

509. Thomée, V., Xu, J.C., Zhang, N.Y., *Superconvergence of the gradient in piecewise linear finite-element approximation to a parabolic problem*, SIAM J. Numer. Anal. **26** (1989), 553–573. MR 90e:65165.

510. Tonoyan, R.N. (ed.), "Applied mathematics, No. 5," Erevan. Univ., Erevan, 1987, pp. 1–136. MR 90a:00002. (in Russian)

511. Tran, T., *The K-operator and the qualocation method for strongly elliptic equations on smooth curves*, J. Integral Equations Appl. **5** (1993), 405–428. MR 95a:65218.

512. Turner, P.R. (ed.), "Numerical analysis and parallel processing," Lectures from the Third S.E.R.C. Numerical Analysis Summer School, Univ. of Lancaster, 1987, Lecture Notes in Math., 1397, Springer, Berlin, 1989, pp. 1–264. MR 90f:65002.

513. Usmani, R.A., *On quadratic spline interpolation*, BIT **27** (1987), 615–622. MR 88j:65030.

514. Vainikko, G., Uba, P., *A piecewise polynomial approximation to the solution of an integral equation with weakly singular kernel*, J. Austral. Math. Soc. Ser. B **22** (1980/81), 431–438. MR 82h:65100.

515. Van der Houwen, P.J., Sommeijer, B.P., *Butcher-Kuntzmann methods for nonstiff problems on parallel computers*, Internat. Conf. on Scientific Computation and Differential Equations (Auckland, 1993), Appl. Numer. Math. **15** (1994), 357–374. MR 95h:65049.

516. Volk, W., *Global superconvergence in the solution of linear boundary value problems*, Numer. Math. **48** (1986), 617–625. MR 87i:65123. (in German)

517. Volk, W., *The iterated Galerkin method for linear integro-differential equations*, J. Comput. Appl. Math. **21** (1988), 63–74. MR 89a:65201.

518. Volle, M., *Convergence by epigraph and level set methods*, C. R. Acad. Sci. Paris Sér. I Math. **299** (1984), 295–298. MR 85h:90135. (in French)

519. Von Seggern, R., *The finite-element method in the numerical treatment of linear integro-differential equations*, Report 1585, Kernforschungsanlage Jülich, Zentralinstitut für Angewandte Math., Jülich, 1979, 1–104. MR 83e:65215. (in German)

520. Von Seggern, R., *A superconvergence result in the application of the finite element method to linear integro-differential equations*, Z. Angew. Math. Mech. **61** (1981), 320–321. MR 84e:45009. (in German)

521. Wahlbin, L.B., *Local behaviour in finite element methods*, in "Handbook of Numer. Anal.,"

vol. II.," P.G. Ciarlet, J.L. Lions (eds), North-Holland, Amsterdam, 1991, pp. 353–522. MR 92f:65001.

522. Wahlbin, L.B., *On superconvergence up to boundaries in finite element methods: a counterexample*, SIAM J. Numer. Anal. **29** (1992), 937–946. MR 94d:65045.

523. Wahlbin, L.B., "Lecture notes on superconvergence in Galerkin finite element methods," Cornell Univ., 1994, pp. 1–243.

524. Wahlbin, L.B., "Superconvergence in Galerkin finite element methods," Lecture Notes in Math., vol. 1605, Springer, Berlin, 1995, pp. 1–166.

525. Wahlbin, L.B., *General principles of superconvergence in Galerkin finite element methods*, in [328], 1997.

526. Wang, H., *Superconvergence of finite element solutions and their derivatives for a class of quasilinear hyperbolic equations*, Hunan Ann. Math. **5** (1985), 38–45. MR 88a:65135. (in Chinese)

527. Wang, H., *Superconvergence of fully discretized finite element solutions and their derivatives for a class of quasilinear hyperbolic equations*, Dongbei Shuxue **2** (1986), 205–214. MR 88h:65217. (in Chinese)

528. Wang, H., *The convergence of BEM for the stationary Stokes problem in three dimensions*, in "Boundary elements," (Beijing, 1986), Pergamon, Oxford, 1986, pp. 143–150. MR 89i:76018.

529. Wang, H., *A boundary element method for steady-state two-dimensional Stokes flows and its asymptotic error estimates*, in "Boundary elements in fluid dynamics," Comput. Mech., Southampton, 1992, pp. 143–154. MR 93k:76079.

530. Wang, H., Ewing, R.E., *Optimal-order convergence rates for Eulerian-Lagrangian localized adjoint methods for reactive transport and contamination in groundwater*, Numer. Methods Partial Differential Equations **11** (1995), 1–31. MR 95m:65153.

531. Wang, J.P., *Superconvergence and extrapolation for mixed finite element methods on rectangular domains*, Math. Comp. **56** (1991), 477–503. MR 91m:65280.

532. Wendland, W.L., *Boundary element methods and their asymptotic convergence*, in "Theoretical acoustics and numerical techniques," CISM Courses and Lectures, 277, Springer, Vienna, 1983, pp. 135–216. MR 86f:65201.

533. Wendland, W.L., Stephan, E., Hsiao, G.C., *On the integral equation method for the plane mixed boundary value problem of the Laplacian*, Math. Methods Appl. Sci. **1** (1979), 265–321. MR 82e:31003.

534. Werner, H., *Spline functions and the numerical solution of differential equations*, in "Special topics of applied mathematics," Proc. Sem., Ges. Math. Datenverarb., Bonn, 1979, North-Holland, Amsterdam, 1980, pp. 173–192. MR 82b:65064.

535. Wheeler, M.F., *Galerkin procedure for estimating the flux for two-point boundary value problems*, SIAM J. Numer. Anal. **11** (1974), 765–768. MR 52#4644.

536. Wheeler, M.F., Whiteman, J.R., *Superconvergent recovery of gradients on subdomains from piecewise linear finite-element approximations*, Numer. Methods Partial Differential Equations **3** (1987), 65–82. MR 90h:65192.

537. Wheeler, M.F., Whiteman, J.R., *Superconvergent recovery of gradients on subdomains from piecewise linear finite-element approximations*, Numer. Methods Partial Differential Equations **3** (1987), 65–82, 357–374. MR 92a:65306.

538. Wheeler, M.F., Whiteman, J.R., *Superconvergence of recovered gradients of discrete time/piecewise linear Galerkin approximations for linear and nonlinear parabolic problems*, Numer. Methods Partial Differential Equations **10** (1994), 271–294. MR 95b:65122.

539. Whiteman, J.R., Beagles, A.E., Warby, M.K., *Theoretical and practical aspects of finite elements in the context of some problems of solid mechanics*, Jahresber. Deutsch. Math.-Verein. **92** (1990), 77–88. MR 91c:73084.

540. Whiteman, J.R., Goodsell, G., *Superconvergent recovery for stresses from finite element approximations on subdomains for planar problems of linear elasticity*, in "The mathematics of finite elements and applications, VI," (Uxbridge, 1987), Academic Press, London, 1988, pp. 29–53. MR 89h:65195.

541. Whiteman, J.R., Goodsell, G., *On some finite element error estimates for stress intensity factors in mode I linear elastic fracture problems*, Transactions of the Fifth Army Conf. on Appl. Math. and Computing (West Point, NY, 1987), ARO Rep., 88-01, U.S. Army Res. Office, Research Triangle Park, NC, 1988, 541–548. MR 89d:73046.

542. Whiteman, J.R., Goodsell, G., *Some gradient superconvergence results in the finite element method*, in "Numerical analysis and parallel processing," (Lancaster, 1987), Lecture Notes in Math., 1397, Springer, Berlin, 1989, pp. 182–260. MR 91a:65234.

543. Whiteman, J.R., Goodsell, G., *Some features of the nodal recovery of gradients from finite element approximations which produces superconvergence*, in "Proc. Conf. EQUADIFF 7," (ed. J. Kurzweil), Prague, 1989, Teubner-Texte zur Mathematik, Band 118, Teubner, Leipzig, 1990, pp. 305–308.

544. Whiteman, J.R., Goodsell, G., *A survey of gradient superconvergence for finite element approximations to second order elliptic problems on triangular and tetrahedral meshes*, in "The mathematics of finite elements and applications, VII," (Uxbridge, 1990), Academic Press, London, 1991, pp. 55–74. MR 92d:65007.

545. Wiberg, N.E., Abdulwahab, F., *An efficient postprocessing technique for stress problems based on superconvergent derivatives and equilibrium*, Numer. Methods in Engineering '92, Elsevier, 1992, 25–32.

546. Wiberg, N.E., Abdulwahab, F., *Superconvergent patch recovery in problems of mixed form*, Comm. Numer. Methods Engrg. **13** (1997), 207–217.

547. Wiberg, N.E., Abdulwahab, F., *Patch recovery based on superconvergent derivatives and equilibrium*, Internat. J. Numer. Methods Engrg. **36** (1993), 2703–2724. MR 94c:73065.

548. Wiberg, N.E., Li, X.D., *Superconvergence patch recovery of finite-element solution and a posteriori L_2-norm error estimate*, Comm. Numer. Methods Engrg. **10** (1994), 313–320.

549. Winther, R., *A finite element method for a version of the Boussinesq equation*, SIAM J. Numer. Anal. **19** (1982), 561–570. MR 83f:65184.

550. Wohlgemuth, R., *Superkonvergenz des Gradienten im Postprocessing von Finite-Elemente-Methoden*, Preprint Nr. 94, Tech. Univ. Chemnitz, 1989, 1–15.

551. Wu, D.G., Huang, D.R., *Superconvergence of quadratic interpolation splines and quadratic spline finite element solutions*, Numer. Math. J. Chinese Univ. **10** (1988), 224–229. MR 90a:65032. (in Chinese)

552. Xiang, C.J., *Superconvergence of the spline-Bounds method for solving Volterra integral equations*, Numer. Math. J. Chinese Univ. **13** (1991), 191–196. MR 92j:65202. (in Chinese)

553. Xu, C.F., *The split iteration method at successive levels for solving finite element equations and its convergence rate*, Math. Appl. **1** (1988), 59–66. MR 90f:65212. (in Chinese)

554. Xu, D., *Finite element methods of the two nonlinear integro-differential equations*, Appl. Math. Comput. **58** (1993), 241–273. MR 94k:65188.

555. Xu, J.C., *The error analysis and the improved algorithms for the infinite element method*, in "Proc. of the 1984 Beijing symposium on differential geometry and differential equations," Science Press, Beijing, 1985, pp. 326–331. MR 87b:65213.

556. Yan, Y., *Cosine change of variable for Symm's integral equation on open arcs*, IMA J. Numer. Anal. **10** (1990), 521–535. MR 91j:65202.

557. Yang, B., Huang, M.Y., *An asymptotic expression for the error of approximate solutions by the mixed finite element method*, Lanzhou Daxue Xuebao **22** (1986), 5–9. MR 88e:65151. (in Chinese)

558. Yang, Y.D., *L^p-estimates and superconvergence of finite element approximations for eigenvalue problems*, Numer. Math. J. Chinese Univ. **9** (1987), 83–87. MR 88h:65197. (in Chinese)

559. Yang, Y.D., *A theorem on superconvergence of shifts of finite element approximations for eigenvalue problems*, J. Math. (Wuhan) **10** (1990), 229–234. MR 91h:65166. (in Chinese)

560. Yang, Y.D., *Superconvergence phenomena for finite difference methods*, Guizhou Shifan Daxue Xuebao Ziran Kexue Ban **13** (1995), 9–12. MR 97c:65176. (in Chinese)

561. Ying, L., Guo, B.Y. (eds), "Numerical methods for partial differential equations," Proceedings of the Second Conference held at Nankai University, Tianjin, 1991, World Scientific Publishing Co., River Edge, NJ, 1992, pp. 1–190. MR 93b:65006.

562. Ženíšek, A., "Nonlinear elliptic and evolution problems and their finite element approximation," Academic Press, 1990. MR 92bc65003.

563. Zennaro, M., *One-step collocation: uniform superconvergence, predictor-corrector method, local error estimate*, SIAM J. Numer. Anal. **22** (1985), 1135–1152. MR 86m:65084.

564. Zhang, L.Q., *Spline collocation approximation to periodic solutions of ordinary differential equations*, J. Comput. Math. **10** (1992), 147–154. MR 93a:65106.

565. Zhang, L., Li, L., *On superconvergence of isoparametric bilinear finite elements*, Comm. Numer. Methods Engrg. **12** (1996), 849–862. MR 97g:65250.

566. Zhang, T., *Superconvergence of a finite element for a singular two-point boundary value problem*, Numer. Math. J. Chinese Univ. **10** (1988), 185–188. MR 90b:65162. (in Chinese)

567. Zhang, T., *The L_p norm approximation properties of generalized elliptic projections and their gradient superconvergence*, J. Northeast. Univ. Nat. Sci. **15** (1994), 15–19. MR 95f:65187. (in Chinese)

568. Zhang, Z., *Arch beam models: finite element analysis and superconvergence*, Numer. Math. **61** (1992), 117–143. MR 93b:65180.

569. Zhang, Z., Zhang, S., *Derivative superconvergence of rectangular finite elements for the Reissner-Mindlin plate.*, Comput. Methods Appl. Mech. Engrg. **134** (1996), 1–16. MR 97h:73083.

570. Zhang, Z., Zhu, J.Z., *Superconvergence of the derivative patch recovery technique and a posteriori error estimation*, in "Modeling, mesh generation, and adaptive numerical methods for partial differential equations," (Minneapolis, MN, 1993), IMA Vol. Math. Appl., 75, Springer, New York, 1995, pp. 431–450. MR 97i:65170.

571. Zhou, A., *Global superconvergence approximations of the mixed finite element method for the Stokes problem and the linear elasticity equation*, RAIRO Math. Modél. Math. Anal. Numér. **30** (1996), 401–411.

572. Zhou, A., Li, J.C., *The full approximation accuracy for the stream function-vorticity-pressure method*, Numer. Math. **68** (1994), 427–435. MR 95k:76094.

573. Zhou, A., Lin, Q., *Optimal and superconvergence estimates of the finite element method for a scalar hyperbolic equation*, Acta Math. Sci. (English Ed.) **14** (1994), 90–94. MR 95c:65161.

574. Zhou, G. H., Rannacher, R., *Pointwise superconvergence of the streamline diffusion finite-element method*, Numer. Methods Partial Differential Equations **12** (1996), 123–145. MR 96m:65112.

575. Zhou, Q.H., *A mathematical representation of the condensation method*, Natur. Sci. J. Xiangtan Univ. **14** (1992), 24–27. MR 93g:65100. (in Chinese)

576. Zhu, J.Z., Zienkiewicz, O.C., *Superconvergence recovery technique and a posteriori error estimators*, Internat. J. Numer. Methods Engrg. **30** (1990), 1321–1339. MR 91i:65141.

577. Zhu, Q.D., *The derivative good points for the finite element method with 2-degree triangular element*, Natur. Sci. J. Xiangtan Univ. **4** (1981), 36–45. (in Chinese)

578. Zhu, Q.D., *A superconvergence result for the finite element method*, Numer. Math. J. Chinese Univ. **3** (1981), 50–55. MR 82f:65121. (in Chinese)

579. Zhu, Q.D., *Natural inner superconvergence for the finite element method*, in "Proc. of the China–France symposium on finite element methods," (Beijing, 1982), Science Press, Beijing, 1983, pp. 935–960. MR 85h:65253.

580. Zhu, Q.D., *Uniform superconvergence estimates of derivatives for the finite element method*, Numer. Math. J. Chinese Univ. **5** (1983), 311–318. MR 85k:65096. (in Chinese)

581. Zhu, Q.D., *Uniform superconvergence estimates for the finite-element method*, Natur. Sci. J. Xiangtan Univ. (1985), 10–26. MR 88h:65223. (in Chinese)

582. Zhu, Q.D., *Some L^∞ estimates and interior superconvergence estimates for piecewise linear finite element approximations*, Natur. Sci. J. Xiangtan Univ. (1987), 9–16. MR 88g:65124. (in Chinese)

583. Zhu, Q.D., *A fast high-accuracy method for finite element*, Chinese J. Numer. Math. Appl. **14** (1992), 52–57. MR 94j:65123.

584. Zhu, Q.D., *The superconvergence and postprocessing for the finite element method*, in "Proc. of Prague Math. Conf.," 1996, pp. 371–376.

585. Zhu, Q.D., *A survey of superconvergence techniques in finite element methods*, in [328], 1997.

586. Zhu, Q.D., Chen, H.S., *High accuracy error analysis of a linear finite element method for general boundary value problems*, Natur. Sci. J. Xiangtan Univ. **10** (1988), 1–10. MR 90j:65136. (in Chinese)

587. Zhu, Q.D., Lin, Q., *Local asymptotic expansion for finite elements*, J. Comput. Math. **3** (1986), 263-265.

588. Zhu, Q.D., Lin, Q., "The superconvergence theory of the finite element method," Hunan Science and Technology Publishing House, Changsha, 1989, pp. 1–314. MR 93j:65191. (in Chinese)

589. Zhu, Q.D., Lin, Q., *Asymptotically exact error estimates and local superconvergence for the finite element method*, Math. Numer. Sinica **15** (1993), 219–224. MR 97a:65096. (in Chinese)

590. Zienkiewicz, O.C., "The finite element method in engineering science," McGraw-Hill, London, 1971. MR 47#4518.

591. Zienkiewicz, O.C., Cheung, Y.K., "The finite element method in structural and continuum mechanics," McGraw-Hill, London, 1967. MR 47#4518.

592. Zienkiewicz, O.C., Zhu, J.Z., *Adaptivity and mesh generation*, Internat. J. Numer. Methods Engrg. **32** (1991), 783–810.

593. Zienkiewicz, O.C., Zhu, J.Z., *The superconvergent patch recovery and a posteriori error estimates, Parts 1,2*, Internat. J. Numer. Methods Engrg. **33** (1992), 1331–1364, 1365–1382. MR 93c:73098-9.

594. Zienkiewicz, O.C., Zhu, J.Z., *The superconvergent patch recovery (SPR) and adaptive finite element refinement*, Comput. Methods Appl. Mech. Engrg. **101** (1992), 207–224. MR 93h:73042.

595. Zlámal, M., *Some superconvergence results in the finite element method. Mathematical aspects of finite element methods*, in "Proc. Conf., Consiglio Naz. delle Ricerche (C.N.R.)," Rome, 1975, Lecture Notes in Math., Vol. 606, Springer, Berlin, 1977, pp. 353–362. MR 58#8365.

596. Zlámal, M., *Superconvergence of gradients in the finite element method*, in "Variational-difference methods in mathematical physics," Proc. All-Union Conf., Novosibirsk, 1977, Akad. Nauk SSSR Sibirsk. Otdel., Vychisl. Tsentr, Novosibirsk, 1978, pp. 15–22. MR 82e:65116. (in Russian)

597. Zlámal, M., *Superconvergence and reduced integration in the finite element method*, Math. Comp. **32** (1978), 663–685. MR 58#13794.

598. Zlámal, M., *Superconvergence of the gradient of finite element solutions*, Wiss. Z. Hochschule Architektur Bauwesen Weimar, 1979, 375–380.

599. Zlámal, M., *A box finite element method giving solution gradients with a high order accuracy*, in [327], 1994, pp. 501–504. MR 95d:65008.

600. Zlotnik, A.A., *Some finite-element and finite-difference methods for solving mathematical physics problems with non-smooth data in an n-dimensional cube. Part I*, Soviet J. Numer. Anal. Math. Modelling **6** (1991), 421–451. MR 92k:65172.